# SURVIVING THE DESIGN OF MICROPROCESSOR AND MULTIMICROPROCESSOR SYSTEMS

# WILEY SERIES ON PARALLEL AND DISTRIBUTED COMPUTING
## Series Editor: Albert Y. Zomaya

---

**Introduction to Parallel Algorithms** / C. Xavier and S. S. Iyengar

**Mobile Processing in Distributed and Open Environments** / Peter Sapaty

**Surviving the Design of Microprocessor and Multiprocessor Systems** / Veljko Milutinović

**Parallel and Distributed Simulation Systems** / Richard Fujimoto

**Solutions to Parallel and Distributed Computing Problems: Lesson From Biological Sciences** / Albert Y. Zomaya, Fikret Ercal, and Stephan Olariu (*Editors*)

**New Parallel Algorithms for Direct Solution of Linear Equations** / C. Siva Ram Murthy, K. N. Balasubramanya Murthy, and Srinivas Aluru

# SURVIVING THE DESIGN OF MICROPROCESSOR AND MULTIMICROPROCESSOR SYSTEMS

## LESSONS LEARNED

Veljko Milutinović

A WILEY-INTERSCIENCE PUBLICATION

**JOHN WILEY & SONS, INC.**

New York • Chichester • Weinheim • Brisbane • Singapore • Toronto

*Library of Congress Cataloging in Publication Data:*

Milutinović, Veljko
Surviving the design of microprocessor and multimicroprocessor systems / Veljko Milutinović.
p. cm.-- (Wiley series on parallel and distributed computing)
Includes bibliographical references and index.
ISBN 0-471-35728-6 (alk. paper)
1. Microprocessors--Design and construction. 2. Computer architecture. I. Title. II.
Series

TK7895.M5 M565 2000
004.2'2--dc21                                                                                            99-042185

Printed in the United States of America

10 9 8 7 6 5 4 3 2 1

There are several different styles in technical texts and monographs. The most familiar is the review style of the basic textbook. This style simply considers the technical literature and represents the data in a more orderly or useful way. Another style appears most commonly in monographs. This reviews either a particular aspect of a technology or all technical aspects of a single complex engineering system. A third style, represented by this book, is an integration of the first two styles, coupled with a personal reconciliation of important trends and movements in technology.

The author, Professor Milutinović, has been one of the most productive leaders in the computer architecture field. Few readers will not have encountered his name on an array of publications involving the important issues of the day. His publications and books span almost all areas of computer architecture and computer engineering. It would be easy, then, but inaccurate, to imagine this work as a restatement of his earlier ideas. This book is different, as it uniquely synthesizes Professor Milutinović's thinking on the important issues in computer architecture.

The issues themselves are presented quite concisely: cache, instruction level parallelism, prediction strategies, the I/O (input/output) bottleneck, multithreading, and multiprocessors. These are currently the principal research areas in computer architecture. Each one of these topics is presented in a crisp way, highlighting the important issues in the field together with Professor Milutinović's special viewpoints on these issues, closing each section with a statement about his own group's research in this area. This statement of issues is coupled with three important case studies of fundamentally different computer systems implementations. The case studies use details of actual engineering implementations to help synthesize the issues presented. The result is a necessarily somewhat eclectic, personal statement by one of the leaders of the field about the important issues that face us at this time.

This work should prove invaluable to the serious student.

MICHAEL J. FLYNN

*Stanford University*

Design of microprocessor and/or multimicroprocessor systems represents a continuous struggle; success (if achieved) lasts infinitesimally long and disappears forever, unless a new struggle (with unpredictable results) starts immediately. In other words, it is a continuous survival process, which is the main motto of this book.

This book is about survival of those who have contributed to the state of the art in the rapidly changing field of microprocessing and multimicroprocessing on a single chip, and about the concepts that yet have to find their way into the next generation microprocessors and multimicroprocessors on a chip, in order to enable these products to stay on the competitive edge.

This book is based on the assumption that one of the ultimate goals of the single-chip design is to have an entire distributed shared memory system on a single silicon die, together with numerous specialized accelerators, including the complex ones of the single-instruction, multiple-data processing (SIMD) and/or multiple-instruction, single-data processing (MISD) type.* (Such an opinion is based on the author's experiences, and on a number of important references,[†] such as Hammond et al. (1997). Consequently, the book concentrates on the major problems to be solved on the way to this ultimate goal (distributed shared memory on a single chip) and summarizes the author's experiences that led to such a conclusion (in other words, the problem is "how to invest one billion transistors on a single chip").

---

*For other possible goals, see *IEEE Computer*, September 1997. (Special Issue on how to use one billion transistors on a single chip).

[†]The essence of the work by Hammond et al. (1997) (see References list at end of book preceding Appendix A) is described here briefly. Researchers at Stanford University [Hammond et al. 1997] have conducted a simulation study to compare three different single-chip architectures of the same transistor count (one billion): (1) superscalar, (2) simultaneous multithreading, and (3) shared memory multiprocessor. They looked both into the state-of-the-art compiler technology and the state-of-the-art semiconductor technology. When the minimum feature size decreases, the transistor gate delay decreases linearly, while the wire delay stays nearly constant or increases. Consequently, as the feature size improves, on-chip wire delays improve 2–4 times more slowly than the gate delay. This means that resources on a processor chip that must communicate with each other must be physically close to each other. For this reason, the simulation study has shown (as was expected) that the shared memory multiprocessor architecture exhibits (on average) the best performance, for the same overall transistor count. In addition, because of its modular structure, the complexity of the shared memory multiprocessor architecture is much easier to manage at design time and test time.

The internal appendixes of this book (at the end of the book) cover the details of one important DSM concept [reflective memory system (RMS)], as well as the details of the tools that can be used to evaluate new architectural ideas, or to characterize the applications of interest.

Appendix D of this book (on the WWW) is an "external" appendix listing the microprocessor and multimicroprocessor-based designs of the author himself, and about the lessons that he has learned through his own professional survival process, which has lasted for about two decades now; concepts from microprocessor and multimicroprocessor boards of the past represent potential solutions for the microprocessor and multimicroprocessor chips of the future, and (which is more important) represent the ground for the authors belief that an ultimate goal is to have an entire distributed shared memory on a single chip, together with numerous specialized accelerators.

At first, distributed shared memory on a single chip may sound as a contradiction; however, it is not. As the dimensions of chips become larger, their full utilization can be obtained only with multimicroprocessor architectures. After the number of microprocessors reaches 16, the SMP architecture is no longer a viable solution since bus becomes a bottleneck; consequently, designers will be forced, in a number of cases, to move to the distributed shared memory paradigm (implemented in hardware, or partially in hardware and partially in software).

In this book, the issues of importance for current onboard microprocessor and multimicroprocessor-based designs, as well as for future on-chip microprocessor and multimicroprocessor designs, have been divided into eight different topics. The first one is about the general microprocessor architecture, and the remaining seven are about seven different problem areas of importance for the "ultimate goal": *distributed shared memory on a single chip, together with numerous specialized accelerators.*

After long discussions with the most experienced colleagues (see the list in the acknowledgment section), and the most enthusiastic students (they always have excellent suggestions), the major topics have been selected, as follows: (1) microprocessor systems on a chip, (2) cache and cache hierarchy, (3) instruction-level parallelism, (4) prediction strategies, (5) input/output bottleneck, (6) multithreaded processing, (7) shared memory multiprocessing systems, and (8) distributed shared memory systems.

Topics related to uniprocessing are of importance for microprocessor based designs of today and microprocessor on-chip designs of immediate future. Topics related to multiprocessing are of importance for multimicroprocessor-based designs of today and multimicroprocessor on-chip designs of the not-so-immediate future.

As already indicated, the author is one of the believers in the prediction that future on-chip machines, even if not of the multimicroprocessor or multimicrocomputer type, will include strong support for multiprocessing (single logical address space) and multicomputing (multiple logical address spaces). Consequently, as far as multiprocessing and multicomputing are concerned, only the issues of importance for future on-chip machines have been selected.

Efficient implementation of a sophisticated machine implies numerous efforts in domains like process technology, system design (e.g., bus design, system integrity checking, fault tolerance), and compiler design. In other words, the ultimate goal of this book cannot be reached without efforts in process technology, computer design, and system software. However, these issues are outside the scope of this book.

This book also includes a prologue section, which explains the roots of the idea behind it: combining synergistically the general body of knowledge and the particular experiences of an individual who has survived several pioneering design efforts of which some were relatively successful commercially.

As already indicated, this book includes both internal (Appendixes A–C) and external (Appendix D) appendixes with case studies. Internal appendixes are a part of this book, and are not further discussed here. External appendixes are available on the World Wide Web (WWW), and are briefly discussed here (only the initial three case studies, developed before this book was published; the list of case studies in the external appendix (D) is expected to grow over time).

The author was deeply engaged in all designs presented in the external appendix. Each project, in the field which is the subject of this book, includes three major activities: (1) envisioning of the strategy (project directions and milestones), (2) consulting on the tactics (product architecture and organization), and (3) engaging in the battle (design and almost exhaustive testing at all logical levels, until the product is ready for production).

The first case study on the WWW is a multimicroprocessor implementation of a data modem receiver for high-frequency (HF) radio. This design has often been quoted as the world's first multimicroprocessor based high frequency data modem. The work was done in 1970s; however, the interest in the results reincarnated in both the 1980s (due to technology impacts that enabled miniaturization) and in 1990s (due to application impacts of wireless communications). The author, absolutely alone, took all three activities (roles) defined above (one technician only helped with wirewrapping, using the list prepared by the author), and brought the prototype to a performance success (the HF modem receiver provided better performance on a real HF medium, compared to the chosen competitor product), and to a market success (after the preparation for production was done by others; wirewrap boards and older-date components were turned, by others, into the printed-circuit boards and newer-date components) in less than two years (possible only with the enthusiasm of a novice). See the references in the reference section, as a pointer to details (these references are not the earliest ones, but the ones that convey most information of interest for this book).

The second case study on the WWW is on a multimicroprocessor implementation of a GaAs systolic array for Gram–Schmidt orthogonalization (GSO). This design has been often quoted as the world's first GaAs systolic array. The work was done in the 1980s; the interest in the results did not reincarnate in the 1990s. The author took only the first two roles; the third one was taken by the others (see the acknowledgment section), but never really completed, since the project was canceled before its full completion, due to enormous cost (total of 8192 microprocessor nodes, each one running at the speed of 200 MHz). See the reference in the reference section, as a pointer to details (these references are not the earliest ones, but the ones that convey most information of interest for this book).

The third case study is on the implementation of a board (and the preceding research) which enables a personal computer (PC) to become a node in distributed shared memory (DSM) multiprocessor of the reflective memory system (RMS) type. This design has been often quoted as the world's first DSM plugin board for PC technology (some efforts with larger visibility came later; one of them, with probably the highest visibility [Gillett 1996], as an indirect consequence of this one). The work was done in the 1990s. The author took only the first role and was responsible for the project (details were taken care of by graduate students); fortunately, the project was completed successfully (and what is more important for a professor, papers were published with timestamps prior to those of the competition). See the references in the reference section, as a pointer to details (these references are not the earliest ones, but the ones that convey most information of interest for this book).

All three case studies have been presented with enough details, so the interested readers (typically undergraduate students) can redesign the same product using a state-of-the-art technology. Throughout the book, the concepts or ideas and lessons or experiences are in the foreground; the technology characteristics and implementation details are in the background, and can be modified (updated) by the reader, if so desired. This book, V. Milutinović, *Surviving the Design of Microprocessor and Multimicroprocessor Systems: Lessons Learned*, is nicely complemented with other books of the same author. One of them is V. Milutinović, *Surviving the Design of a 200 MHz RISC Microprocessor: Lessons Learned*, IEEE Computer Society Press, Los Alamitos, CA, 1997. These two books together (in various forms) have been used for about a decade now, by the author himself, as the coursework material for two undergraduate courses that he has taught at numerous universities worldwide. Also, there are three books on the more advanced topics, that have been used in graduate teaching on the follow-up subjects:

Protić, J., Tomašević, M., Milutinović, V., *Tutorial on Distributed Shared Memory: Concepts and Systems*, IEEE Computer Society Press, Los Alamitos, CA, 1998.

Tartalja, I., Milutinović, V., *Tutorial on Cache Consistency Problem in Shared Memory Multiprocessors: Software Solutions*, IEEE Computer Society Press, Los Alamitos, CA, 1997.

Tomašević, M., Milutinović, V., Tutorial on Cache Coherence Problem in *Shared Memory Multiprocessors: Hardware Solutions*, IEEE Computer Society Press, Los Alamitos, CA, 1993.

This book covers only the issues that are, in the opinion of the author, of strong interest for future design of microprocessors and multimicroprocessors on the chip, or the issues that have impacted his opinion about future trends in microprocessor and multimicroprocessor design.

All issues presented here have been treated selectively, with more attention paid to topics that are believed to be of more importance. This explains the difference in the breadth and depth of coverage throughout the book.

Also, the selected issues have been treated at the various levels of detail. This was done intentionally, in order to create room for creativeness of the students. Typical homework requires that the missing details be completed, and the inventiveness with which the students fulfill the requirement is sometimes unbelievable (the best student projects can be found on the author's Web page). Consequently, one of the major educational goals of this book, if not the major one, is to help create the inventiveness among the students. Suggestions on how to achieve this goal more efficiently are more than welcome.

Finally, a few words on the educational approach used in this book. It is well known that "one picture is worth of one thousand words." Consequently, the stress in this book has been placed on the utilization of modern presentation methodologies in general, as well as figures and figure captions, in particular. All necessary explanations have been put into the figures and figure captions. The main body of the text has been kept to its minimum—only the issues of interest for the global understanding of the topic and/or the thoughts on experiences gained and lessons learned. Consequently, students claim that this book is fast to read and easy to comprehend.

Important prerequisites for reading this book are Flynn (1995), Hennessy and Patterson (1996), and Patterson (1994)—the three most important textbooks in the fields of computer architecture, organization, and design.

In conclusion, this book teaches the concepts of importance for putting together a DSM system on a single chip. It is well suited for graduate and advanced undergraduate students, and as already mentioned, has been used as such at numerous universities worldwide. It is also well suited for practitioners from industry (for innovation of their knowledge) and for managers from industry (for a better understanding of the future trends).

VELJKO MILUTINOVIĆ

*University of Belgrade*
*vm@etf.bg.ac.yu*
`http://galeb.etf.bg.ac.yu/~vm/`

## ACKNOWLEDGMENTS

This book would not be possible without the help of numerous individuals; some of them helped the author to master the knowledge and to gather the experiences necessary to write this book; others have helped to create the structure or to improve the details. Since the book of this sort would not be possible if the author did not take place in the three large projects defined in the preface, the acknowledgment will start from those involved in the same projects, directly or indirectly.

In relation to the first project (MISD for DFT), the author is thankful to professor Georgije Lukatela from whom he has learned a lot, and also to his colleagues who worked on the similar problems in the same or other companies (Radenko Paunovic, Slobodan Nedic, Milenko Ostojic, David Monsen, Philip Leifer, and John Harris).

In relation to the second project (SIMD for GSO), the author is thankful to professor Jose Fortes who had an important role in the project, and also to his colleagues who were involved with the project in the same team or within the sponsor team (David Fura, Gihjung Jung, Salim Lakhani, Ronald Andrews, Wayne Moyers, and Walter Helbig).

In relation to the third project (MIMD for RMS), the author is thankful to professor Milo Tomašević who has contributed significantly, and also to colleagues who were involved in the same project, within his own team or within the sponsor team (Savo Savic, Milan Jovanovic, Aleksandra Grujic, Ziya Aral, Ilya Gertner, and Mark Natale).

The list of colleagues and professors who have helped with the overall structure and contents of the book through formal or informal discussions and direct or indirect advice, on one or more elements of the book, during the seminars presented at their universities or during friendly chats between conference sessions, or have influenced the author in other ways, includes but is not limited to the following individuals: Tihomir Aleksic, Vidojko Ciric, Jack Dennis, Hank Dietz, Jovan Djordjevic, Jozo Dujmovic, Milos Ercegovac, Michael Flynn, Borko Furht, Jean-Luc Gaudiot, Anoop Gupta, Reiner Hartenstein, John Hennessy, Kai Hwang, Liviu Iftode, Emil Jovanov, Zoran Jovanovic, Borivoj Lazic, Bozidar Levi, Kai Li, Oskar Mencer, Aleksandar Milenković, Srdjan Mitrovic, Trevor Mudge, Vojin Oklobdzija, Milutin Ostojic, Yale Patt, Branislava Perunicic, Antonio Prete, Jelica Protić, Bozidar Radenkovic, Jasna Ristic, Eduardo Sanchez, Richard Schwartz, H. J. Siegel,

Alan Jay Smith, Ljubisa Stankovic, Dusan Starcevic, Per Stenstrom, Daniel Tabak, Igor Tartalja, Jacques Tiberghien, Mateo Valero, Dusan Velasevic, and Dejan Zivkovic.

The list also includes numerous individuals from industry worldwide who have provided support or have helped clarify details on a number of issues of importance: Tom Brumett, Roger Denton, Charles Gimarc, Gordana Hadzic, Hans Hilbrink, Lee Hoevel, Petar Kocovic, Oleg Panfilov, Lazar Radicevic, Charles Rose, Djordje Rosic, Gad Sheaffer, Mark Tremblay, Helmut Weber, and Maurice Wilkes.

Students have helped a lot to maximize the overall educational quality of the book. Several generations of students have used the book before it went to press. Their comments and suggestions were of extreme value. Those who deserve special credit are listed here: Aleksandar Bakic, Jovanka Ciric, Dragana Cvetkovic, Goran Davidovic, Zoran Dimitrijevic, Dusan Dingaric, Vladan Dugaric, Danko Djuric, Ilija Ekmecic, Damir Horvat, Zoran Horvat, Igor Ikodinovic, Milan Jovanovic, Predrag Knezevic, Dejan Krivokuca, Dusko Krsmanovic, Petar Lazarevic, Davor Magdic, Darko Marinov, Gvozden Marinkovic, Boris Markovic, Predrag Markovic, Marjan Mihanovic, Jelena Mirkovic, Milan Milicevic, Nenad Nikolic, Milja Pesic, Dejan Petkovic, Zvezdan Petkovic, Milena Petrovic, Milos Prvulovic, Bozidar Radunovic, Dejan Raskovic, Nenad Ristic, Andrej Skorc, Ivan Sokic, Miodrag Stefanovic, Milan Trajkovic, Miljan Vuletic, Slavisa Zigic, and Srdjan Zgonjanin.

VELJKO MILUTINOVIĆ

*vm@etf. bg.ac.yu*
```
http:// galeb.etf.bg.ac.yu/~ vm/
```

| | |
|---|---|
| ALU | arithmetic logic unit |
| AST | active sector table |
| AURC | automatic update release consistency |
| BHR | branch history register |
| BHSR | branch history shift register |
| BHT | branch history table |
| BIPS | billion instructions per second |
| BIST | built-in self test |
| BPB | branch prediction buffer |
| BPS | branch prediction strategy |
| BPT | branch prediction table |
| BRL | bus, ring, or LAN |
| BTB | branch target buffer |
| CU | compilation unit |
| DCS | distributed computer system |
| DDA | data decoupled architecture |
| DHLF | dynamic history length fitting |
| DIB | dual independent bus |
| DMM | data memory module |
| DSM | distributed shared memory |
| EIDE | enhanced integrated drive electronics |
| EPIC | explicitly parallel instruction computing |
| FFT | fast Fourier transform |
| FIFO | first in, first out |
| FPGA | field programmable gate array |
| FPU | floating-point unit |
| FRC | functional redundancy checking |
| GBH | global branch history |
| GMC | grid, mesh, or crossbar |
| HEP | heterogeneous element processor |
| HLRC | home-based LRC |
| HPI | host-port interface |
| ICN | interconnection network |
| IDU | instruction decoding unit |
| IFU | instruction fetch unit |
| ILP | instruction-level parallelism |

| | |
|---|---|
| IO | in-order |
| IOC | input/output controller |
| IPC | instructions per clock |
| ISA | instruction-level architecture |
| LAP | local access path |
| LRC | lazy release consistency |
| LSU | load/store unit |
| LVC | local variable cache |
| MAF | miss address file |
| MCM | memory consistency model |
| MCS | multicomputer system |
| MESI | Modified/Exclusive/Shared/Invalid |
| MH-BPS | multihybrid BPS |
| MIMD | multiple instruction, multiple data |
| MLL | machine-level language |
| MLP | machine-level parallelism |
| MMM | mirror memory multiprocessor |
| MMU | memory management unit |
| MOESI | Modified/Owned/Exclusive/Shared/Invalid |
| MOESIA | MOESI-Advanced |
| MOSE | Modified/Owned/Shared/Exclusive |
| MOSI | Modified/Owned/Shared/Invalid |
| MRMW | multiple reader, multiple writer |
| MRSW | multiple reader, single writer |
| MSHR | miss status handling registers |
| NOPC | network of personal computers |
| NOW | network of workstations |
| NTS | nontemporal streaming |
| OC | optimizing compiler |
| OLTP | online transaction processing |
| OoO | out-of-order |
| OOP | object-oriented programming |
| OS | operating system |
| PEM | processing element module |
| PGA | pin grid array |
| PHT | pattern history table |
| PSC | predictor selection counter |
| PSU | packet switching unit |
| PTC | parity testing and checking |
| RAC | remote access cache |
| RAID | redundant array of independent disks |
| RMS | reflective memory system |
| ROP | RISC operation |
| RST | reduced state transition |
| SCI | Scalable Coherent Interface |

| | |
|---|---|
| SEC | single edge contact |
| SEE | selective eager execution |
| SFP | spatial footprint predictor |
| SHT | spatial footprint history table |
| SI | Shared/Invalid |
| SIMD | single instruction, multiple data |
| SISD | single instruction, single data |
| SLH | spatial locality hint |
| SMP | shared memory processing |
| SMT | simultaneous multithreading |
| SP | superpipelined |
| SRSW | single reader, single writer |
| SS | superscalar |
| SSE | streamlining SIMD execution |
| SSMT | simultaneous subordinate microthreading |
| SVM | share virtual memory |
| TLB | translation look-aside buffer |
| TMI | transition module interface |
| UMA | uniform memory access |
| UPCR | useless private copies removal |
| VHP | value history pattern |
| VHT | value history table |
| VLIW | very long instruction word |
| WI | write-invalidate |
| WIN | word invalidate |
| WTI | writethrough invalidate |
| WU | write-update |

# PART I

# IMPORTANT FACTS

As already indicated, this author believes that one efficient solution for the "one billion transistor chip" of the future is a complete distributed shared memory machine on a single chip, together with a number of specialized on-chip accelerators.

The eight sections that follow cover (1) essential facts about the current microprocessor architectures and (2) the seven major problem areas, to be resolved on the way to the final goal stated above.

# CHAPTER 1

# Microprocessor Systems

This chapter includes two sections. The section on basic issues covers the past trends in microprocessor technology and characteristics of some contemporary microprocessor machines from the workstation market, namely, Intel Pentium, Pentium MMX, Pentium Pro, and Pentium II/III, as the main driving forces of today's personal computing market. The section on advanced issues covers future trends in state of the art microprocessors.

## 1.1. BASIC ISSUES

It is interesting to compare current Intel complex instruction set computing (CISC)-type products (which drive the personal computer market today) with the reduced instruction set computing (RISC) products of Intel and of the other companies. At the present time, DEC (Digital Equipment Corporation) Alpha family includes three representatives: 21064, 21164, and 21264. The PowerPC family was initially devised by IBM, Motorola, and Apple and includes a series of microprocessors starting at PPC 601 (IBM name) or MPC 601 (Motorola name); the follow-up projects have been referred to as 602, 603, 604, 620, and 750. The Sun SPARC family follows two lines: V.8 (32-bit machines) and V.9 (64-bit machines). The MIPS Rx000 series started with R2000/3000, followed by R4000, R6000, R8000, R10000, and R12000. Intel has introduced two different RISC machines: i960 and i860 (Pentium II has a number of RISC features included at the microarchitecture level). The traditional Motorola RISC line includes MC88100 and MC88110. The Hewlett-Packard series of RISC machines is referred to as PA (Precision Architecture).

All comparative data both for modern microprocessors and for microprocessors that are sitting on our desks for years now have been presented in the form of tables (manufacturer names and Internet Universal Resource Locators (URLs)* are given in Table 1.1). One has to be aware of the past, before starting to look into the future.

Tables 1.2–1.6 include only data for microprocessors declared by their manufacturers as being (predominantly or partially) of the RISC type. Conse-

---

*Note that URLs of manufacturers are subject to change.

**TABLE 1.1. Microprocessor Family Home Pages**

| Company | Internet URL[a] of Microprocessor Family Home Page |
|---|---|
| IBM | `http://www.chips.ibm.com/products/powerpc/` |
| Motorola | `http://www.mot.com/SPS/PowerPC` |
| DEC | `http://www.europe.digital.com/semiconductor/alpha/alpha.htm` |
| Sun | `http://www.sun.com/microelectronics/products/microproc.html` |
| MIPS | `http://www.mips.com/products/index.html` |
| Hewlett-Packard | `http://hpcc920.external.hp.com/computing/framed/technology/micropro` |
| AMD | `http://www.amd.com/K6` |
| Intel | `http://www.intel.com/pentiumii/home.htm` |

*Source:* Prvulovic (1997).

[a]Listed URL addresses and their contents do change over time.

**TABLE 1.2. Microprocessor Technology**[a]

| Microprocessor | Company | Technology | Transistors | Frequency (mHz) | Package |
|---|---|---|---|---|---|
| PowerPC 601 | IBM, Motorola | $0.6\,\mu$m, 4 L, CMOS | 2,800,000 | 80 | 304 PGA |
| PowerPC 604e | IBM, Motorola | $0.35\,\mu$m, 5 L, CMOS | 5,100,000 | 225 | 255 BGA |
| PowerPC 620[b] | IBM, Motorola | $0.35\,\mu$m, 4 L, CMOS | 7,000,000 | 200 | 625 BGA |
| PowerPC 750[b] | IBM, Motorola | $0.29\,\mu$m, 5 L, CMOS | 6,350,000 | 300 | 360 CBGA |
| Alpha 21064[b] | DEC | $0.7\,\mu$m, 3 L, CMOS | 1,680,000 | 300 | 431 PGA |
| Alpha 21164[b] | DEC | $0.35\,\mu$m, 4 L, CMOS | 9,300,000 | 500 | 499 PGA |
| Alpha 21264[b] | DEC | $0.35\,\mu$m, 6 L, CMOS | 15,200,000 | 500 | 588 PGA |
| SuperSPARC | Sun Microelectronics | $0.8\,\mu$m, 3 L, CMOS | 3,100,000 | 60 | 293 PGA |
| UltraSPARC-I[b] | Sun Microelectronics | $0.4\,\mu$m, 4 L, CMOS | 5,200,000 | 200 | 521 BGA |
| UltraSPARC-II[b] | Sun Microelectronics | $0.35\,\mu$m, 5 L, CMOS | 5,400,000 | 250 | 521 BGA |
| R4400[b] | MIPS Technologies | $0.6\,\mu$m, 2 L, CMOS | 2,200,000 | 150 | 447 PGA |
| R10000[b] | MIPS Technologies | $0.35\,\mu$m, 4 L, CMOS | 6,700,000 | 200 | 599 LGA |
| PA7100 | Hewlett-Packard | $0.8\,\mu$m, 3 L, CMOS | 850,000 | 100 | 504 PGA |
| PA8000[b] | Hewlett-Packard | $0.35\,\mu$m, 5 L, CMOS | 3,800,000 | 180 | 1085 LGA |
| PA8500[b] | Hewlett-Packard | $0.25\,\mu$m, ? L, CMOS | > 120,000,000 | 250 | ? ? |
| MC88110 | Motorola | $0.8\,\mu$m, 3 L, CMOS | 1,300,000 | 50 | 361 CBGA |
| AMD K6 | AMD | $0.35\,\mu$m, 5 L, CMOS | 8,800,000 | 233 | 321 PGA |
| i860 XP | Intel | $0.8\,\mu$m, 3 L, CHMOS | 2,550,000 | 50 | 262 PGA |
| Pentium II | Intel | $0.35\,\mu$m, ? L, CMOS | 7,500,000 | 300 | 242 SEC |

*Sources:* Prvulovic (1997) and Stojanovic (1995).

[a]*Key:* $x$L—$x$-layer metal ($x = 2, 3, 4, 5, 6$); PGA—pin grid array; CMOS—complementary metal oxide semiconductor; BGA—ball grid array; CBGA —ceramic ball grid array; LGA—land grid array; SEC—single edge contact; DEC —Digital Equipment Corporation; AMD —Advanced Micro Devices.

[b]64-bit microprocessors, all others are 32-bit microprocessors.

**TABLE 1.3. Microprocessor Architecture**[a]

| Microprocessor | IU Registers | FPU Registers | VA | PA | EC Dbus | SYS Dbus |
|---|---|---|---|---|---|---|
| PowerPC 601 | $32 \times 32$ | $32 \times 64$ | 52 | 32 | None | 64 |
| PowerPC 604e | $32 \times 32 + RB(12)$ | $32 \times 64 + RB(8)$ | 52 | 32 | None | 64 |
| PowerPC 620 | $32 \times 64 + RB(8)$ | $32 \times 64 + RB(8)$ | 80 | 40 | 128 | 128 |
| PowerPC 750 | $32 \times 32 + RB(12)$ | $32 \times 64 + RB(6)$ | 52 | 32? | 64, 128 | 64 |
| Alpha 21064 | $32 \times 64$ | $32 \times 64$ | 43 | 34 | 128 | 128 |
| Alpha 21164 | $32 \times 64 + RB(8)$ | $32 \times 64$ | 43 | 40 | 128 | 128 |
| Alpha 21264 | $32 \times 64 + RB(48)$ | $32 \times 64 + RB(40)$ | ? | 44 | 128 | 64 |
| SuperSPARC | $136 \times 32$ | $32 \times 32^{c}$ | 32 | 36 | None | 64 |
| UltraSPARC-I | $136 \times 64$ | $32 \times 64$ | 44 | 36 | 128 | 128 |
| UltraSPARC-II | $136 \times 64$ | $32 \times 64$ | 44 | 36 | 128 | 128 |
| R4400 | $32 \times 64$ | $32 \times 64$ | 40 | 36 | 128 | 64 |
| R10000 | $32 \times 64 + RB(32)$ | $32 \times 64 + RB(32)$ | 44 | 40 | 128 | 64 |
| PA7100 | $32 \times 32$ | $32 \times 64$ | 64 | 32 | ? | ? |
| PA8000 | $32 \times 64 + RB(56)$ | $32 \times 64$ | 48 | 40 | 64 | 64 |
| PA8500 | $32 \times 64 + RB(56)$ | $32 \times 64$ | 48 | 40 | 64 | 64 |
| MC88110 | $32 \times 32$ | $32 \times 80$ | 32 | 32 | None | ? |
| AMD K6 | $8 \times 32 + RB(40)$ | $8 \times 80$ | 48 | 32 | 64 | 64 |
| i860 XP | $32 \times 32$ | $32 \times 32^{c}$ | 32 | 32 | None | ? |
| Pentium II | ? | $8 \times 80$ | 48 | 36 | 64 | 64 |

*Sources:* Prvulovic (1997) and Stojanovic (1995).

[a]*Key*: IU—integer unit; FPU—floating-point unit; VA—virtual address (bits); PA—physical address (bits); EC Dbus—external cache databus width (bits); SYS Dbus—system bus width (bits); RB—rename buffer (size expressed in the number of registers).

[b]The number of integer unit registers shows the impact of initial RISC research, on the designers of a specific microprocessor. Only Sun Microsystems have opted for the extremely large register file, which is a sign of a direct or indirect impact of Berkeley RISC research. In the other cases, smaller register files indicate the preferences corresponding directly or indirectly to the Stanford MIPS research.

[c]Can also be used as a $16 \times 64$ register file.

quently, these tables do not include data on Pentium and Pentium Pro. However, Pentium and Pentium Pro are presented in greater detail later in the text. Therefore, the tables can serve as a basis for comparison of Pentium and Pentium Pro with other relevant products.

Table 1.2 compares the chip technologies. Table 1.3 compares selected architectural issues. Table 1.4 compares the instruction level parallelism and the count of the related processing units. Table 1.5 is related to cache memory, and Table 1.6 includes miscellaneous issues, like translation look-aside buffer (TLB) structures and branch prediction solutions.

The main reason for including Table 1.2 is to give readers more insight into the major technological aspects of the design and their impact on the transistor count and clock speed. When the feature size goes down, the number of transistors goes up, at best quadratically, if the same or similar speed is to be maintained. That is the case with DEC Alpha and HP

**TABLE 1.4. Microprocessor ILP Features**[a]

| Microprocessor | ILP Issue | LSU Units | IU Units | FPU Units | GU Units |
|---|---|---|---|---|---|
| PowerPC 601 | 3 | 1 | 1 | 1 | 0 |
| PowerPC 604e | 4 | 1 | 3 | 1 | 0 |
| PowerPC 620 | 4 | 1 | 3 | 1 | 0 |
| PowerPC 750 | 4 | 1 | 2 | 1 | 0 |
| Alpha 21064 | 2 | 1 | 1 | 1 | 0 |
| Alpha 21164 | 4 | 1 | 2 | 2 | 0 |
| Alpha 21264 | 4 | 1 | 4 | 2 | 0 |
| SuperSPARC | 3 | 0 | 2 | 2 | 0 |
| UltraSPARC-I | 4 | 1 | 4 | 3 | 2 |
| UltraSPARC-II | 4 | 1 | 4 | 3 | 2 |
| R4400 | 1[b] | 0 | 1 | 1 | 0 |
| R10000 | 4 | 1 | 2 | 2 | 0 |
| PA7100 | 2 | 1 | 1 | 3 | 0 |
| PA8000 | 4 | 2 | 2 | 4 | 0 |
| PA8500 | 4 | 2 | 2 | 4 | 0 |
| MC88110 | 2 | 1 | 3 | 3 | 2 |
| AMD K6 | 6[c] | 2 | 2 | 1 | 1[d] |
| i860 XP | 2 | 1 | 1 | 2 | 1 |
| Pentium II | 5[c] | ? | ? | ? | ? |

*Sources:* Prvulovic (1997) and Stojanovic (1995).

[a]*Key*: ILP—instruction-level parallelism; LSU—load/store or address calculation unit; IU—integer unit; FPU—floating point unit; GU—graphics unit.

[b]Superpipelined.

[c]RISC instructions, one or more of them are needed to emulate an $80 \times 86$ instruction.

[d]MMX (multimedia extensions) unit.

Precision Architecture microprocessors. However, if the goal is to increase the speed dramatically, the transistor count increase will be affected. That is the case with Sun SPARC (Sun Microsystems Scalable Process Architecture) microprocessors. Until recently, pin count was one of the major bottlenecks of microprocessor technology, but recently the pin count problem has improved considerably. Another dramatic improvement is noticed in the number of metal layers. The greater the number of metal layers, the larger the transistor count increase when the feature size decreases. Consequently, the most dramatic transistor count increases are in the cases characterized (among other things) by a dramatic improvement in the number of metal layers (e.g., DEC Alpha 21264 and HP Precision Architecture 8500).

The main reason for including Table 1.3 is to give readers more insight into the number and type of on-chip resources for register storage and bus communications. The number of integer unit registers can be either medium (if flat organizations are used) or relatively large (if structured organizations are used). Flat organization means that all registers of the CPU are visible to the currently running program. Structured organizations mean that registers are organized into specific structures (often called "windows"), with only one

**TABLE 1.5. Microprocessor Cache Memory**[a]

| Microprocessor | L1 Icache, kB | L1 Dcache, kB | L2 cache, kB |
|---|---|---|---|
| PowerPC 601 | 32, 8WSA, UNI | | — |
| PowerPC 604e | 32, 4WSA | 32, 4WSA | — |
| PowerPC 620 | 32, 8WSA | 32, 8WSA | —[b] |
| PowerPC 750 | 32, 8WSA | 32, 8WSA | —[b] |
| Alpha 21064 | 8, DIR | 8, DIR | —[b] |
| Alpha 21164 | 8, DIR | 8, DIR | 96, 3WSA[b] |
| Alpha 21264 | 64, 2WSA | 64, DIR | —[b] |
| SuperSPARC | 20, 5WSA | 16, 4WSA | — |
| UltraSPARC-I | 16, 2WSA | 16, DIR | —[b] |
| UltraSPARC-II | 16, 2WSA | 16, DIR | —[b] |
| R4400 | 16, DIR | 16, DIR | —[b] |
| R10000 | 32, 2WSA | 32, 2WSA | —[b] |
| PA7100 | 0 | | —[c] |
| PA8000 | 0 | | —[c] |
| PA8500 | 512, 4WSA | 1024, 4WSA | — |
| MC88110 | 8, 2WSA | 8, 2WSA | — |
| AMD K6 | 32, 2WSA | 32, 2WSA | —[b] |
| i860 XP | 16, 4WSA | 16, 4WSA | — |
| Pentium II | 16, ? | 16, ? | 512, ?[d] |

*Sources:* Prvulovic (1997) and Stojanovic (1995).

[a]*Key:* Icache—on-chip instruction cache; kB—kilobytes; Dcache—on-chip data cache; L2 cache—on-chip L2 cache; DIR—direct mapped; $x$WSA—$x$-way set associative ($x = 2, 3, 4, 5, 8$); UNI—unified L1 instruction and data cache.

[b]On-chip cache controller for external L2 cache.

[c]On-chip cache controller for external L1 cache.

[d]L2 cache is in the same package, but on a different silicon die.

*Comment:* It is only an illusion that early HP microprocessors were lagging behind the others, as far as the on-chip cache memory support is concerned; they were using the so-called on-chip assist cache, which can be treated as a zero-level cache memory, and works on slightly different principles, compared to the traditional cache (as will be explained later). On the other hand, DEC was the first to place both level 1 and level 2 caches on the same chip with the CPU.

structure element visible to the currently running program. All advanced microprocessors include a rename buffer. Rename buffers enable runtime resolution of data dependencies. The larger the rename buffer, the better the execution speed of programs. Consequently, the newer microprocessors feature considerable increases in the sizes of their rename buffers. The contribution of rename buffers in the floating-point execution stream is not as dramatic, so they are less often found in the floating-point units. System buses (and external cache data buses, if implemented) are 64 or 128 bits wide.

Actually, Table 1.2 shows the strong and the not-so-strong sides of different manufacturers, as well as their basic development strategies. Some manufacturers generate large transistor count chips that are not very fast, and vice versa. Also, the pin count of chip packages differs, as do the number

**TABLE 1.6. Miscellaneous Microprocessor Features**[a]

| Microprocessor | ITLB | DTLB | BPS |
|---|---|---|---|
| PowerPC 601 | 256, 2WSA, UNI | | —[b] |
| PowerPC 604e | 128, 2WSA | 128, 2WSA | $512 \times 2BC$ |
| PowerPC 620 | 128, 2WSA | 128, 2WSA | $2048 \times 2BC$ |
| PowerPC 750 | 128, 2WSA | 128, 2WSA | $512 \times 2BC$ |
| Alpha 21064 | 12 | 32 | $4096 \times 2BC$ |
| Alpha 211644 | 8 ASSOC | 64 ASSOC | $ICS \times 2BC$ |
| Alpha 21264 | 128 ASSOC | 128 ASSOC | 2LMH, $32 \times RAS$ |
| SuperSPARC | 64 ASSOC, UNI | | ? |
| UltraSPARC-I | 64 ASSOC | 64 ASSOC | $ICS \times 2BC$ |
| UltraSPARC-II | 64 ASSOC | 64 ASSOC | $ICS \times 2BC$ |
| R4400 | 48 ASSOC | 48 ASSOC | — |
| R10000 | 64 ASSOC | 64 ASSOC | $512 \times 2BC$ |
| PA7100 | 16 | 120 | ? |
| PA8000 | 4 | 96 | $256 \times 3BSR$ |
| PA8500 | 160, UNI | | $> 256 \times 2BC$ |
| MC88110 | 40 | 40 | ? |
| AMD K6 | 64 | 64 | $8192 \times 2BC$, $16 \times RAS$ |
| i860 XP | 64, UNI | | ? |
| Pentium II | ? | ? | ? |

*Source:* Prvulovic (1997).

[a]*Key:* ITLB—translation look-aside buffer for code (entries); DTLB—translation look-aside buffer for data (entries); 2WSA—two-way set associative; ASSOC—fully associative; UNI—unified TLB for code and data; BPS—branch prediction strategy; 2BC—2-bit counter; 3BSR—3-bit shift register; RAS—return-address stack; 2LMH—two-level multihybrid (gshare for the last 12 branch outcomes and pshare for the last 10 branch outcomes); ICS—instruction cache size (2BC for every instruction in the instruction cache).

[b]Hinted instructions available for static branch prediction.

*Comment:* The great variety in TLB design numerics is a consequence of the fact that different manufacturers see differently the real benefits of having a TLB of a given size. Grouping of pages, in order to use one TLB entry for a number of pages, has been used by DEC and viewed as a viable price/performance tradeoff. Variable page size has been first used by MIPS Technologies machines.

of on-chip levels of metal or the minimal feature size. Note that the number of companies manufacturing general-purpose microprocessors is relatively small.

The main reason for including Table 1.4 is to give readers more insight into the ways in which the features related to instruction-level parallelism (ILP) are implemented in popular microprocessors. The issue width goes up to 6 (AMD K6 represents a sophisticated attempt to get maximum out of ILP). In order to be able to execute all fetched operations in parallel, in all scenarios of interest, a relatively large number of execution units are needed. On average, this number is bigger than the issue width. One exception to this is AMD K6 with a relatively small number of execution units, which are used efficiently, due to the sophisticated internal design of the AMD K6 [Shriver and Smith 1998].

From Table 1.4 one can see that the total number of units (integer, floating-point, and graphics) is always larger than or equal to the issue width. Intel and Motorola had a head start in the hardware acceleration of the graphics function, which is the strategy adopted later by the follow-up machines of most other manufacturers. Zero in the LSU column indicates no independent load/store units.

The main reason for including Table 1.5 is to give readers more insight into the cache memory design of popular microprocessors. L1 refers to level 1 cache memory, and L2 refers to level 2 cache memory. L1 is closer to the CPU and smaller than L2. L1 is typically smaller in capacity for one order of magnitude. Since it is smaller, it is faster, and fewer clock cycles are needed to access it. In commercial designs, typically the principle of inclusion is satisfied, which means that all data in L1 are also present in L2. Often, L1 is on the same chip as the CPU. In more recent microprocessors, both L1 and L2 are on the same chip as the CPU. DEC Alpha is the first microprocessor to include both L1 and L2 on the same chip as the CPU. In the case of Pentium II, L1 and L2 are on two different chips, but on the same package. L1 cache is typically divided into separate instruction and data caches. L2 typically includes both instructions and data in one cache.

The main reason for including Table 1.6 is to give readers more insight into the design of the TLB (translation look-aside buffer) and BPS (branch prediction strategies) in popular microprocessors. The number of TLB entries goes up to 256. In most microprocessors, there are separate TLBs for instructions and data. Exceptions are Intel i860, Sun SPARC, and Hewlett-Packard PA 8500. As far as BPS, typically only the less sophisticated techniques are used (2-bit counter and return address stack). The only exception is DEC Alpha 21264 (two-level multihybrid). Detailed explanations of these and more sophisticated techniques are given later in this book, in the section on branch prediction strategies.

The following sections give a closer look into the Intel Pentium, Pentium MMX, Pentium Pro, and Pentium II/III machines.

### 1.1.1. Pentium

The major characteristics of the Pentium processor include the features that make it different in comparison with the i486 processor. The Pentium processor is built out of 3.1 MTr (million transistors) using the Intel $0.8\text{-}\mu\text{m}$ BiCMOS silicon technology. It is packed into a 273-pin PGA (pin grid array) package, as indicated in Figure 1.1.

Power and ground pins are marked on Figure 1.1; They are the main reason for showing this figure (to emphasize the fact that signal departure from power and ground is a major issue). With more power and ground pins, the logic on the chip attains power and ground levels with less departure from the ideal.

Pentium pin functions are shown in Table 1.7. The major reason for showing pin functions is to emphasize the fact that, as far as the main pin

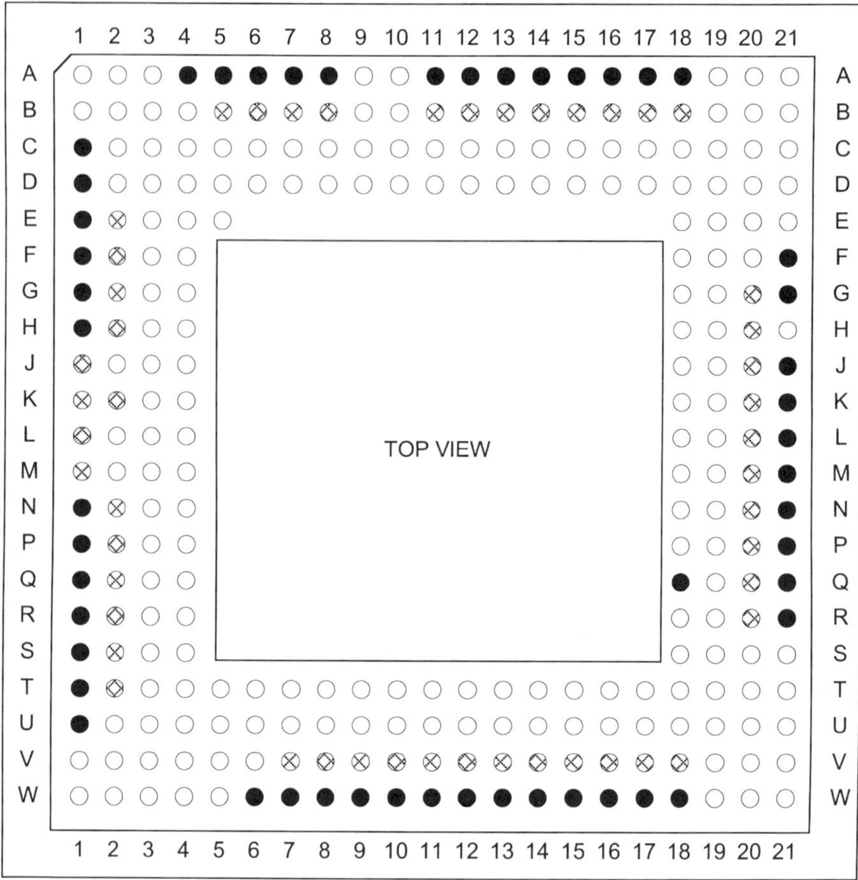



**Figure 1.1.** Pentium pin layout. (black dots—$V_{CC}$, crossed pattern dots—$V_{SS}$). [*Source:* Intel (1993).]

*Comment:* The total pin count is 273. As it is typical for packages with a relatively high pin count, a large percentage of pins is dedicated to power and ground (in the case of Pentium, 50 and 49, respectively). However, note that much larger packages are in use today, with DEC and HP among companies leading in the packaging technology.

functions are concerned, there is no major difference between Pentium and the earliest members of the x86 family.

Pentium is fully binary-compatible with previous Intel machines in the x86 family. Some of the above-mentioned enhancements are supported with new instructions. The memory management unit (MMU) is fully compatible with i486, while the floating-point unit (FPU) has been redesigned for better performance.

A block diagram of the Pentium processor is shown in Figure 1.2. The figure is important because it shows elements that are expected to be further

**TABLE 1.7. Pentium Pin Functions**[a]

| Function | Pins |
|---|---|
| Clock | CLK |
| Initialization | RESET, INIT |
| Address bus | A31−A3, BE7#−BE0# |
| Address mask | A20M# |
| Databus | D63−D0 |
| Address parity | AP, APCHK# |
| Data parity | DP7−DP0, PCHK#, PEN# |
| Internal parity error | IERR# |
| System error | BUSCHK# |
| Bus cycle definition | M/IO#, D/C#, W/R#, CACHE#, SCYC, LOCK# |
| Bus control | ADS#, BRDY, NA# |
| Page cacheability | PCD, PWT |
| Cache control | KEN#, WB/WT# |
| Cache snooping/consistency | AHOLD, EADS#, HIT#, HITM#, INV |
| Cache flush | FLUSH# |
| Write ordering | EWBE# |
| Bus arbitration | BOFF#, BREQ, HOLD, HLDA |
| Interrupts | INTR, NMI |
| Floating-point error reporting | FERR#, IGNNE# |
| System management mode | SMI#, SMIACT# |
| Functional redundancy checking | FRCMC# (IERR#) |
| TAP port | TCK, TMS, TDI, TDO, TRST# |
| Breakpoint/performance monitoring | PM0/BP0, PM1/BP1, BP32 |
| Execution tracing | BT3BT0, IU, IV, IBT |
| Probe mode | R/S#, PRDY |

*Source:* Intel (1993).

[a]Traditional pin functions are clock, initialization, addressing, data, bus control, bus arbitration, and interrupts. These or similar functions can be found all the way back to the earliest x86 machines. Pin functions like parity, error control, cache control, tracing, breakpoint, and performance monitoring can be found in immediate predecessors, in a somewhat reduced form.
[b]TAP—processor boundary scan.

explored in future microprocessors. For example, caches are the kernel of the later support for SMP and DSM. U and V pipelines are the kernel of the later support for ILP, and so on.

The core of the processor is the pipeline structure, which is shown in Figure 1.3 in comparison with the pipeline structure of the i486. A precise description of activities in each pipeline stage can be found in Intel (1993). As can be seen from Figure 1.3, Pentium is a superscalar processor (which is an important departure from i486). However, the depth of the Pentium pipeline has not changed (compared to i486).

Internal error detection is based on functional redundancy checking (FRC), built-in self-test (BIST), and parity testing and checking (PTC). Constructs for performance monitoring count occurrences of selected internal events and trace execution through the internal pipelines.

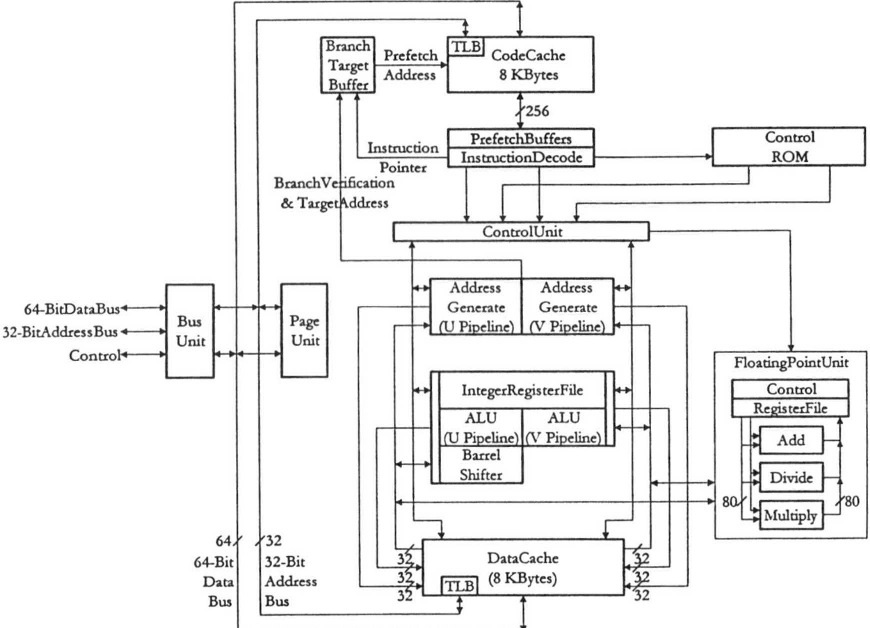


**Figure 1.2.** Pentium block diagram. [*Source:* Intel (1993).]
*Comment:* This block diagram sheds light on the superscaling of Pentium processor. Two pipelines (U and V) are responsible for the execution efficiency. Note that advanced implementations also include blocks such as MMX to accelerate multimedia code.

### 1.1.1.1. *Cache and Cache Hierarchy*

Internal cache organization follows the two-way set associative approach. Each of the two caches (data cache and instruction cache) includes 128 sets. Each set includes two entries of 32 bytes each (one set includes 64 bytes). This means that each of the two caches is 8 kB large. Both caches use the LRU replacement policy. One can add second-level caches off the chip, if needed by the application.

One bit has been assigned to control the cacheability on a page-by-page basis: PCD (1 = CachingDisabled; 0 = CachingEnabled). One bit has been assigned to control the write policy for the second level caches: PWT (1 = WriteThrough; 0 = WriteBack). The circuitry used to generate signals PCD and PWT is shown in Figure 1.4. The states of these 2 bits appear at the pins PWT and PCD during the memory access cycle. Signals PCD and KEN# are internally ANDed in order to control the cacheability on a cycle by cycle basis (see Fig. 1.4).

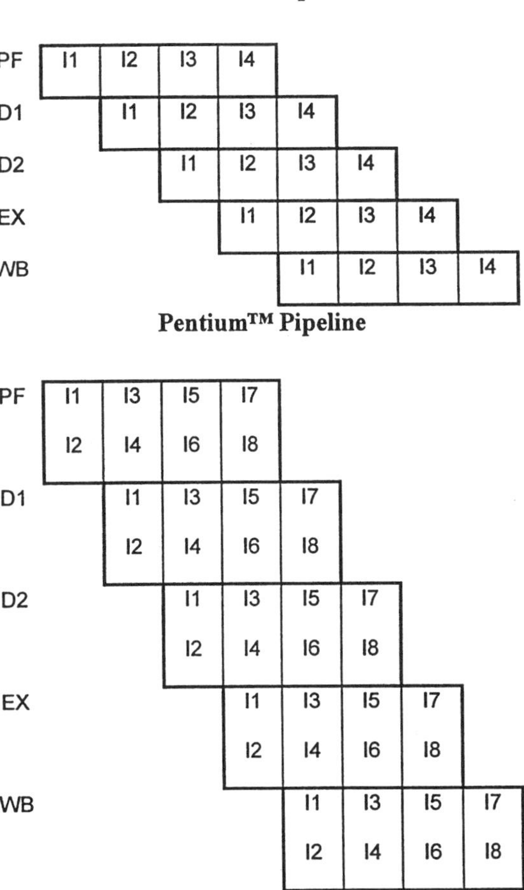


**Figure 1.3.** Intel 486 pipeline versus Pentium pipeline (PF—prefetch; D1/2—decoding 1/2; EX—execution; WB—writeback). [*Source:* Intel (1993).]
*Comment:* This block diagram sheds light on the superscaling of the Pentium processor, in comparison with the i486 processor. The depth of the pipeline has not changed; only the width. This is a consequence of the fact that technology has changed drastically in the sense of on-chip transistor count, and minimally in the sense of off-chip-to/on-chip delay ratio.

### 1.1.1.2. Instruction-Level Parallelism

Pentium is a 32-bit microprocessor based on 32-bit addressing; however, it includes a 64-bit databus. The internal architecture is superscalar with two pipelined integer units; consequently, it can execute, on average, more than one instruction per clock cycle.

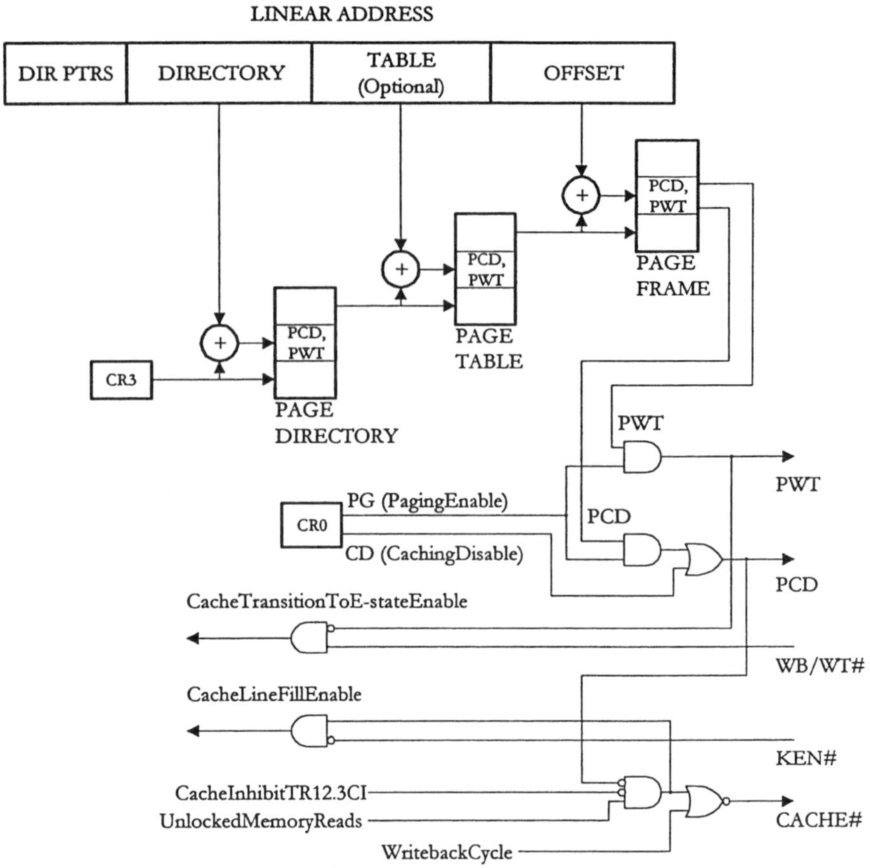

LINEAR ADDRESS

| DIR PTRS | DIRECTORY | TABLE (Optional) | OFFSET |

PAGE DIRECTORY

PAGE TABLE

PAGE FRAME

CR3

CR0

PG (PagingEnable)

CD (CachingDisable)

PWT

PCD

PWT

PCD

CacheTransitionToE-stateEnable

WB/WT#

CacheLineFillEnable

KEN#

CacheInhibitTR12.3CI

UnlockedMemoryReads

WritebackCycle

CACHE#

**Figure 1.4.** Generation of PCD and PWT (PCD—a bit that controls cacheability on a page-by-page basis; PWT—a bit that controls write policy for the second level caches; PTRS—pointers; CR$i$—control register bit $i$; $i = 0, 1, 2, 3, \ldots$). [*Source:* Intel (1993).] *Comment:* The Pentium processor enables the cacheability to be controlled on a page-by-page basis, as well as the write policy of the second level cache, which is useful for a number of applications, including DSM. Figure 1.4 also sheds light on the method used to transform linear address elements into a relevant control signals.

### 1.1.1.3. Branch Prediction

Branch prediction is based on a branch target buffer (BTB). The address of the instruction that is currently in the D1 stage is applied to BTB. If a BTB hit occurs, the assumption is made that the branch will be taken, and the execution continues with the target instruction without any pipeline stalls and/or flushes (because the decision has been made early enough, i.e., during the D1 stage). If a BTB miss occurs, it is assumed that the branch will not be taken, and the execution continues with the next instruction, also without pipeline stalls and/or flushes. The flushing of the pipeline occurs if

- Loop for computing prime numbers:

```
    for(k=i+prime; k<=SIZE; k+=prime)
        flags[k]=FALSE;
```

- Allocation:

```
    prime-ecx
    k-edx
    FALSE-al
```

- Assembly code:

```
    inner_loop:
        mov byte ptr flags[edx], al
        add edx, ecx
        cmp edx, SIZE
        jle inner_loop
```

- Pairing: mov+add  and  cmp+jle
- One loop iteration execution time:

$$T_{exe}[\text{Pentium (with branch prediction)}]=2$$

$$Texe[i486]=6$$

**Algorithm 1.1.** Benefits of branch prediction. [*Source:* Intel (1993).]
*Comment: Pairing* refers to the two-issue superscaling capability of the Pentium processor. If the speed is compared using the number of clock cycles per loop iteration, the technological differences (conditionally speaking) can be expressed through the duration of a single clock cycle.

the branch is mispredicted (one way or the other), or if it was predicted correctly, but the target address did not match. The number of cycles wasted on misprediction depends on the branch type.

The code in Algorithm 1.1 is an example in which branch prediction reduces the Pentium execution time by three times, compared to i486.

### 1.1.1.4. Input / Output

Organization of the interrupt mechanism is a feature that is of importance for incorporation of an off-the-shelf microprocessor into microprocessor and multimicroprocessor systems. Interrupts inform the processor or the multiprocessor of the occurrence of external asynchronous events. External interrupt related details are specified in Algorithm 1.2.

### 1.1.1.5. A Comment About Multithreading

Multithreading on the fine-grain level is not supported in the Pentium processor. Of course, the Pentium processor can be used in the coarse-grain

- Pentium recognizes 7 external interrupts with the following priority:

      BUSCHK#

      R/S#

      FLUSH#

      SMI#

      INIT

      NMI

      INTR

- Interrupts are recognized at instruction boundaries.

- In Pentium,
  the instruction boundary is at the first clock in the execution stage of the instruction pipeline.

- Before an instruction is executed, Pentium checks if any interrupts are pending.
  If yes, it flushes the instruction pipeline, and services the interrupt.

**Algorithm 1.2.** External interrupt (BUSCHK#—pin T03; R/S#—pin R18; FLUSH#—pin U02; SMI#—pin P18; INIT—pin T20; NMI—pin N19; INTR—pin N18). [*Source:* Intel (1993).]

*Comment:* The bus check input (BUSCHK#) allows the system to signal an unsuccessful completion of a bus cycle; if this input is active, address bus signals and selected control bus signals are latched into the machine check registers, and the Pentium processor vectors to the machine check exception. The R/S# interrupt input is asynchronous and edge-sensitive; it is used to force the processor into the idle state, at the next instruction boundary, which means that it is well suited for debugging related work. The cache flush (FLUSH#) interrupt input forces the Pentium processor to write back to memory all modified lines of the data cache; after that, it invalidates both internal caches (data cache and instruction cache). The system management interrupt (SMI#) forces the Pentium processor to enter the system management mode at the next instruction boundary. The initialization (INIT) interrupt input pin forces the Pentium processor to restart the execution, in the same way as after the RESET signal, except that internal caches, write buffers, and floating-point registers preserve their initial values (those existing prior to INIT). The earliest x86 machines include only NMI and INTR, that is, the two lowest-priority interrupts of the Pentium processor (nonmaskable interrupts typically used for power failure and maskable interrupts typically used in conjunction with priority logic).

multiprocessor systems, in which case the multithreading paradigm is achieved through appropriate software and additional hardware constructs off the processor chip.

### 1.1.1.6. Support for Shared Memory Multiprocessing

Multiprocessor support exists in the form of special instructions, constructs for easy incorporation of the second level cache, and the implementation of the Modified/Exclusive/Shared/Invalid (MESI) and the Shared/Invalid (SI) protocols. The MESI and the SI cache consistency maintenance protocols are supported in a way that is completely transparent to the software.

The 8-kB data cache is reconfigurable on a line-by-line basis, as a writeback or writethrough cache. In the writeback mode, it fully supports the MESI cache consistency protocol. Parts of the data memory can be made noncacheable, either by software action or by external hardware. The 8-kB instruction cache is inherently write protected and supports the SI protocol.

The data cache includes two state bits to support the MESI protocol. The instruction cache includes one state bit to support the SI protocol. Operating modes of the two caches are controlled with 2 bits in the register CR0: CD (code disable) and NW (not writethrough). System reset makes CD = NW = 1. The best performance is potentially obtained with CD = NW = 0. Organization of code and data caches is shown in Figure 1.5.

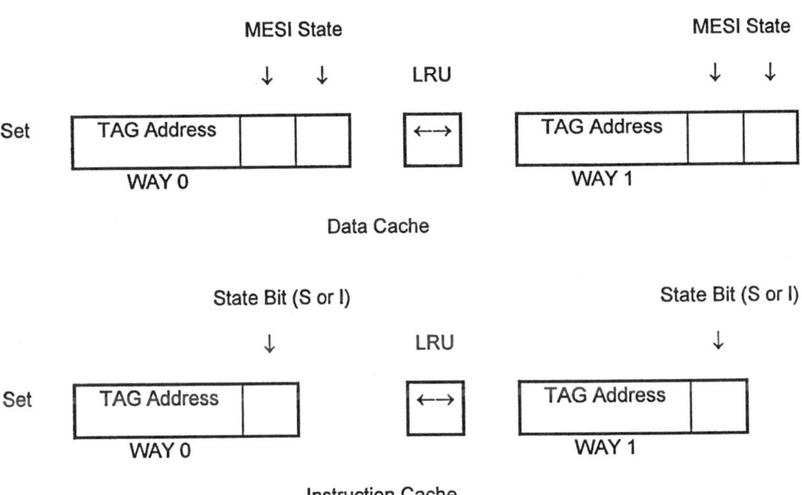


**Figure 1.5.** Organization of instruction and data caches (MESI—Modified/Exclusive/Shared/Invalid protocol; LRU—least recently used). [*Source:* Intel (1993).]
*Comment:* This figure stresses the fact that the Pentium processor uses a two-way set-associative cache memory. In addition to tag address bits, two more bits are needed to specify the MESI state (in the data cache), while one more bit is needed to specify the SI state (in the instruction cache).

A special snoop (inquire) cycle is used to determine whether a line (with a specific address) is present in the code or data cache. If the line is present and is in the M (modified) state, the processor (rather than the memory) has the most recent information and must supply it.

The on-chip caches can be flushed by external hardware (input pin FLUSH# low) or by internal software (instructions INVD and WBINVD). The WBINVD causes the modified lines in the internal data cache to be

```
M—Modified:

An M-state line is available in ONLY one cache,
and it is also MODIFIED (different from main  memory).
An M-state line can be accessed (read/written to)
without sending a cycle out on the bus.

E—Exclusive:

An E-state line is also available in only one cache in the
system,
but the line is not MODIFIED (i.e., it is the same as main
memory).
An E-state line can be accessed (read/written to) without
generating a bus cycle.
A write to an E-state line will cause the line to become
MODIFIED.

S—Shared:

This state indicates that the line is potentially shared with
other caches
(i.e., the same line may exist in more that one cache).
A read to an S-state line will not generate bus activity,
but a write to a SHARED line will generate a writethrough cycle
on the bus.
The write-through cycle may invalidate this line in other caches.
A write to an S-state line will update the cache.

I—Invalid:

This state indicates that the line is not available in the cache.
A read to this line will be a MISS,
and may cause the Pentium processor to execute LINE FILL.
A write to an INVALID line causes the Pentium processor
to execute a write-through cycle on the bus.
```

**Algorithm 1.3.** Definition of states for the MESI and the SI protocols (LINE FILL—fetching the whole line into the cache from main memory). [*Source:* Intel (1993).]
*Comment:* This algorithm gives only the precise description of the MESI and the SI protocols. Detailed explanations of the rationales behind, and more, are given later on in this book (section on caching in shared memory multiprocessors).

written back, and all lines in both caches are to be marked invalid. The INVD causes all lines in both data and code caches to be invalidated, without any writeback of modified lines in the data cache.

As already indicated, each line in the Pentium processor data cache is assigned a state, according to a set of rules defined by the MESI protocol. These states tell whether a line is valid ($I$ = Invalid), available to other caches ($E$ = Exclusive or $S$ = Shared), and has been modified in comparison to memory or not ($M$ = Modified). An explanation of the MESI protocol is given in Algorithm 1.3. The data cache state transitions on read, write, and snoop (inquire) cycles are defined in Tables 1.8, 1.9, and 1.10, respectively.

**TABLE 1.8.  Data Cache State Transitions for UNLOCKED Pentium Processor Initiated Read Cycles**[a]

| Present State | Pin Activity | Next State | Description |
|---|---|---|---|
| M | N/A | M | Read hit; data are provided to processor core by cache. No bus cycle is generated. |
| E | N/A | E | Read hit; data are provided to processor core by cache. No bus cycle is generated. |
| S | N/A | S | Read hit; data are provided to processor core by cache. No bus cycle is generated. |
| I | CACHE# low AND KEN# low AND WB/WT# high AND PWT low | E | Data item does not exist in cache (MISS). A bus cycle (read) will be generated by the Pentium processor. This state transition will happen if WB/WT# is sampled high with first BRDY# or NA#. |
| I | CACHE# low AND KEN# low AND (WB/WT# low OR PWT high) | S | Same as previous read miss case except that WB/WT# is sampled low with first BRDY# or NA#. |
| I | CACHE# high AND KEN# high | I | KEN# pin inactive; the line is not intended to be cached in the Pentium processor. |

*Source:* Intel (1993).

[a]Locked accesses to data cache will cause the accessed line to transition to Invalid state.
*Comment:* For more details see the section on caching in shared memory multiprocessors and Tomašević and Milutinović (1993). In comparison with a "theoretical" case, this "practical" case includes a number of additional state transition conditions, related to pin signals.

**TABLE 1.9. Data Cache State Transitions for UNLOCKED Pentium Processor Initiated Write Cycles**[a]

| Present State | Pin Activity | Next State | Description |
|---|---|---|---|
| M | N/A | M | Write hit; update data cache. No bus cycle generated to update memory. |
| E | N/A | M | Write hit; update cache only. No bus cycle generated; line is now MODIFIED. |
| S | PWT low AND WB/WT#[b] high | E | Write hit; data cache updated with write data item. A writethrough cycle is generated on bus to update memory and/or invalidate contents of other caches. The state transition occurs after the writethrough cycle completes on the bus (with the last BRDY#). |
| S | PWT low AND WB/WT# low | S | Same as above case of write to S-state line except that WB/WT# is sampled low. |
| S | PWT high | S | Same as above cases of writes to S state lines except that this is a write hit to a line in a writethrough page; status of WB/WT# pin is ignored. |
| I | N/A | I | Write MISS; a writethrough cycle is generated on the bus to update external memory. No allocation done. |

*Source:* Intel (1993).

[a]Locked accesses to data cache will cause the accessed line to transition to Invalid state.

[b]WB/WT writeback/writethrough.

*Comment:* For more details see the section on caching in shared memory multiprocessors and Tomašević and Milutinović (1993). In comparison with a "theoretical" case, this "practical" case includes a number of additional state transition conditions, related to pin signals.

**TABLE 1.10. Data Cache State Transitions During Inquire Cycles**

| Present State | Next State INV = 1[a] | Next State INV = 0 | Description |
|---|---|---|---|
| M | I | S | Snoop hit to a MODIFIED line indicated by HIT# and HITM# pins low. Pentium processor schedules the writing back of the modified line to memory. |
| E | I | S | Snoop hit indicated by HIT# pin low; no bus cycle generated. |
| S | I | S | Snoop hit indicated by HIT# pin low; no bus cycle generated. |
| I | I | I | Address not in cache; HIT# pin high. |

*Source:* Intel (1993).

[a]INV—invalid bit.

*Comment:* For more details see the section on caching in shared memory multiprocessors and Tomašević and Milutinović (1993). A good exercise for the reader is to create a state diagram, using the data from this table and from Tables 1.8 and 1.9.

### *1.1.1.7. Support for Distributed Shared Memory*

There is no special support for distributed shared memory. However, the Pentium architecture includes two write buffers (one per pipe), which enhances the performance of consecutive writes to memory. Writes into these two buffers are driven out onto the external bus to memory, using the strong ordering approach. This means that the writes cannot bypass each other. Consequently, the system supports sequential memory consistency in hardware. More sophisticated memory consistency models can be achieved in software.

### 1.1.2. Pentium MMX

The Pentium MMX MultiMedia eXtensions microprocessor is a regular Pentium with 57 additional instructions for fast execution of typical primitives in multimedia processing, such as (1) vector manipulations, (2) matrix manipulations, and (3) bit-block moves.

For typical multimedia applications, Pentium MMX is about 25% faster compared to Pentium, which makes it better suited for Internet server and workstation applications. See Algorithm 1.4 for a list of MMX instructions.

Appearance of the MMX can be treated as the proof of validity of the opinion that accelerators will play an important role in future machines. The MMX subsystem can be treated as an accelerator on the chip.

### 1.1.3. Pentium Pro

The major highlights of Pentium Pro include the features that make it different in comparison with the Pentium processor. It was first announced in 1995. It is based on a 5.5-MTr design, the Intel 0.6-$\mu$m BiCMOS silicon technology, and runs at 150 MHz. A newer 0.3-$\mu$m version runs at 200 MHz. The slower version achieves 6.1 SPECint95 and 5.5 SPECfp95. The faster version achieves 8.1 SPECint95 and 6.8 SPECfp95.

Both the internal and the external buses are 64 bits wide. The processor supports the split transactions approach, which means that address and data cycles are decoupled (another independent activity can happen between the address cycle and the data cycle). The processor includes 40 registers visible to the programmer, each one 32 bits wide. Also, there is a relatively large number of registers not visible to the programmer (e.g., the reorder buffer). Branch prediction is based on a branch target buffer (BTB).

Pentium Pro is superpipelined and includes 14 stages. Pentium Pro is also superscalar, and the instruction fetch unit (IFU) fetches 16 bytes per clock cycle, while the instruction decoding unit (IDU) decodes 3 instructions per clock cycle. Processor supports the speculative and the out-of-order execution.

```
EMMS—Empty MMX state
MOVD—Move doubleword
MOVQ—Move quadword
PACKSSDW—Pack doubleword to word data (signed with saturation)
PACKSSWB—Pack word to byte data (signed with saturation)
PACKUSWB—Pack word to byte data (unsigned with saturation)
PADD—Add with wraparound
PADDS—Add signed with saturation
PADDUS—Add unsigned with saturation
PAND—Bitwise And
PANDN—Bitwise AndNot
PCMPEQ—Packed compare for equality
PCMPGT—Packed compare greater (signed)
PMADD—Packed multiply add
PMULH—Packed multiplication
PMULL—Packed multiplication
POR—Bitwise Or
PSLL—Packed shift left logical
PSRA—Packed shift right arithmetic
PSRL—Packed shift right logical
PSUB—Subtract with wrap-around
PSUBS—Subtract signed with saturation
PSUBUS—Subtract unsigned with saturation
PUNPCKH—Unpack high data to next larger type
PUNPCKL—Unpack low data to next larger type
PXOR—Bitwise Xor
```

**Algorithm 1.4.** New instructions of the Pentium MMX processor (MMX—Multi-Media eXtension). The most logical meaning of the mnemonic MMX is MultiMedia eXtensions. However, that is not the case, for legal reasons. [*Source:* Intel (1997a).] *Comment:* These 26 code types expand to 57 instructions, when different data types are taken into consideration (one quad, two doublewords, 4 words, or 8 bytes). Some multimedia applications have reported a 20% performance increase with Pentium MMX, compared to Pentium. Note that real application performance increase is always smaller than the benchmark performance increase.

The first-level caches are on the processor chip; both of them are two-way set-associative 8-kB caches with buffers that handle four outstanding misses. The second-level cache includes both data and code. It is four-way set-associative, is 256 kB large, and includes 8 error correction code (ECC) bits per 64 data bits.

Support for shared memory (multi)processing (SMP) and distributed shared memory (DSM) is in the form of the MESI protocol support. Some primitive support for multicomputer systems (MCSs) and distributed computer systems (DCSs) is in the form of two dedicated ports, one for input and one for output.

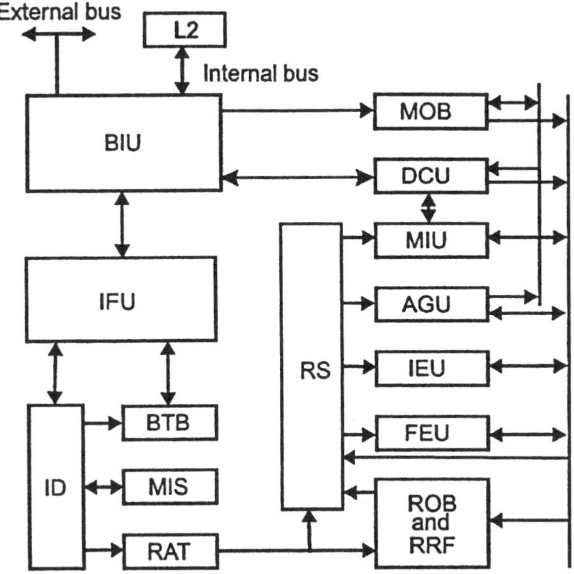


**Figure 1.6.** Pentium Pro block diagram [AGU—address generation unit; BIU—bus interface unit; BTB—branch target buffer; DCU—data cache unit; FEU—floating-point execution unit; ID—instruction decoder; IEU—integer execution unit; IFU—instruction fetch unit (with I-cache); L2—level 2 cache; MIS—microInstruction sequencer; MIU—memory interface unit; MOB—memory (re)order buffer; RAT—register alias table; ROB—reorder buffer; RRF—retirement register file; RS—reservation station]. [*Source:* Papworth (1996).]

*Comment:* The L2 cache is on the same package as the CPU, but on another chip (the Pentium Pro package includes two chips). Major additions, compared to the Pentium processor, are those related to the exploitation of the instruction level parallelism (RAT, ROB, RRF, and RS).

A block diagram of the Pentium Pro machine is given in Figure 1.6, together with brief explanations. Detailed explanations of underlying issues in instruction-level parallelism (ILP) and branch prediction strategies (BPSs) are given later in this book.

### 1.1.4. Pentium II and Pentium III

To a first approximation, Pentium II can be treated as a regular Pentium Pro with additional MMX instructions to accelerate multimedia applications. However, it includes other innovations aimed at decreasing the speed gap between the CPU and the rest of the system.

Also, a new packaging technology is used in Pentium II for the first time by Intel. It is the single edge contact (SEC) connector, which enables the CPU to operate at greater speeds. For the most recent speed comparison

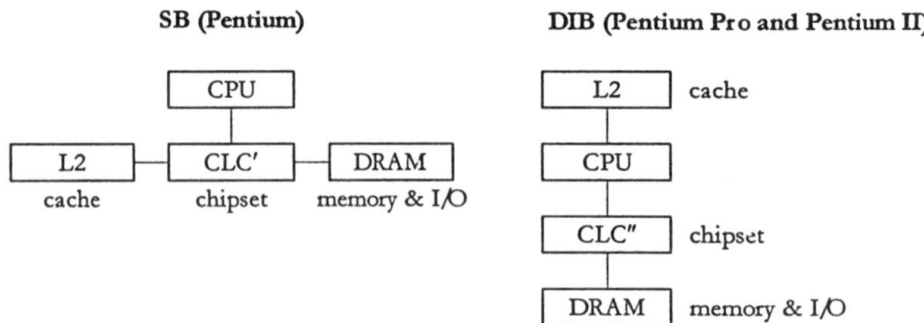


**Figure 1.7.** SB and DIB bus structures (SB—single independent bus; DIB—dual independent bus; CLC—control logic chipset; L2—second level cache; SEC—single edge contact connector). [*Source:* Intel (1997a).]
*Comment:* The dual-bus structure (DIB) eliminates a critical system bottleneck of the single-bus structure (SB).

data, the interested reader is referred to Intel's WWW presentation (http://www.intel.com/).

An important element of Pentium II and Pentium Pro is the dual independent bus (DIB) structure. As indicated in Figure 1.7, one of the two DIB buses goes toward L2 cache; the other one goes toward DRAM (dynamic RAM) main memory, and I/O, and other parts of the system. The only difference between the Pentium Pro and the Pentium II processors, with respect to their L2 caches, is that the Pentium Pro implemented the L2 in the same package as the CPU, while the Pentium II implements its L2 on a cartridge with a single-edge connector. In fact, the Pentium Pro's L2 is accessible at the full CPU speed of 200 MHz; the Pentium II's L2 runs at only half the speed of the CPU (300 MHz divided by 2, or 150 MHz) and is therefore more of a bottleneck than the original design. (This was done to enable the use of commodity SRAMs instead of the custom SRAMs needed by the Pentium Pro, and to obviate the need for the Pentium Pro's expensive ceramic two-die package.)

During Q1/99 Intel presented a new microprocessor: the Pentium III (codename "Katmai"). It operates at 450 and 500 MHz. It is compatible with the previous generation of Intel Pentium II processors, and the major improvements are (1) streaming SIMD execution (SSE) and (2) chip ID.

The SSE implies 70 new instructions intended mainly for 3D graphics, digital image and sound processing, voice recognition, and so on. According to Intel, as the result of an efficient hardware implementation, these new instructions improve the execution time of application code by more than 50%.

The main idea behind chip ID is, according to Intel, improving e-commerce over the Internet by avoiding unnecessary manually entered passwords. Instead, the processor introduces itself with the unique chip ID on e-commerce sites. Discussion about this new feature is still going on; Intel has an option to disable this feature, unless a customer explicitly requests it.

## 1.2. ADVANCED ISSUES

One of the basic issues of importance in microprocessing is the correct prediction of future development trends. The facts presented here have been adapted from a talk by an Intel researcher [Sheaffer 1996] and represent a possible way for Intel to proceed in future developments after the year 2000 (it summarizes the views of several Intel VIPs).

Quite soon after the year 2000, the on-chip transistor count will reach about 1 GTr (one gigatransistor) using the minimum lithographic dimension of less than 0.1 $\mu$m and a gate oxide thickness of less than 40 Å. Consequently, microprocessors on a single chip will be faster. Their clock rate is expected to be about 4 GHz, to enable an average speed of about 100 BIPS (billion instructions per second). This means that the microprocessor speed, measured using standard benchmarks, may reach about 100 SPECint95 (which is about 3500 SPECint92).

Under such conditions, the basic issue is how to design future microprocessors, in order to maintain the existing trend of approximately doubling the performance in every 18 months. Hopefully, some of the answers will be clear, once the reading of this book is completed.

With regard to process technology, the delay trends are given in Figure 1.8 and the area trends in Figure 1.9 for the case of Intel products. The position of an ongoing project  can be estimated using extrapolation. Frequency of

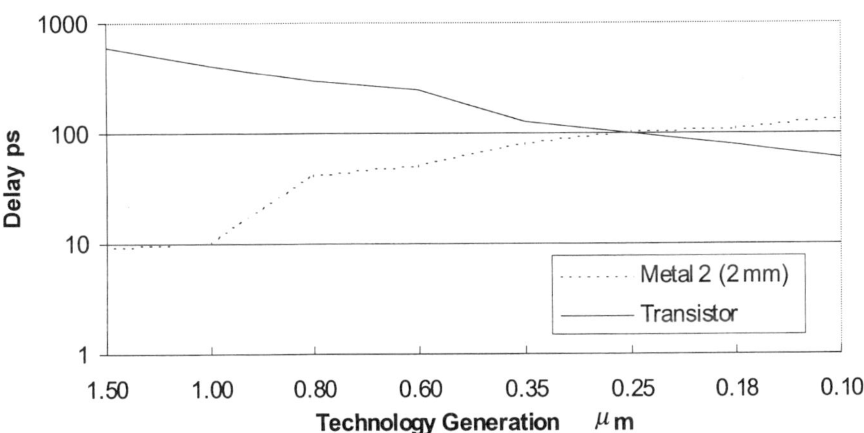


**Figure 1.8.** Microprocessor chip delay trends (metal 2 (2 mm)—two-level metal). [*Source:* Sheaffer (1996).]

*Comment:* As the basic geometry becomes smaller, gates become faster (because of the smaller distances that carriers have to cross), but wires become slower (because of the increased parasitic capacitances and resistances). Architectures that are suitable for basic geometries larger than 0.25 $\mu$m have to be reconsidered before an attempt is made to reimplement them in basic geometries smaller than 0.25 $\mu$m.

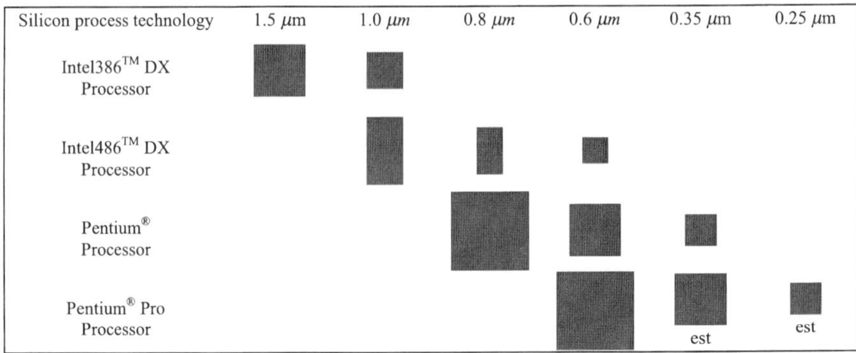


**Figure 1.9.** Microprocessor chip area trends (est—estimated). [*Source:* Sheaffer (1996).]

*Comment:* Once a new architecture is created, it is typically reimplemented once or twice in newer technologies, which results in smaller chips and better speeds. In some companies, the decision to go into the next generation architecture comes after the technology has advanced enough so that the next generation architecture can fit into a chip of approximately the same area as the first attempt of the previous generation. Consequently, the first implementations of i386 and i486 are of approximately the same chip area. However, in some cases, the move to the next generation is possible only after a major technological advancement is achieved; consequently, the chip area of the initial implementation of Pentium is considerably larger than the chip area of the initial implementation of i486. In conclusion, development of microprocessor technology includes both evolutionary phases, as well as revolutionary steps.

operation and transistor count represent alternative ways to express performance and complexity. Related data for the line of Intel products are given in Figures 1.10 and 1.11.

Figure 1.10 can be used to extrapolate future trends at Intel, concerning clock frequency. However, the ultimate goal is not a fast clock, but a fast application. Therefore, the speed of the clock has to be carefully examined against how much work the microprocessor can do in one clock cycle, which is an architectural issue. Note that the dots for PPro-225 and PPro-300 are located above the position that would be obtained by linear extrapolation. In other words, the trend at Intel shows the elements of superlinearity. On the other hand, in reality, most of the time it is the opposite—the curves like the one in Figure 1.10 typically show saturation, as indicated in Becker et al. (1998) and Hartenstein (1997).

Figure 1.11 can be used to extrapolate future trends at Intel, concerning the transistor count per die. Note that in some cases the microprocessor package includes two or more dies, while Figure 1.11 refers only to the number of transistors per single die. Again, as in the case of Figure 1.10, the trend at Intel shows elements of superlinearity.

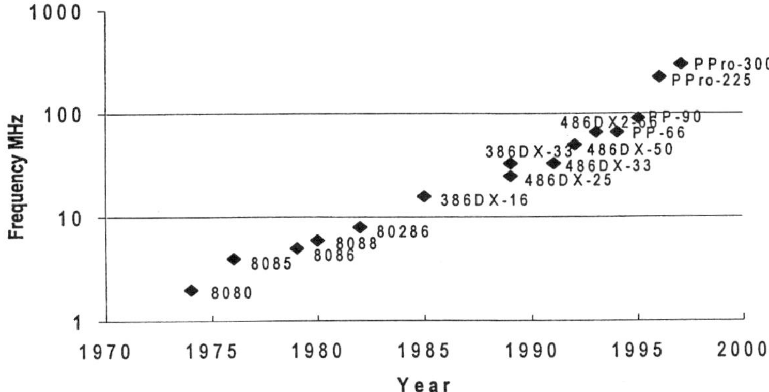



**Figure 1.10.** Microprocessor chip operation frequency (PP—Pentium processor; PPro —Pentium Pro processor). [*Source:* Sheaffer (1996).]
*Comment:* The raw speed of microprocessors increases at the steady speed of about one order of magnitude per decade. Note, however, that what is important is not the speed of the clock (raw speed), but the time to execute a useful application (semantic speed).

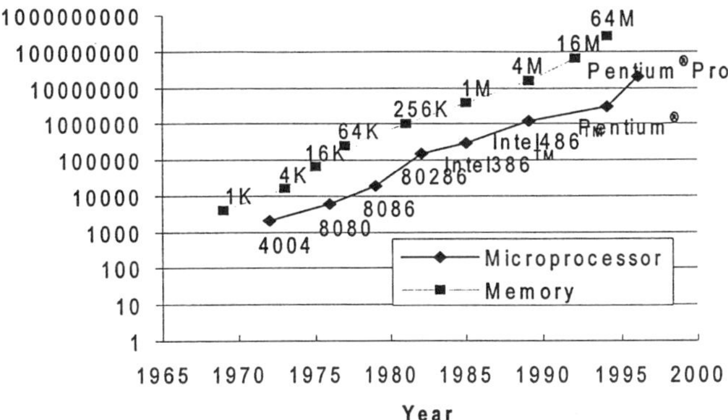



**Figure 1.11.** Microprocessor and memory complexity. [*Source:* Sheaffer (1996).]
*Comment:* The fact that the memory curve is above the processor curve tells that highly repetitive structures pack more transistors onto the same area die. The distance between the two curves has a slightly increasing trend, which means that the higher the complexity, the more difficult it is to fit a given irregular structure onto the same area die.

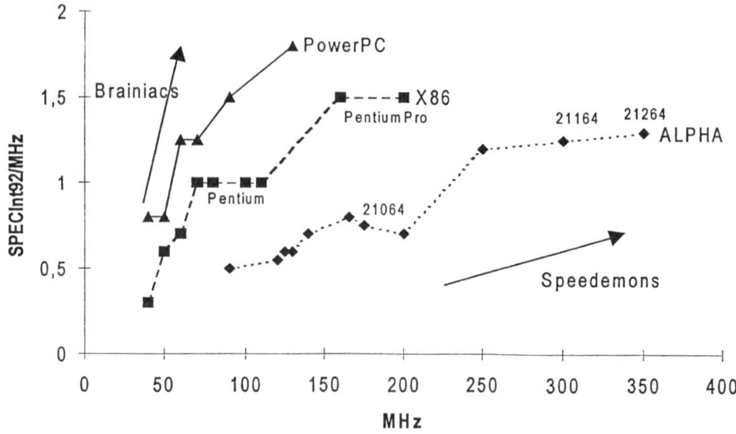


**Figure 1.12.** Microprocessor sophistication. [*Source:* Sheaffer (1996).]
*Comment:* Brainiacs tend to have a slower clock; however, on average, they accomplish more during a single clock period. On the other hand, speedemons are characterized by an extremely fast clock and little work done during each clock period. Of course, what counts is the speed of the compiled high-level language code (semantic speed). Consequently, an architecture is superior if it has been designed wisely, not if it has been designed as a brainiac or a speedemon. Typically, the design path in between the two extremes has the highest success expectations. Maybe that explains the decision of Intel to select the medium approach in between the two extremes.

Generally, modern microprocessors have been classified in two basic groups: (1) brainiacs (lots of work accomplished during a single clock cycle, but slow clock), and (2) speedemons (fast clock, but not much work accomplished during a single clock cycle). Figure 1.12 can be treated as a proof of the statement that Intel microprocessors belong in between the two extreme groups. It appears that Intel processors have started as brainiacs; however, over time they have taken a course that brings them closer to speedemons. Figure 1.13 compares the time budget of brainiacs and speedemons.

From Figure 1.13 one can see that, in the case of brainiacs, most of the clock cycles are used to perform arithmetical, logical, or related operations. On the other hand, in the case of speedemons, a good portion of time is used to manage the memory more efficiently.

In principle, there are two educational approaches: (1) from general to specific (analysis) and (2) from specific to general (synthesis). Both approaches have their merits. In this book, the second of the two approaches is used: the Intel case study is presented first (already presented in the text so far), and then the future architectural developments are discussed next (in the text to follow, from now on).

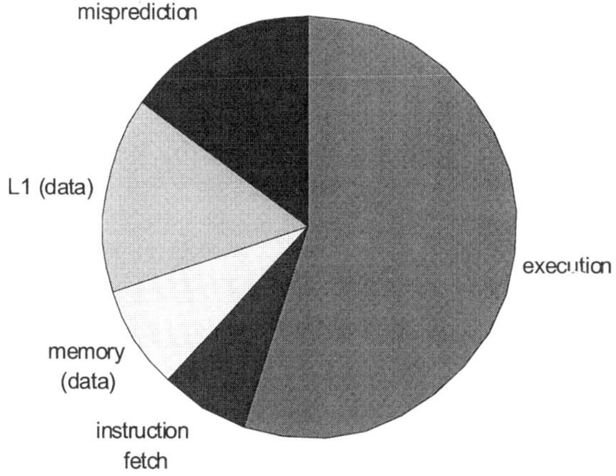


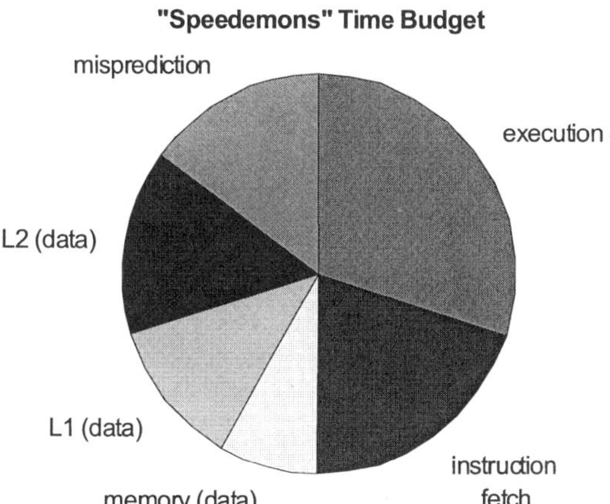


**Figure 1.13.** Microprocessor time budget (L1/2—first/second-level cache). [*Source:* Sheaffer (1996).]

*Comment:* Brainiacs use more on-chip resources to enrich the execution unit; consequently, often there is no room for L2 cache on the CPU chip. On the other hand, speedemons are the first ones to include L2 on the same chip as the CPU, which is only partially a consequence of the fact that the execution unit is simpler.

Finally, a frequent question is how to select the optional microprocessor for a given application. There are two answers to that question. One answer implies that one can select only among the existing microprocessors on the market. In such a case, nontechnical reasons for selection may be more relevant than technical reasons. The other answer to the same question implies that one can design and implement a new microprocessor, in which case the question translates into "What is the best architecture to select for the new design?" This book gives an answer when the goal is to design a DSM on a single chip.

The author and his associates were active in the microprocessor design and also in several designs of potential accelerators for mission critical applications. For details, see Davidovic et al. (1997), Fortes et al. (1986), Helbig and Milutinović (1989), Milutinović (1978, 1979, 1980a, 1980b, 1980c, 1984, 1985a, 1985b, 1985c, 1986, 1987a, 1987b, 1988a), Milutinović et al. (1986a, 1986b), Milutinović and Petkovic (1995), Raskovic et al. (1995), and Vuletic et al. (1997).

## PROBLEMS

**1.1.** Explain why a higher number of metal layers means a larger transistor count increase when the feature size decreases. Explain the impacts.

**1.2.** Compare the transistor count invested (by various manufacturers) into the IU and FPU registers of Table 1.3. Interpret the differences.

**1.3.** Calculate the ratio of the number of execution units and the issue width for all microprocessors listed in Table 1.4. Interpret the differences.

**1.4.** Study the details of AMD K6 and explain the essence of the features that you believe are the major contributors to its excellent performance. Suggest further improvements.

**1.5.** When you compare DEC and HP microprocessors, how many transistors were saved in HP processors, by having only the assist cache, rather than both L1 and L2 caches on the same chip as the CPU? Create a small application example (a short routine) showing the effectiveness of the assist cache of HP microprocessors compared with DEC microprocessors.

**1.6.** Compare the L2/L1 capacity ratio (Rc) for DEC Alpha 21264, Intel Pentium II, and a few commercial board designs (that one can find on the Internet) using other microprocessors listed in Table 1.2. How does Rc for DEC Alpha 21264 and Intel Pentium II compare with Rc for commercial boards that you have analyzed?

**1.7.** What are the pros and cons of having separate TLBs for instructions and data? Compare with unified TLBs.

**1.8.** In Figure 1.2, explain the possible ways to integrate the MMX accelerator. Explain the pros and cons of different schemes.

**1.9.** Consult the open literature and describe precisely (in the time domain) all seven external interrupts listed in Algorithm 1.2.

**1.10.** Suggest architectural add-ons that would make Pentium more efficient when used in SMP systems. Do the same for the case of DSM systems.

# IMPORTANT ISSUES

The seven chapters to follow cover the seven problem areas of importance for future microprocessors on a single VLSI chip, keeping in mind the ultimate goal advocated in this book, which is an entire DSM on a single chip, together with a number of simple or complex accelerators.

# CHAPTER 2

# Cache and Cache Hierarchy

This chapter includes two main sections. The section on basic issues covers three simple examples (associative, direct-mapped, and set-associative cache), and gives the basic definitions of cache hierarchy. The section on advanced issues covers selected state-of-the-art contributions.

## 2.1. BASIC ISSUES

This section gives only three examples and the elementary definitions, since the basic cache concepts are widely known (the text to follow assumes that the basic cache concepts are well understood, which is one of the prerequisites for reading this book).

The major issue in cache design is how to design a close-to-the-CPU memory that is both fast enough and large enough, but cheap enough. In other words, the major problem here is not technology, but technoeconomics, under conditions of constantly changing technology costs and characteristics (which is what some researchers sometimes forget).

The solution to the problem defined above, which includes the elements of both technology and economics, is cache memory. This solution was made possible after it was recognized that programs do not access their data in a completely random way—the principle of locality is inherent to data and code access patterns. The discussion that follows implies the processor-memory organization from Figure 2.1 and the related access patterns.

Three cache organizations are possible: (1) associative, as one extreme; (2) set-associative, as a combined solution; and (3) direct-mapped, as another extreme. In all these cases, the memory capacity is assumed to be equal to $2^m$ ($m$ is a relatively large integer) and divided into consecutive blocks of $b$ bytes ($b$ is also a power of 2).

### 2.1.1. Fully Associated Cache

An example using the associative approach is shown in Figure 2.2. Each cache line contains one block, together with the block number and a validity bit (showing whether the corresponding block contains a valid data item).

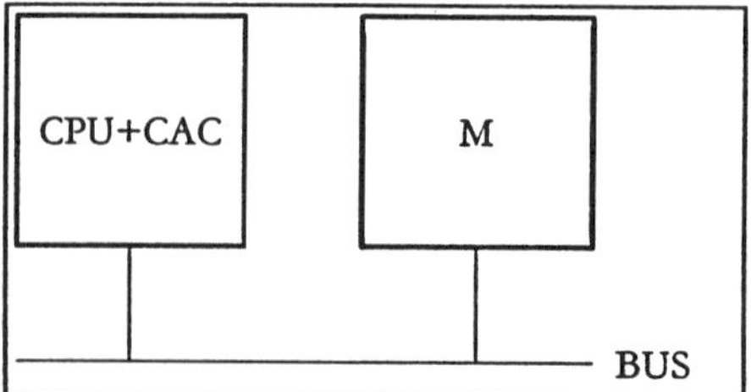

**Figure 2.1.** On-chip cache (CAC—cache and cache hierarchy). [*Source:* Tanenbaum (1990).]

*Comment:* This figure presents a simple processor/memory model used to explain the caching approaches in the figures to follow. Note that some new microprocessor system architectures make a departure from this simple processor/memory model (like the dual independent bus structure of Intel Pentium II). However, the difference does not affect the essence of caching.

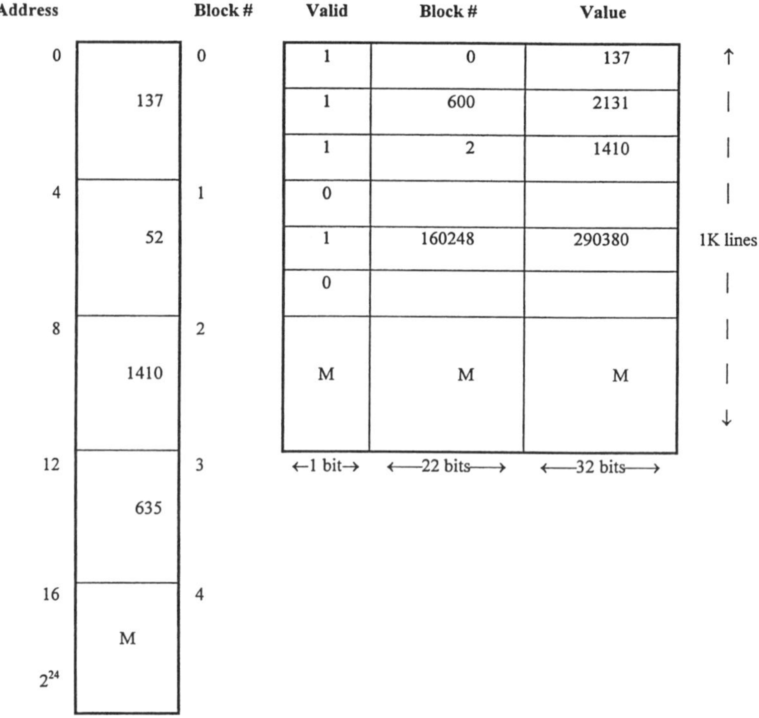

**Figure 2.2.** An associative cache with 1024 lines and 4-byte blocks. [*Source:* Tanenbaum (1990).]

*Comment:* In the case of a fully associative cache it happens that a block cannot be purged out of the cache only because a subset of its address bits matches the same subset of address bits of the newly referenced block. Consequently, if the replacement algorithm is properly designed, the block to be purged is always the block that is the least likely to be referenced again.

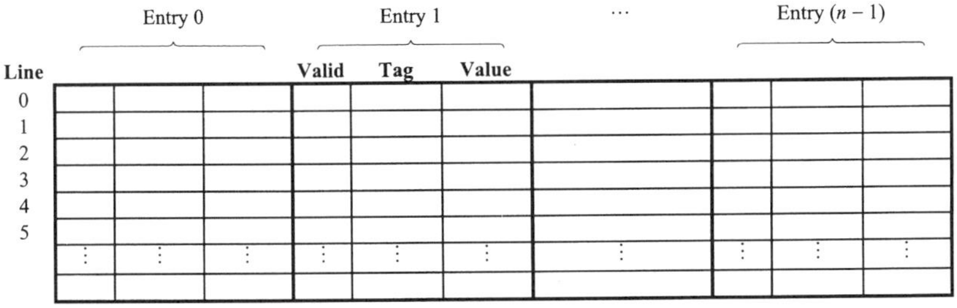

Entry 0          Entry 1          ...          Entry $(n-1)$

Line          Valid    Tag    Value

0
1
2
3
4
5

**Figure 2.3.** A set-associative cache with $n$ entries per line. [*Source:* Tanenbaum (1990).]

*Comment:* In the case of a set-associative cache, the benefits of full associativity are present within a set. Since the block count of a set is considerably smaller than the block count of the entire cache, nonoptimal block replacement is a reality that impacts the performance.

In the example of Figure 2.2, memory is byte-addressable, and one block occupies 4 bytes. Consequently, the *Block#* field is 22 bits wide, because the memory addresses are 24 bits wide. If the *Valid* bit is zero, the contents of the fields *Block#* and *Value* are irrelevant. The *Value* field is 32 bits wide.

### 2.1.2. Set-Associative Cache

An example using the set-associative approach is shown in Figure 2.3. Each cache line contains several entries, together with the block *Tag* and a *Valid* bit.

Full associativity is present within a set. The larger the set, the closer the performance of the set-associative and the fully associative caches. Of course, the set-associative cache is simpler to implement.

### 2.1.3. Direct-Mapped Cache

An example using the direct-mapped approach is shown in Figure 2.4. Each cache line contains one block, together with the *Tag* and *Valid* bits. The direct-mapped cache is the simplest to implement; however, it is characterized by the worst performance of all three cache organizations.

Note that a relatively large number of addresses map onto the same cache line. Consequently, addresses that map on the same line also compete for the same line, which is a potential source of unnecessary replacment of data to be used in the near future.

Cache memory can be organized hierarchically. If that is the case, typically the inclusion principle is satisfied. This means that data in lower levels (closer to CPU) are also present in higher levels (closer to memory). The larger the speed gap between the CPU and the memory, the larger the optimal number of cache hierarchy levels.

| Line | Valid | Tag | Value | Addresses that use this line: |
|---|---|---|---|---|
| 0 | 1 | 2 | 12130 | 0, 4096, 8192, 12288, ... |
| 1 | 1 | 1 | 170 | 4, 4100, 8196, 12292, ... |
| 2 | 1 | 3 | 2142 | 8, 4104, 8200, 12296, ... |
| 3 | 0 | | | 12, 4108, 8204, 12300, ... |
| 4 | 0 | | | 16, 4112, 8208, 12304, ... |
| 5 | 0 | | | 20, 4116, 8212, 12308, ... |
| ⋮ | ⋮ | ⋮ | ⋮ | |
| 1023 | 1 | | | 4092, 8188, 12284, ... |

| | 12 | 10 | 2 |
|---|---|---|---|
| **Address bits** | Tag | Line | 00 |

**Figure 2.4.** A direct-mapped cache with 1024 4-byte lines and a 24-bit address. [*Source:* Tanenbaum (1990).]

*Comment:* In the case of the direct-mapped cache, the benefits of associativity are nonexistent. Consequently, a match of a subset of address bits of a block will force block replacement, even for a block that may be reused later. Some compiled vector processing code may result in patterns that periodically destroy the item needed next. The interested reader is encouraged to develop such a code, to demonstrate the issue.

According to Peir et al. (1999), techniques for achieving fast cache access, while maintaining high hit ratios, can broadly be classified into the following four categories:

1. *Decoupled Cache.* Data array access and line selection are done independently of the tag array access and comparison, which reduces the delay imbalance between the paths through the tag and the data arrays; this improves the cache clock rate.

2. *The Multiple Access Cache.* The direct-mapped cache (which is characterized with fast access) is accessed more than once; this results in a hit ratio typical of set-associative caches, but with an access time that is (in spite of multiple accesses) faster than that of the set-associative cache.

3. *Augmented Cache.* In addition to the direct-mapped cache, a small fully associative cache is added to improve performance; overall, this improves the hit ratio, while the access time slowdown is practically negligible.

4. *Multilevel Cache.* The lowest-level cache is the smallest and the fastest, while the upper-level caches are bigger and slower; this enables large capacity for the highest-level cache and a good speed for the lowest-level cache, which is the reasoning that led to the introduction of the cache memory concept in the first place.

In the section on advanced issues, which follows next, several enhancements are discussed. In each of the enhancements to be discussed, some elements of the above-mentioned four categories could be recognized.

## 2.2. ADVANCED ISSUES

This section gives a short overview of selected research efforts in the field of uniprocessor cache memory. Some of the solutions presented here are also applicable in the field of multiprocessing.

An idea that deserves attention is *to use pointers to page numbers rather than page numbers*, in cache architectures of modern microprocessors and multimicroprocessors [Seznec 1996]. Figures 2.5–2.8 give the basic explanation.

By using pointers to the page numbers, one saves a relatively large number of transistors in a new microprocessor design. This is because the same data can be found in various microprocessor resources (in the general case, caches, TLB, various tables and directories, etc.).

| AddressTag | V | Data |
|---|---|---|

**Figure 2.5.** Structure of a cache line (V—valid bit). [*Source:* Seznec (1996).]
*Comment:* This structure includes only the fields that are absolutely necessary to explain the concept. Realistic microprocessor system caches include a number of additional tags.

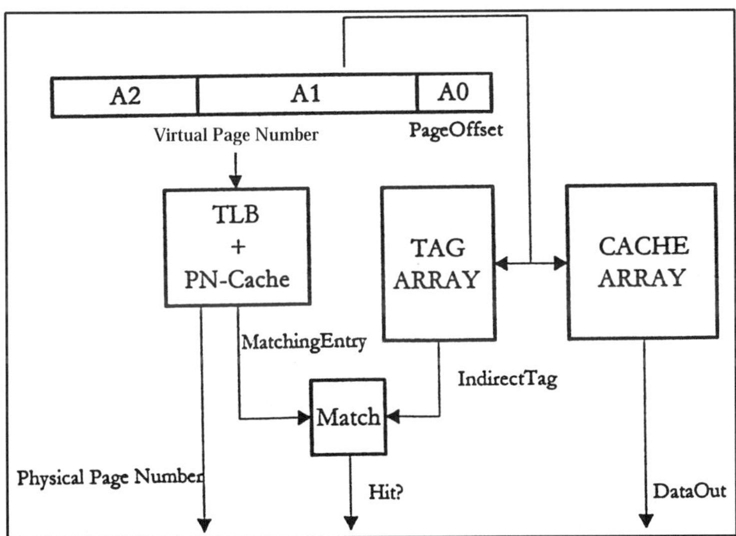


**Figure 2.6.** Indirect-tagged VV cache; TLB serves as a PN-cache (VV—virtually indexed, virtually tagged; PN—page number). [*Source:* Seznec (1996).]
*Comment:* Unlike the next two structures, this one does not include a separate page number cache. The physical page number is obtained directly from the TLB.

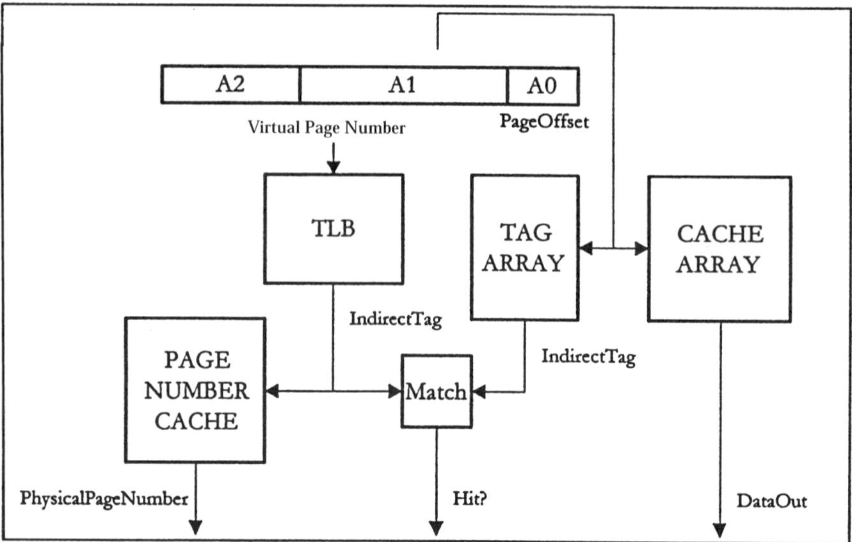


**Figure 2.7.** Indirect-tagged VP cache with physical page numbers stored in the PN-cache (VP—virtually indexed, physically tagged; PN—page number). [*Source:* Seznec (1996).]

*Comment:* Note that the address fields A0 and A1 are applied at the same time to both the tag array and the cache array.

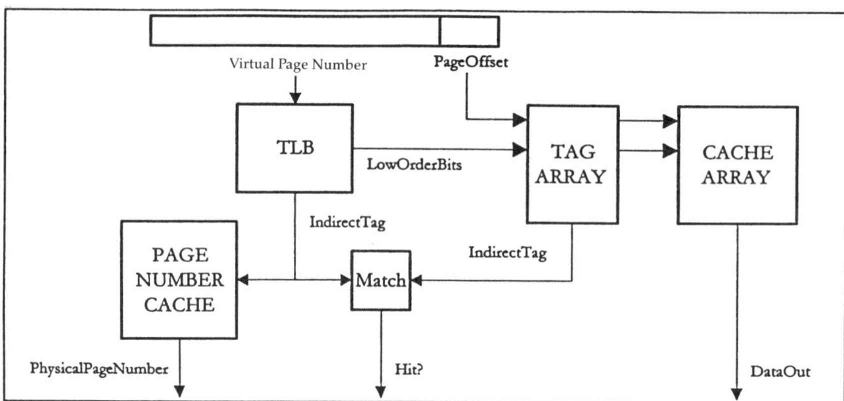


**Figure 2.8.** Indirect-tagged PP cache (PP—physically indexed, physically tagged; PN—page number). [*Source:* Seznec (1996).]

*Comment:* Note that the page offset is applied in conjunction with the low order bits from the TLB.

This is of special interest for object-oriented environments where the pointers to page numbers may exhibit a higher level of locality, which is crucial for cache efficiency. Variations of this idea are expected to be seen in future machines for object-oriented environments.

Another contribution that deserves attention is the *difference-bit cache* [Juan et al. 1996]. The difference-bit cache is a two-way set-associative cache with a smaller access time, which is important, since the cycle time of a pipelined processor is usually determined by the cache access time.

The smaller access time was achieved by noticing that the two tags for a set have to differ by at least one bit, and this bit can be used to select the way in a two-way set-associative cache. Consequently, all hits of a difference-bit cache are one cycle long.

Recent research at Hewlett-Packard [Chan et al. 1996] introduces the concept of *assist cache*, which can be treated as a zero-level cache, in systems in which only the zero-level cache is on the microprocessor chip, and L1/2 are off the processor chip.

The major idea behind the assist cache is to treat spatial data (elements of complex data structures) in a different way compared to other data, such as temporal data (single variables). Spatial data are likely to be used only once, and consequently after being used, on the next replacement, they are not placed back into the off-chip L1 cache (and off-chip L2 cache, if present); this cache is bypassed and spatial data go directly to memory.

An important research track is related to the *victim cache*. First mention of the underlying concept dates back to DeJuan et al. (1987, 1988). The concept was introduced under the name victim cache in Jouppi (1990). An improvement referred to as *selective victim caching* is described in Stiliadis and Varma (1997).

This research starts from the design trade-off between direct-mapped and set-associative or fully associative caches. Direct-mapped caches are faster (which means a shorter cycle time); however, their miss ratio is higher (which means more cycles for execution of a program) because semantically related data can map into the same cache block. Fully associative caches represent the opposite extreme (longer access time, but higher hit ratio). Set-associative caches are in between.

One question is what to do in order to achieve an access time typical of the direct-mapped cache and, at the same time, a hit ratio comparable to that of the set-associative approach. Victim cache is a solution. It is a small fully associative cache. Its place in the cache/memory hierarchy is one level above the first-level cache (L1), which is direct-mapped. Blocks replaced from the direct-mapped L1 cache are placed into the fully associative victim (V) cache (kind of level $1\frac{1}{2}$). If this block is referenced again, before being replaced from the V cache, the L1 miss penalty will be relatively small (because the victim cache is small and physically close). Since the V cache is fully associative, many blocks that conflict with each other in the direct-mapped L1 cache coexist without problems in the V cache. Consequently, the speed of the directed-mapped cache makes L1 faster, and the major drawback of the direct-mapped cache (mapping of two or more useful data items into the same cache block) is eliminated by the victim cache.

In the case of the *selective victim cache*, incoming blocks to L1 are placed selectively into either the L1 or V cache, based on their past behavior. This implies the existence of a prediction mechanism, to determine the access likelihood of the incoming block. Blocks that are more likely to be accessed are placed in the L1 cache; blocks that are less likely to be accessed are placed in the V cache.

Similarly, when a miss is being serviced from the V cache, the prediction mechanism is applied again, to determine whether an interchange of the conflicting blocks is required (which is mandatory in the case of the conventional V cache). For 10 different SPEC92 benchmarks, the selective victim cache provides a cache miss rate improvement of up to about 20%, and an interchange traffic improvement between L1 and V of up to about 70%. The selective victim cache approach can be based on a number of different prediction mechanisms.

Another question is: What is better—(1) fewer cycles that are longer (fully associative cache), or (2) more cycles that are shorter (direct-mapped)? Hill (1988) compares the direct-mapped cache and the set-associative cache (fully associative cache has no practical value in a number of applications) and concludes that the direct-mapped cache gives a shorter overall execution time for a set of important system and application programs. Milutinović (1986c) comes to the same conclusion for GaAs technology, through a side study leading to an early GaAs RISC [Helbig and Milutinović 1989].

An important new trend in cache design uses techniques that optimize the cache design and performance under conditions typical of superscalar microprocessors [Vajapeyam and Mitra 1997, Hopkins 1997].

The *spatial footprint predictor* (SFP) cache predicts which portions of a cache block will be used before eviction [Kumar and Wilkerson 1998]. In traditional cache memories, more than 50% of cached data is evicted before use, which means that the potential for such a prediction is relatively high. Efficient prediction enables smaller cache blocks (combined with *multiple block fetch on a miss*) to achieve a better ratio of used-before-evicted data versus evicted-before-used data, and consequently a better performance. The predictor from Kumar and Wilkerson (1998) improves the cache miss rate, on average, by about 18% for SPEC applications. The miss rate improvement potential is estimated to be about 35%. The SFP approach requires no ISA modifications.

Prediction is based on two tables: the spatial footprint history table (SHT) and the active sector table (AST). The SHT stores some of the previously recorded footprints. The AST records the footprint while a sector is active in the cache. Sectors are aligned groups of adjacent cache blocks and can be active or inactive (initially, all sectors are inactive). Upon sector activation, an AST entry is allocated and the predicted spatial footprint is used to fetch new blocks. Upon sector deactivation, the recorded footprint is used to update the predictor. Cache blocks within a sector that are referenced while the sector is active define the spatial footprint of that sector. A spatial footprint is stored as a bit vector, with a bit for every block in the sector. Consequently, blocks from a sector are fetched only if they are very likely to be used in the near future.

An alternative approach, referred to as *trace cache* [Peleg and Wiser 1994, Rotenberg et al. 1997] places logically contiguous instructions into physically contiguous storage. Consequently, multiple cache blocks can efficiently be fetched on each cycle. The maximal number of fetch blocks supplied on each cycle is limited by the capabilities of the branch predictor.

In general, instruction fetching past a taken branch is a serious problem for instruction cache design. One possible solution for the problem is the trace cache. Like the classical instruction cache, the trace cache is accessed using the starting address of the next block. Unlike the classical instruction cache, it stores logically consecutive instructions in physically consecutive storage.

Actually, a trace cache line stores a segment of the dynamic instruction trace, across multiple potentially taken branches. In other words, the block contains dynamically consecutive instructions, not statically consecutive instructions.

As a group of instructions is processed, it is latched into the fill unit. Consecutive groups of instructions are concatenated until the block is completely filled. One possible size of the block is equal to the issue width of instructions. The fill unit is off the critical execution path, so its latency does not affect the CPU clock.

One important characteristic of the trace cache is that it continues to contribute to the overall system performance as its size increases. This is not the case with a classical instruction cache; if the instruction cache size grows beyond the working set of larger applications, a larger instruction cache may not bring additional performance improvements. With the growing size of the trace cache, it can be partitioned into a smaller one-cycle component and a larger multiple-cycle component. Multiple-level hierarchies of trace caches can also be built.

Patel (1998) introduces two mechanisms to improve the efficiency of the trace cache: branch promotion (dynamically converting strongly biased branches into branches with static predictions—statically predicted branches do not consume the branch predictor bandwidth when they are fetched) and trace packing (dynamically packing trace segments into as many instructions as possible—the fill unit will not divide a block of newly retired instructions across trace segments). Both techniques together improve the effective fetch rate of the traditional trace cache by up to about 17% [Patel et al. 1998].

Further enhancements of the trace cache concept can be found in Rotenberg et al. (1999), Patel et al. (1999), and Black et al. (1999). The first paper [Rotenberg et al. (1999)] gives a more accurate performance analysis. The second paper [Patel et al. 1999] evaluates a number of trace cache design choices. The third paper [Black et al. 1999] introduces the concept of the block-based trace cache, which is aimed at using the storage more efficiently; instead of storing blocks of traces, it stores pointers to blocks of traces.

Several research efforts try to improve the cache subsystem performance by treating spatial and temporal data in different ways, with different mechanisms. The major approaches will now be briefly summarized from Prvulovic et al. (1999a).

### 2.2.1. Exploitation of Temporal Locality

Much work has been done in exploiting the temporal locality of the data accesses, using cache bypassing [Tyson et al. 1995, Johnson 1997a]. The main idea is not to cache nontemporal data. In this way, the entire cache capacity is dedicated to caching the data that will be reused.

Still, nontemporal data can exhibit a high degree of spatial locality, so it is useful to cache the corresponding block for some time. Besides, if some data item is wrongly presumed as being nontemporal, caching it in a small buffer reduces the penalty of such a mistake. For both of these reasons, a small buffer is often added that caches such nontemporal data. The result is a cache split according to the temporal locality, consisting of a larger temporal subcache (main cache) and a smaller nontemporal subcache (buffer).

The previously mentioned assist cache [Chan et al. 1996] is another example in this category. The first-level cache consists of a large, external, direct-mapped main cache and a smaller, on-chip, fully associative assist cache. Besides reducing conflicts in the main cache, the assist cache also serves as a "nontemporal" subcache, while the main cache serves as a "temporal cache." At compile time, some memory access instructions are marked as "spatial locality only."

Data accessed by these instructions are not cached in the main cache, but only in the assist cache. Other data items are considered as having both temporal and spatial locality and can be cached by the main cache.

The concept of nontemporal streaming (NTS) cache [Rivers et al. 1996] is similar to the concept of assist cache. The cache is split into a larger, direct-mapped, temporal subcache, and a smaller, fully associative, nontemporal subcache. The decision on which subcache should be used is made at runtime. Each entry in the temporal subcache is associated with a bit vector containing one bit for each word used. If no reuse is detected while a particular cache block was in the temporal subcache, it is marked as nontemporal (on eviction from temporal subcache). Future accesses to this memory block would cause a fetch into a nontemporal subcache. If reuse is detected in a nontemporal subcache, the appropriate block is marked back to temporal.

### 2.2.2. Exploitation of Spatial Locality

The spatial locality of data accesses has traditionally been exploited by "sequential" prefetching techniques and sectored caches. "Sequential" prefetching techniques, such as "always," "tagged," "miss," and "bidirectional" [Tse and Smith 1998], prefetch blocks that are neighbors to the block being currently accessed. In this way, prefetching utilizes spatial locality across cache block boundaries. Unfortunately, as was found in Tse and Smith (1998), these prefetching techniques increase the bus traffic, make cache tag and data ports busy, and introduce pollution into the cache.

Sectored caches exploit the fact that in many cases there is not enough spatial locality to justify bringing the entire block into the cache.

In sectored caches, each cache block is divided into several (usually two or four) subblocks that share the common tag; however, each one has its own validity bit. When a nonspatial access occurs, only one of the subblocks is brought into the cache, therefore reducing bus traffic. However, this approach still wastes cache space in the case of nonspatial blocks, since the unused subblocks cannot be used to cache other data.

One interesting approach that avoids this waste of cache space is presented in Johnson et al. (1997b). In this design, the entire cache contains single-word blocks; however, for data that are detected as being spatial, several blocks (corresponding to one larger block in split designs) are fetched. In effect, several consecutive blocks containing spatial data behave as if they together form one larger cache block. The main drawback of this approach is that it involves increased tag overhead (more tags for a cache of equal capacity) over the conventional cache, although most of the cache blocks are used by the spatial data.

A multiword cache block is a waste of cache space, if the cached data do not exhibit spatial locality, while a single-word cache block introduces unnecessary tag overhead for data that do exhibit spatial locality. Caches that are split according to the spatial locality exploit these facts. The cache is split into two subcaches. Spatial data are cached in a subcache with a larger block size, while nonspatial data are cached in a subcache that contains smaller blocks (typically one word).

The split temporal/spatial cache was introduced in Milutinović (1995a). In this design, the cache is split into a temporal subcache (with block size of one word) and a spatial subcache (with a block size of four words). Three possibilities are explored for the ratio between capacities of spatial and temporal primary subcaches: 1:1, 1:2, and 1:4. A secondary cache is included only for the temporal subcache.

The decision as to which subcache to use is done at run time, profile time, or both, using the same counter-based heuristic. Two counters per cache block are present in the spatial subcache. One of those counters, the $Y$ counter, is incremented at each access to a particular cache block. At the same time, the other counter ($X$ counter) is incremented when the upper half of the cache block is accessed and decremented at access to the lower half. When the $Y$ counter saturates, and the absolute value of the $X$ counter is larger than a specified threshold, the cache block is marked as temporal and removed from the spatial subcache. The "always" prefetch [Tse and

Smith 1998] strategy is also employed in this design, but only if both the block causing the prefetch and the prefetched block are marked as spatial. In this prefetching scheme, each spatial cache access causes one consecutive cache block to be prefetched.

The major drawback of this design is the assumption that most data are either predominantly temporal or predominantly spatial. In fact, most data exhibit both types of locality. Two issues result from this. The first is that a secondary cache would be useful in the spatial part, since it caches the data that exhibit both types of locality as well as spatial only data. The other issue is the counter-based heuristic. If data are accessed so that one word is used several times before the next word is accessed, the heuristic may mark that block as temporal if the $Y$ counter saturates while only one-half of the block is used. Thereafter, this block is always cached in the temporal part and causes four misses (one per word used) instead of only one, which it would cause in the spatial subcache.

An interesting piece of research is presented in Sahuquillo and Pont (1999). It is an extension of the split temporal/spatial cache for the shared memory multiprocessor systems.

The dual-data cache was introduced in Gonzalez et al. (1995). In this design, the primary data cache is split into a smaller (33% of total cache capacity) temporal subcache and a larger (66% of the total cache capacity) spatial subcache. The temporal subcache has a block size of 8 bytes, while the spatial subcache has a larger block size (16 or 32 bytes). The decision as to which subcache to use is made at run time, using a variant of the stride directed prefetch predictor. The predictor, in essence, tries to detect if a particular load/store instruction accesses memory at addresses that differ by a constant stride. Using this information, at each cache miss the cache controller decides if the data should be cached in temporal subcache or the spatial subcache, or not cached at all. Prefetching the next block on a cache miss is also done in this design, but only in the spatial subcache. The most important drawback of this design is that its locality detection is based on instructions rather than on data. This means that if the CPU accesses a data block in a spatial manner, but words of that block are accessed by different instructions, the detection logic could still decide that a block is nonspatial. Moreover, one instruction could see the block as nonspatial while the other may see it as spatial. This means that a particular word may be cached in both the spatial and nonspatial subcaches. While the authors have shown that this does not lead to inconsistencies if properly implemented, the performance may still suffer.

A substantially modified design of the dual data cache is proposed in Sanchez et al. (1997). The temporal subcache in this new design is only a small fully associative buffer (up to 16 single word 8-byte entries) while the spatial subcache remains large with a block size of 32 bytes.

The decision as to which subcache to use for each data access is made at compile time, by performing data locality analysis. Moreover, the compiler can mark a particular load/store instruction as noncached (i.e., bypass), if no locality for that particular data access is expected. The main drawback of this approach is that the changes to the instruction set are required to implement different flavors of load/store instructions. Also, locality information is still based on instructions rather than on data, and the locality of many data accesses is difficult to determine at compile time.

Research on array cache [Tomasko et al. 1997] design splits the primary data cache into a smaller (25% of the total cache capacity) scalar subcache and a larger (75% of the total cache capacity) array subcache. Block size in the scalar subcache is 32 bytes, while in the array subcache it is larger (experimentally varied from 64 to 512 bytes). The decision as to which subcache to use for a particular data access is done at compile time, by marking the scalar variable accesses to use the scalar subcache, while array accesses go to the array subcache. The main drawback of this approach is that changes to the instruction set are required to implement different flavors of load/store instructions. In addition, the scalar/array heuristic may be widely inaccurate. Many arrays are accessed in a "random" manner that exhibits almost no spatial locality, while scalars that are used in a particular part of the program are usually stored in the neighboring memory words.

### 2.2.3. The Most Recent Contributions to Locality Exploitation

Prvulovic et al. (1999b) categorize data into two categories: spatial and nonspatial. Such a data treatment enables the performance to be considerably improved over the approach of Milutinović (1995a).

Kalamatianos et al. (1999) stress temporal locality and different analysis methods aimed at the elimination of conflicts between different program units in the uniprocessor cache.

Temam (1999) moves the concept into the upper memory levels, and introduces an algorithm that can be used for the efficient exploitation of spatial and temporal locality in upper memory levels.

Kandemir et al. (1999) improve cache locality by combining loop and data transformations, using a novel compiler-based algorithm that improves the overall performance considerably.

The February 1999 issue of *IEEE Transactions on Computers* is completely devoted to research activities oriented to enhancing and exploiting different types of localities in both uniprocessor and multiprocessor systems. The importance of the problem and various aspects of it are treated in Milutinović and Valero (1999).

The author and his associates were active in the cache architecture research and design. For more information, see Milutinović et al. (1995a, 1996a, 1996b) and Prvulovic et al. (1999a, 1999b).

## PROBLEMS

**2.1.** For the case of a direct-mapped cache, develop a piece of code that destroys the data item to be needed next. Show what happens, for the same example, in the case of a set-associative or a fully associative cache.

**2.2.** Compute the transistor count saving, for one of the popular microprocessors, if pointers to page numbers are used rather than page numbers, as in Seznec (1996). What is the impact on the clock cycle time?

**2.3.** Develop the difference-bit approach [Juan et al. 1996] for the case of a four-way set-associative cache memory. What is the problem with such a solution?

**2.4.** Create a piece of code that demonstrates the benefits of the assist cache approach from Chan et al. (1996). Create another piece of code, which includes data with much more spatial locality, and show how many clock cycles have been saved.

**2.5.** Compare the victim cache from Jouppi (1990) and the selective victim cache from Stiliadis and Varma (1997) for a piece of code of your choice. What further improvements would you suggest?

**2.6.** For a piece of code of your choice, show which of the two cases is better: (**a**) fewer cycles that are longer or (**b**) more cycles that are shorter. Also create an example that shows the opposite, and explain what are the code characteristics that make one or the other case superior.

**2.7.** Develop a detailed block scheme of the spatial footprint predictor (SFP) cache from Kumar and Wilkerson (1998). Propose an improvement that takes into consideration the temporal locality, as well.

**2.8.** Explain why the trace cache continues to contribute to the overall system performance when the size of the trace cache grows. Why is that not the case with the conventional instruction cache?

**2.9.** Explain the essence of branch promotion and trace packing. Compare performance and complexity, for these two improvements of the trace cache approach.

**2.10.** Propose alternative ways to measure the level of spatial and temporal locality of data in a given program. Propose ways to visualize the changes, during the execution of a program.

# CHAPTER 3

# Instruction-Level Parallelism

This chapter includes two sections. The first one covers basic issues (background). While the second one discusses advanced issues (state of the art).

## 3.1. BASIC ISSUES

Instruction-level parallelism (ILP) and its efficient exploitation are crucial for maximization of the speed of compiled high-level language (HLL) code. Methods to exploit ILP include superscalar (SS), superpipelined (SP), and very long instruction word (VLIW) processing. In principle, ILP can be viewed as a measure of the average number of instructions that an appropriate SS, SP, or VLIW processor might be able to execute at the same time.

Instruction timing of a RISC processor and an SS processor are given in Figures 3.1 and 3.2, respectively. The ILP is a function of the number of dependencies in relation to other instructions. Unfortunately, an architecture is not always able to support all available ILP. Consequently, another measure—machine-level parallelism (MLP)—is often used, as a measure of the ability of an SS processor to take advantage of the available ILP. For the best performance/complexity ratio, the MLP and the ILP have to be balanced.

Fundamental performance limitations of an SS processor are illustrated in Figures 3.3, 3.4, and 3.5, respectively. Figure 3.3 explains the case of true data dependencies in a superscalar environment (one cycle is lost). Figure 3.4 explains the case of procedural dependencies in the same superscalar environment (the number of lost cycles is much larger). Figure 3.5 explains the case of resource conflicts in the same superscalar case (effects are the same as in Figure 3.3, but the cause is different). Finally, Figure 3.6 shows the impact of resource conflicts for the case of a long operation, in two different scenarios, when the execution unit is internally not pipelined (worst-case scenario, from the performance point of view) and internally fully pipelined (better-case scenario, from the performance point of view).

Issue (fetch and decode), execution (in the more general sense), and completion (writeback that changes the state of the finite-state machine called *microprocessor*) are the major elements of an instruction. In the least

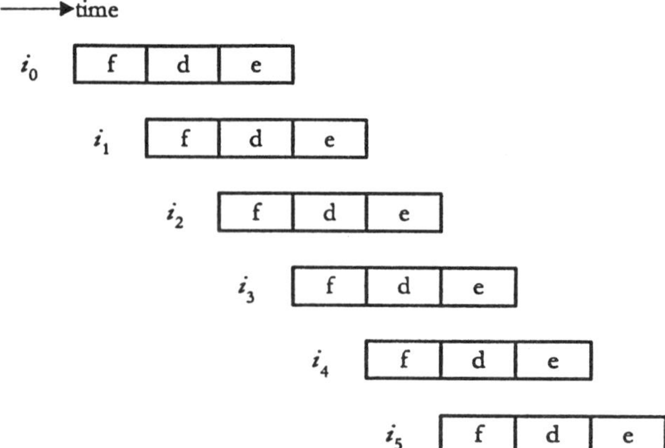

**Figure 3.1.** Instruction timing in a RISC processor without memory limitations ($i$—instruction; f—fetch; d—decode; e—execute). [*Source:* Johnson (1991).]
*Comment:* This is a typical RISC-style instruction execution pipeline with three stages. It represents the baseline for explanation of other cases to follow.

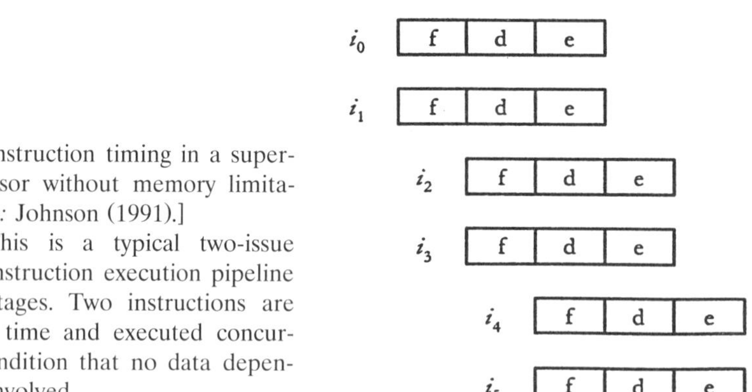

**Figure 3.2.** Instruction timing in a superscalar processor without memory limitations. [*Source:* Johnson (1991).]
*Comment:* This is a typical two-issue superscalar instruction execution pipeline with three stages. Two instructions are fetched at a time and executed concurrently, on condition that no data dependencies are involved.

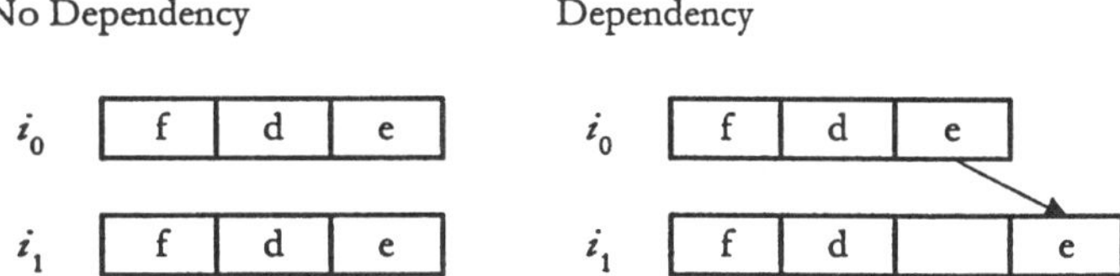

**Figure 3.3.** True data dependencies—effect of true data dependencies on instruction timing. [*Source:* Johnson (1991).]
*Comment:* The same instruction pipeline from Figure 3.2 is presented, in the case when the second of the two concurrently fetched instructions is data dependent on the first one.

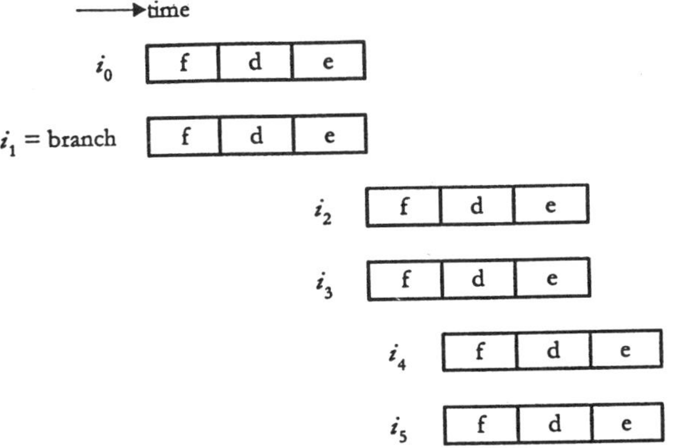



**Figure 3.4.** Procedural dependencies—effect of procedural dependencies on instruction timing. [*Source:* Johnson (1991).]
*Comment:* The same instruction pipeline from Figure 3.2 is presented again, now for the case when the second instruction is a branch.

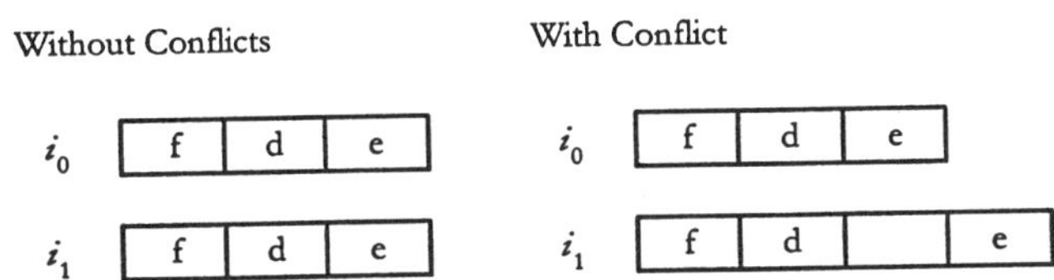



**Figure 3.5.** Resource conflicts—effect of resource conflicts on instruction timing. [*Source:* Johnson (1991).]
*Comment:* This figure, which is related to resource conflict, looks the same as a previous figure that is related to data dependency, except that causes of the one cycle loss are different in the two different cases.

sophisticated microprocessors these elements are in order. In the most sophisticated microprocessors these elements are out of order, which means better performance for more VLSI complexity. The essence will be explained using an example from Johnson (1991).

In this example, a microprocessor that can issue two instructions at a time, execute three at a time, and complete two at a time is assumed. Furthermore, the following program related characteristics are assumed: (1) instruction I1 requires two cycles to execute, (2) instructions I3 and I4 conflict for a functional unit, (3) instruction I5 depends on the datum generated by I4, and (4) instructions I5 and I6 conflict for a functional unit. The case of the in-order issue and in-order completion is given in Figure 3.7. The case of in-order issue and out-of-order completion is given in Figure 3.8. The case of out-of-order issue and out-of-order execution is given in Figure 3.9.

**Case 1**

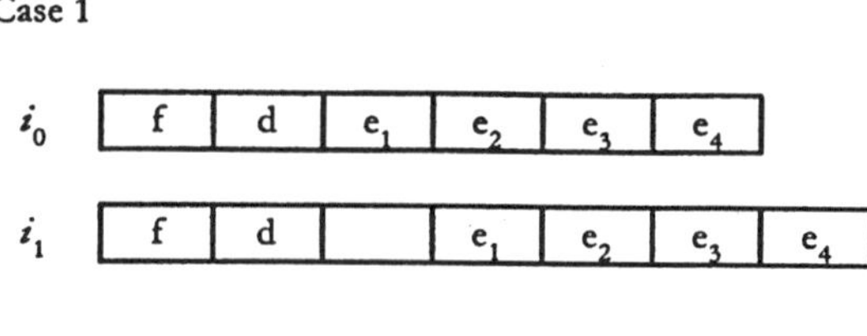

**Case 2**

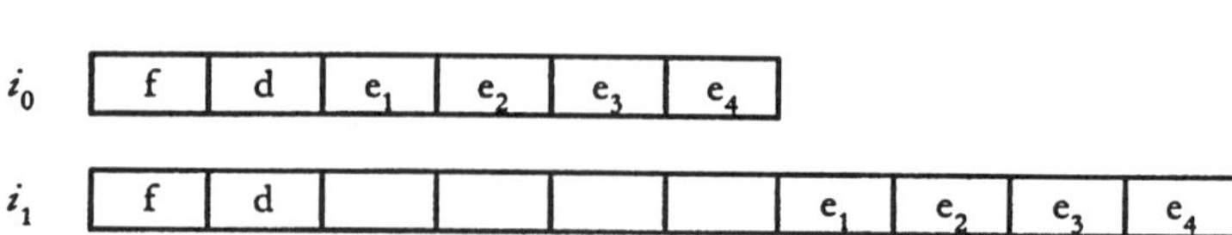

**Figure 3.6.** Resource conflicts—effect of resource conflicts on instruction timing. [*Source:* Johnson (1991).]

*Comment:* The same instruction pipeline from Figure 3.5 is presented, under the condition when the execution takes several pipeline stages to complete. Under such a condition, two alternative solutions represent a price/performance trade-off. The one that wastes less cycles is more expensive to implement.

| Decode | | | Execute | | | Writeback | | Cycle |
|---|---|---|---|---|---|---|---|---|
| $I_1$ | $I_2$ | | | | | | | 1 |
| $I_3$ | $I_4$ | | $I_1$ | $I_2$ | | | | 2 |
| $I_3$ | $I_4$ | | $I_1$ | | | | | 3 |
| | $I_4$ | | | | $I_3$ | $I_1$ | $I_2$ | 4 |
| $I_5$ | $I_6$ | | | | $I_4$ | | | 5 |
| | $I_6$ | | | $I_5$ | | $I_3$ | $I_4$ | 6 |
| | | | | $I_6$ | | | | 7 |
| | | | | | | $I_5$ | $I_6$ | 8 |

**Figure 3.7.** Example II (II—in-order issue, in-order completion). [*Source:* Johnson (1991).]

*Comment:* Instructions I1 and I2 are issued together to the execution unit and have to be written back (i.e., completed) together. The same applies for the pair I3 and I4, as well as the pair I5 and I6. Total execution time is eight cycles.

| Decode | | Execute | | | Writeback | | Cycle |
|---|---|---|---|---|---|---|---|
| $I_1$ | $I_2$ | | | | | | 1 |
| $I_3$ | $I_4$ | $I_1$ | $I_2$ | | | | 2 |
| | $I_4$ | $I_1$ | | $I_3$ | $I_2$ | | 3 |
| $I_5$ | $I_6$ | | | $I_4$ | $I_1$ | $I_3$ | 4 |
| | $I_6$ | $I_5$ | | | $I_4$ | | 5 |
| | | $I_6$ | | | $I_5$ | | 6 |
| | | | | | $I_6$ | | 7 |

**Figure 3.8.** Example IO (IO—in-order issue, out-of-order completion). [*Source:* Johnson (1991).]

*Comment:* Instructions I1 and I2 are issued together to the execution unit; however, they do not have to be written back together. The same applies to I3 and I4, as well as I5 and I6, which saves one clock cycle. Total execution time is seven cycles.

| Decode | | Window | Execute | | | Writeback | | Cycle |
|---|---|---|---|---|---|---|---|---|
| $I_1$ | $I_2$ | | | | | | | 1 |
| $I_3$ | $I_4$ | $I_1, I_2$ | $I_1$ | $I_2$ | | | | 2 |
| $I_5$ | $I_6$ | $I_3, I_4$ | $I_1$ | | $I_3$ | $I_2$ | | 3 |
| | | $I_4, I_5, I_6$ | | $I_6$ | $I_4$ | $I_1$ | $I_3$ | 4 |
| | | $I_5$ | | $I_5$ | | $I_4$ | $I_6$ | 5 |
| | | | | | | $I_5$ | | 6 |

**Figure 3.9.** Example OoO (OoO—out-of-order issue, out-of-order completion). [*Source:* Johnson (1991).]

*Comment:* Instruction pairs do not have to be issued together and can be executed out of order. For this reason, I6 is issued and executed before I5 (which is data dependent on I4) and consequently the execution time takes six clock cycles.

In this particular example, the simplest case (II, in-order issue and in-order completion) takes eight cycles to execute. The medium case (IO, in-order issue and out-of-order completion) takes seven cycles to execute. The most sophisticated case (OoO, out-of-order issue and out-of-order completion) takes six cycles to execute. These values have been obtained after the following machine related characteristics are assumed: (1) instruction is present in the decoding unit until its execution starts, and (2) each instruction is executed in the appropriate execution unit.

In the simplest case (II), the completion can be done only after both paired instructions get fully executed; both are completed together. This approach (II) is typical of scalar microprocessors and rarely used in superscalar microprocessors.

In the medium case (IO), execution of an instruction can start as soon as the related resource is available; also, the completion is done as soon as the execution is finished (I2 completes out of order). There are three basic cases when the issue of an instruction has to be stalled: (1) when such issue could generate a functional unit conflict, (2) when the instruction to be issued depends on the instruction(s) not yet completed, and (3) when the result of the issued instruction could be overwritten by an older instruction that takes longer to execute, or by a following instruction not yet executed. Special-purpose control hardware is responsible for stalling, in all three cases. This approach (IO) was first used in scalar microprocessors; however, its major use is in superscalar microprocessors.

In the most sophisticated approach (OoO), the processor is able to look ahead beyond the instruction that has been stalled (due to one of the reasons listed in the previous paragraph on the IO approach), which is not possible with the IO approach. Fetching and decoding beyond the stalled instruction is made possible by inserting a resource called an *instruction window*, between the decode stage and the execute stage. Decoded instructions are placed into the instruction window (if there is enough space there) and examined for resource conflicts and possible dependencies. The term *instruction window* (or *window of execution*) refers to the full set of instructions that may be considered simultaneously for parallel execution, subject to data dependencies and data conflicts. As soon as an executable instruction is detected (like I6 in Fig. 3.9), it is scheduled for execution, regardless of the program order, namely, out of order (the only condition is that program semantics are preserved). This approach (OoO) introduces one additional type of hazard, when instruction $N + 1$ destroys the input of instruction $N$ (this case has to be watched for by the control hardware). So far, this approach has been used only in superscalar microprocessors.

In this context, important roles are played by the BHT (branch history table) and the BTB (branch target buffer). The BHT helps determine the branch outcome. The BTB helps compute the branch target address. These issues are discussed in more detail later.

A related approach—VLIW (very long instruction word)—is shown in Figure 3.10. Single instruction specifies a larger number of operations to be executed concurrently. Consequently, the number of run-time instructions is smaller, but the amount of compile-time activities is larger (and the code size may increase if the compiler is not adequate for the given architecture or application). The approach is well suited for special-purpose architectures or applications and not well suited for making new VLIW machines, which are binary-compatible with existing machines (binary compatibility is the ability to execute a machine program written for an architecture of an earlier generation).

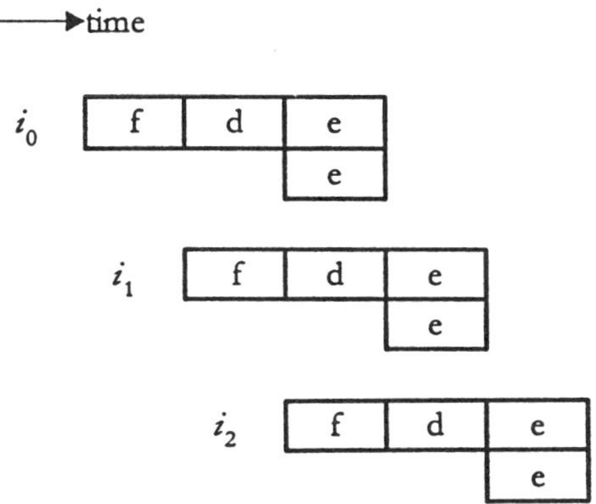

**Figure 3.10.** Instruction timing of a VLIW processor. [*Source:* Johnson (1991).]
*Comment:* The major difference between VLIW and superscalar processing is that VLIW fetches one instruction at the time; however, this instruction includes specifiers for several operations to be executed. On the contrary, superscalar fetches several instructions at the time; however, each one specifies only one operation to execute. Note that the superscalar approach makes it easier to create a microprocessor that is code-compatible with previous generation machines of the same manufacturer.

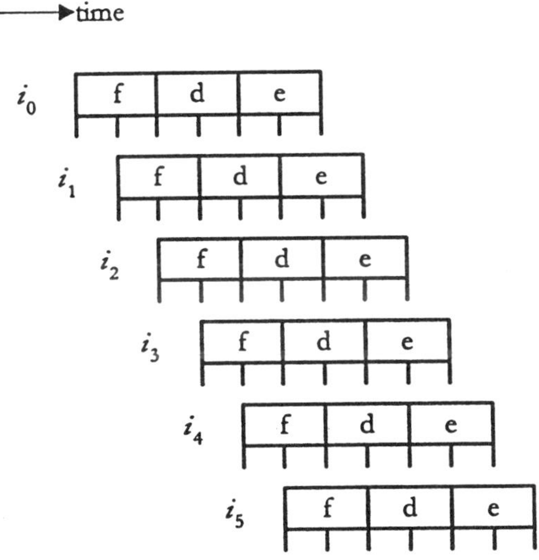

**Figure 3.11.** Instruction timing in a superpipelined processor. [*Source:* Johnson (1991).]
*Comment:* This figure shows the case when the basic clock is subdivided into two subclocks. In theory, the subdivision can go beyond the limit of two. In practice, it rarely goes beyond two, due to the increased complexity of hardware design and increased problems in software utilization.

The related superpipelined approach is shown in Figure 3.11. Stages are divided into substages; the more substages per stage, the deeper the superpipelining. Figure 3.11 refers to the case with the depth of pipelining equal to two. Superpipelining (SP) takes longer to generate a result (1.5 clock cycles for two instructions), compared with superscaling (SS), which takes a shorter time (1 clock cycle for two instructions). On the other hand, SP takes less time for simpler operations (e.g., 0.5 clock cycles), compared with SS, which takes longer (e.g., 1 clock cycle, if and when no clock with finer resolution is available). Latency is shorter with SP (shorter basic clock), and consequently the outcome of a simple branch test is known sooner, but the clock period is shorter with SS (no extra latches necessary for the SP approach). Resources are less duplicated with the SP approach, but its major problem is clock skew. Which is better—SS or SP—depends on the technology and the application.

Of course, the best performance/complexity ratio can be achieved using hybrid techniques that combine SS, SP, and VLIW. Theoretical limits on the ILP exploitation are determined by conditional branches, which affect clock count, and the extra complexity that grows quadratically and affects clock speed. According to several authors, practical limits of ILP exploitation are an eight-way issue (if only hardware employed) and an $N$-operation VLIW (if both hardware and compiler employed); $N > 8$.

Issues discussed so far will briefly be revisited through different examples of different superscalar microprocessors. Each microprocessor will be discussed according to the same template: superscalar features, block diagram, major highlight, branch prediction, execution issues, functional units, and cache memory. For more examples and a more detailed treatment, see Stallings (1996), or the original manufacturer literature.

### 3.1.1. Example: MIPS R10000 and R12000

Microprocessor MIPS R10000 is a four-way superscalar machine. It fetches four instructions at a time from the I cache. Its block diagram is given in Figure 3.12.

Before being put into the I cache, the instructions get precoded, to help the decoding process. While in the cache, code compactness is the issue. While in the decoder, code decodability is the issue. In this context, compactness means less bits per instruction; decodability means less gate delays per decoding. In the case of R10000, the difference between the two forms is equal to 4 bits. Consequently, the extra logic is minimal (since the number of instructions in the decoding unit is much smaller than the number of instructions in the typical size I cache), and the decoding process can be done immediately on the instruction fetch (from the I cache).

The branch prediction unit is placed next to the I cache and shares some features of the cache in order to speed up the branch prediction. The I cache has 512 lines and the branch prediction table (BPT) has 512 entries. The BPT is based on 2-bit counters, as explained later in this book. If a branch is

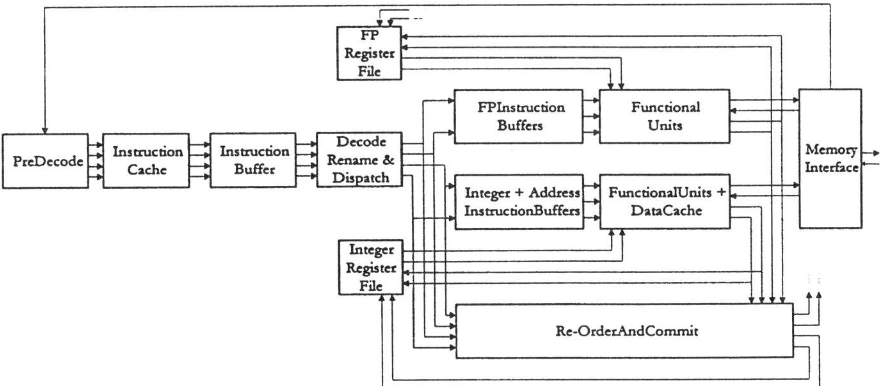


**Figure 3.12.** Organization of MIPS R10000 (FP—floating-point). [*Source:* Smith and Sohi (1995).]

*Comment:* Major units that distinguish microprocessor from its competitors are the predecode unit and the reorder-and-commit unit. The first one assists in compacting the code, which is important for the hit ratio of the L1 cache. The second one helps the out-of-order execution and helps to maintain precise states at the times of exceptions.

predicted as taken, instruction fetching is redirected after a cycle. During that cycle, instructions in the not-taken path continue to be fetched; they get placed into the cache called *resume cache*, to be ready if the prediction happens to be incorrect. The resume cache is large enough for instructions of the not-taken paths related to four consecutive branches.

Major execution phases are instruction fetch, instruction decoding, operand renaming (based on 64 physical and 32 logical registers), and dispatching to appropriate instruction queues (up to four instructions concurrently dispatched into three queues, for later execution in five different functional units).

During the register renaming process, the destination register is assigned a physical register, which is listed as unused in the so-called free list. At the same time, another list is updated to reflect the new logical-to-physical mapping. If needed, an operand is accessed through the list of logical-to-physical mappings.

During the instruction dispatching process, each queue can accept up to four instructions. Also, a reservation bit for each physical result register is set busy. Queues act like reservation stations holding instructions, and physical register designators act as pointers to data. Global register reservation bits for source operands are constantly being tested for availability. As soon as all source operands become available, the instruction is considered free and can be issued.

A reorder buffer is used to maintain a precise state at the time of exception. Exception conditions for noncommitted instructions are held at the reorder buffer. An interrupt occurs whenever an instruction with an exception is ready to be committed.

The five functional units are address adder, two integer arithmetic-logic units (ALUs) (one with a shifter and the other with an integer multiplier, in addition to basic adder and logic unit), floating-point adder, and floating-point multioperation unit (multiplication, division, and square root).

The on-chip primary cache is 32 kB large, two-way set-associative, and includes 32-byte lines. The off-chip secondary cache is typically based on the inclusion principle (if something is in the primary cache, it will also be in the secondary cache).

The architecture of R12000 is basically the same as the architecture of R10000, except that some features are considerably improved (address space, on-chip cache memory, etc.). Also, implementation technology reflects the state of the art at the time of its introduction.

### 3.1.2. Example: DEC Alpha 21164 and 21264

Microprocessor DEC Alpha 21164 is also a four-way superscalar machine. It fetches four instructions at a time from the I cache. Its block diagram is given in Figure 3.13.

Instructions are fetched and placed into one of two instruction buffers (each one is four instructions deep). Instructions are issued (from an instruction buffer) in the program order (not bypassing each other). One instruction buffer is used until emptied, and then the issuing from the next instruction buffer starts (a solution that makes the control logic much less complex).

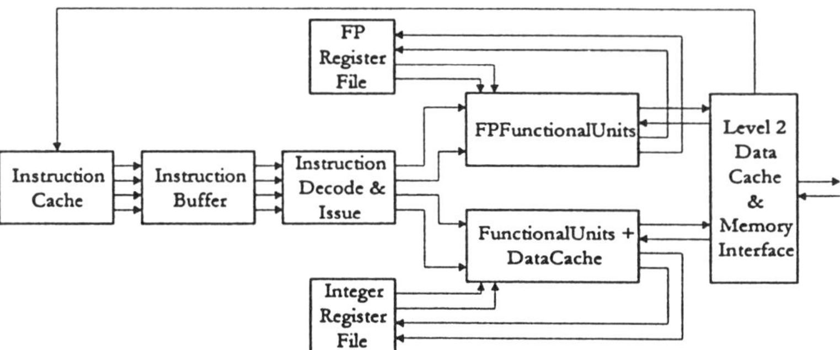


**Figure 3.13.** Organization of DEC Alpha 21164. [*Source:* Smith and Sohi (1995).] *Comment:* This microprocessor belongs to the group of speedemons; consequently, the execution unit includes only the blocks that are absolutely necessary; however, these blocks are designed aggressively (i.e., transistor count is often traded for faster execution), in order to achieve maximal clock speed.

Again, the branch prediction table is associated with the instruction cache. A 2-bit branch history counter is associated with each cache entry. Only one branch can be in the state when it is predicted but not yet resolved. Therefore, the issue is stalled on the second branch if the first one is not yet resolved.

After decoding, instructions are arranged according to the type of functional unit that they are to use. After the operand data are ready, instructions are issued to units at which they are to be executed. Instructions are not allowed to bypass each other.

In order to make the handling of precise interrupts easier, the instruction issue is in order. The final pipeline stage in the integer units updates the destination registers also in order. Bypass registers are included in the data path structure, so that data can be used before they are being written into their destination registers. The final bypass stages of the floating-point units update registers out of order. Consequently, not all floating-point exceptions result in precise interrupts.

This microprocessor includes four functional units: two integer ALUs (one for basic ALU operations plus shift and multiplication; the other for basic ALU operations and evaluation of branches), one floating-point adder, and one floating-point multiplier.

Two levels of cache memory reside on the CPU chip. The first-level cache, direct-mapped for fast one clock access, is split into two parts: an instruction cache and a data cache. Both the instruction cache and data cache are 8 kB large. The second-level cache (three-way set-associative) is shared (joint instruction and data cache). Its capacity is 96 kB.

The primary cache handles up to six outstanding misses. For that purpose, a six-entry MAF (miss address file) is included. Each entry includes the missed memory address and the target register address of the instruction, which exhibits a miss. If the MAF contains two entries with the same memory address, then the two entries will merge into one.

The more recent microprocessor from the same family [Gwennap 1997a], the DEC Alpha 21264, has the following characteristics: (1) four-way issue superscalar, (2) out-of-order execution architecture, (3) speculative execution, and (4) multihybrid branch prediction.

The cycle time is 600 MHz, which causes the microprocessor to deliver an estimated 40 SPECint95 and 60 SPECfp95 performance. This was made possible because both L1 and L2 caches are on the processor chip. Also, the path to memory enables the data transfer rate of over 3 GB/s.


### 3.1.3. Example: PowerPC 620 and 750

Microprocessor PowerPC 620 is a four-issue machine. Its block diagram is given in Figure 3.14.

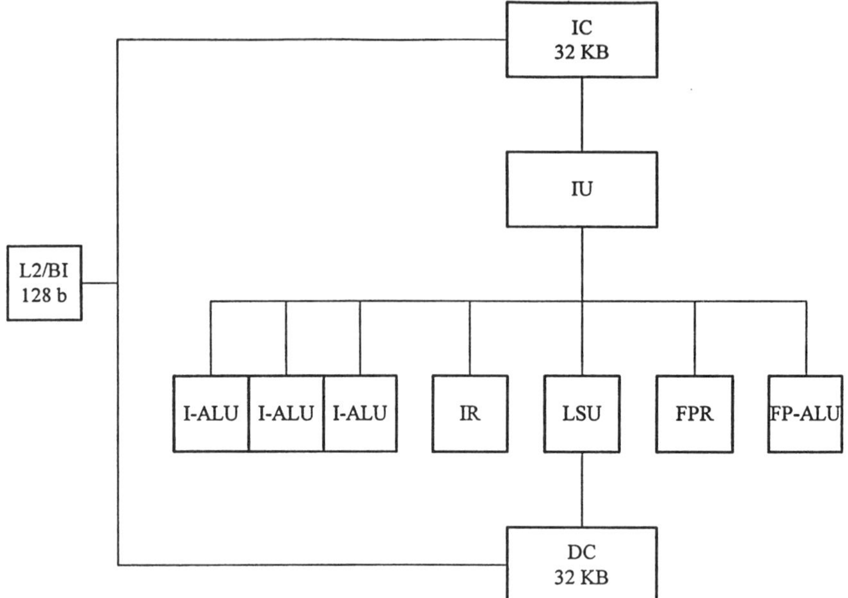


**Figure 3.14.** A block diagram of PowerPC 620 (DC—data cache; IC—instruction cache; IU—instruction unit; FP-ALU—floating-point ALU; FPR—floating-point registers; L2/BI—L2 bus interface; LSU—load/store unit; IR—instruction registers).

*Comment:* The stress in the PowerPC architecture is on good data throughput for advanced Internet applications.

PowerPC 620 is the first 64-bit implementation of the PowerPC architecture. It includes five independent execution units. Four instructions can be issued simultaneously to three integer units and one floating-point unit.

Branch prediction in the PowerPC 620 is based on a branch history table of 2048 entries. According to Thompson and Ryan (1994), it provides a branch prediction rate of about 90%.

The PowerPC 620 implements efficient prediction strategy and speculative execution. If prediction turns out to be correct, data are moved from temporary registers to permanent registers; otherwise, temporary registers are flushed without any damage to the run-time environment.

Each execution unit includes two or more reservation stations that store the dispatched instructions until after the results of other instructions are known. The PowerPC 620 can execute speculatively up to four unresolved branch instructions (the PowerPC 601 can execute speculatively only one unresolved branch instruction).

The on-chip L1 cache memory is split in two parts: an instruction cache memory and a data cache memory. Each one of the two parts is 32 kB in capacity and is implemented as an 8-way set-associative memory.

The PowerPC 750 microprocessor is also designed for desktop and portable markets. It includes an on-chip L2 cache controller supporting the backside L2 caches of up to 1 MB. It has an improved dynamic branch prediction unit, as well as an additional floating-point unit. The block scheme of the PowerPC 750 is similar to that of the PowerPC 620 [Kennedy et al. 1997].

The PowerPC 750 can achieve the SPECint95 performance of over 14 and the SPECfp95 performance of over 10. It is offered as a single-chip package, or as a small daughtercard integrating the processor and the L2 cache SRAMs [Pyron et al. 1998].

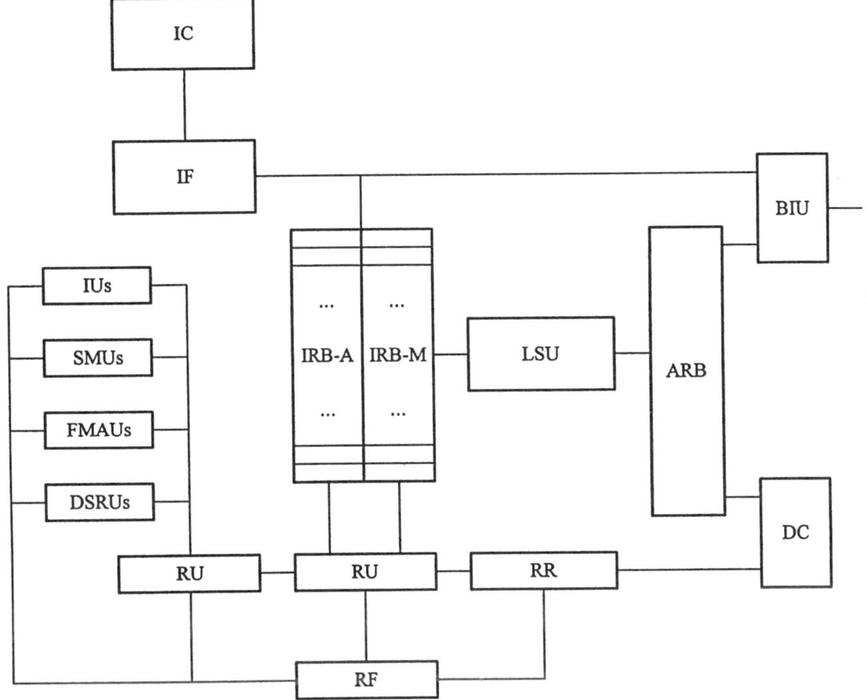



**Figure 3.15.** A block diagram of PA-8000 (ARB—address reorder buffer; BIU—bus interface unit; DC—data cache; DSRUs—divide and square units; FMUs—floating-point units; IC—instruction cache; IRB-A—instruction reorder buffer (ALU buffer); IRB-M—instruction reorder buffer (memory buffer); IUs—integer units; LSU—load/store unit; RF—register file; RR—rename registers; RU—retire unit; SMUs—shift/merge units). [*Source:* Kumar (1997).]
*Comment:* The stress in this architecture is on efficient speculative execution with on-the-fly instruction reordering and an unorthodox cache organization.

### 3.1.4. Example: Precision Architecture PA-8000 and PA-8500

The Hewlett-Packard PA-8000 is a 4-way superscalar microprocessor with efficient speculative execution and on-the-fly instruction reordering. A block diagram of PA-8000 is given in Figure 3.15.

The major architectural goal was to attain industry-leading performance on a broad range of applications. On the way to this goal, some fairly dramatic departures from the mainstream were implemented (especially in the branch prediction and cache domains). The architecture can achieve over 12 SPECint95 and over 20 SPECfp95.

Both dynamic and static branch prediction are used. Software selects the mode of branch prediction on the page-by-page basis. The execution unit supports speculative execution with on-the-fly instruction reordering and relays on a vector of functional units.

The primary cache of PA-8000 is external; PA architectures traditionally include a special on-chip assist cache to better exploit the spatial locality [Chan et al. 1996]. If the so-called SLH (spatial locality hint) bit is set, then, after being purged out of the assist cache, data go directly to memory, rather than going to the off-chip L1. In this way, large amounts of sequential data (data structures like records, vectors, matrices, etc.) are processed without polluting the off-chip caches.

The PA-8500 has a block diagram similar to that of the PA-8000 [Brauch and Fleischman 1998]. One significant difference is the $0.25$-$\mu$m fabrication process that allows a fast L1 cache to be placed on the same chip as the CPU. The cache includes almost 17 Mb of storage on over 100 million transistors. The on-chip L1 data cache is 1 MB, and the on-chip L1 instruction cache is 0.5 MB.

### 3.1.5. Example: AMD K5 and K6

The AMD K5 microprocessor architecture represents a superscalar implementation of the Intel x86 instruction set. Its block diagram is given in Figure 3.16.

Since the Intel x86 instruction set uses variable-length instructions, the fetched instructions are sequentially precoded before being placed into the instruction cache. Instructions in the byte queue wait to be dispatched.

The branch prediction logic is integrated with the instruction cache. There is one prediction entry per cache line. Prediction entry includes a bit to reflect the direction taken by the previous execution of the branch. Each prediction entry also contains a pointer to the target instruction, so it is known where in the cache it can be found.

Decoding takes two cycles (because the x86 instruction set is complex) and creates ROPs (RISC operations). More complex x86 instructions are converted into a sequence of ROPs (essentially a microroutine). After the first decode cycle, instructions get dispatched to reservation stations. Data come to the reservation stations from the register file or the reorder buffer.

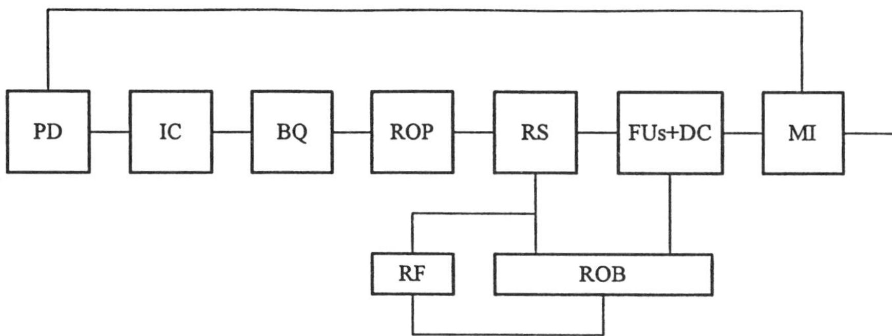

**Figure 3.16.** A block diagram of AMD K5 (BQ—byte queue; DC—data cache; IC—instruction cache; MI—memory interface; PD—predecode; RF—register file; ROP—RISC-like OPeration; ROB—reorder buffer; RS—reservation station).
*Comment:* The main goal of the K5 project was to design an architecture that enables the relatively slow x86 instruction set to execute fast.

Once the data become available, instructions and data enter the functional units. There are six functional units: two integer ALUs, one FP unit, two load/store units, and a branch unit. Result data are kept in the reorder buffer before being placed into the register file [Smith and Sohi 1995].

The core of the AMD K6 processor is the RISC86 microarchitecture, similar to that of the AMD K5. Many resources are larger/longer in size and new resources are added. For example, in comparison with the AMD K5, the AMD K6 architecture includes a multimedia unit [Fetherston et al. 1998].

### 3.1.6. Sun UltraSPARC I and II

The Sun UltraSPARC I is a four-issue superscalar processor based on the SPARC version 9 64-bit RISC architecture. A block diagram of the Sun UltraSPARC I is given in Figure 3.17.

The core instruction set is extended to provide support for graphics and multimedia (2D image processing, 2D and 3D graphics, and image compression). It is especially fast for MPEG-2. The prefetch unit can fetch instructions from all levels of memory hierarchy. The next instruction is fetched according to the result of branch prediction. Instruction execution is based on a 9-stage pipeline. The L1 cache is on the same chip with the CPU and is divided into two parts: an instruction cache (16 kB) and a data cache (also 16 kB).

The Sun UltraSPARC II has the same number of integer unit and floating-point unit registers, and the same width of system buses. It is basically the same architecture; the major differences are technological, resulting in faster speed.

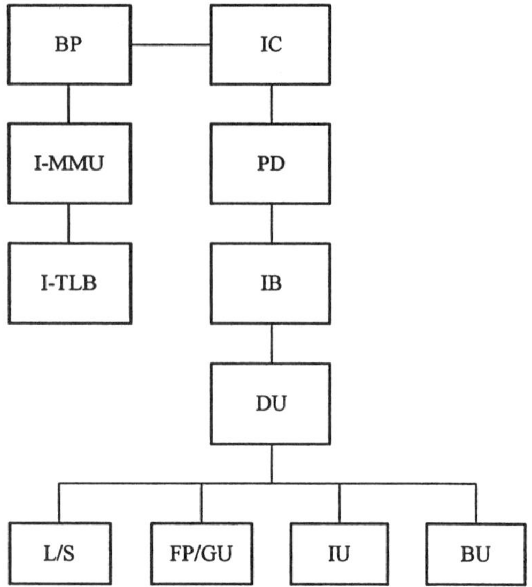


**Figure 3.17.** A block diagram of UltraSPARC I (BP—branch prediction; BU—branch unit; DU—dispatch unit; FP/GU—floating point/graphics unit; IB—instruction buffer; IC—instruction cache; I-MMU—instruction MMU; I-TLB—instruction TLB; IU—integer unit; L/S—load/store unit; PD—predecoding unit). [*Source:* Tremblay and O'Connor (1996).]
*Comment:* The major strength of Sun UltraSPARC I is efficient support for multimedia.

### 3.1.7.  A Summary

An innovative point-of-view on modern microprocessor based computers is given in Tredennick (1996). It includes an interesting diagram that shows differences between the first five generations of microprocessors, as shown in Figure 3.18.

Tredennick's paper (1996) also includes a number of interesting statements. The higher the chip complexity, the larger the design teams, the higher the design costs, and the longer the idea-to-market time. Consequently, current leading-edge design efforts exceed 3 years, while new product developments are required about every 18 months to remain competitive. In other words, 18 months after a company introduces a new product, the competing company introduces a new product, with twice the performance. Therefore, overlapping design teams have to be used, in order to stay on the competitive edge. Also, development costs per microprocessor rise at the rate of 25% per year. This increased development cost can be absorbed only with a larger number of units sold, as indicated in Figure 3.19.

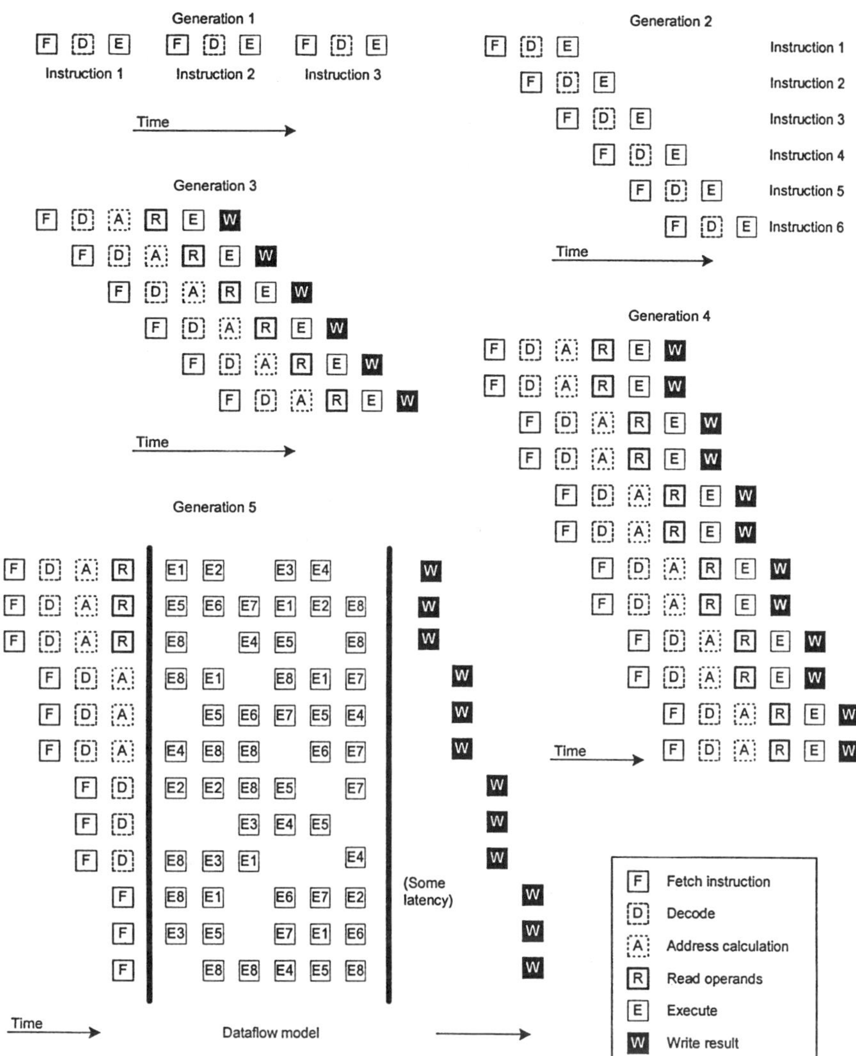



**Figure 3.18.** Differences among the first five generations of microprocessors.
*Comment:* Changes for five generations of microprocessors are shown using an execution "scrabble" diagram.

If the curves of Figure 3.19 are examined more carefully, we see that Intel's products are to the right of the $X$ axis (where the development costs are easy to amortize), which makes it difficult to compete against Intel. Also, since development costs rise at 25% per year, unless the unit volume of an Intel competitor grows at better than 25% per year, its cost per chip will grow with each successive generation, while Intel's chip cost is almost constant

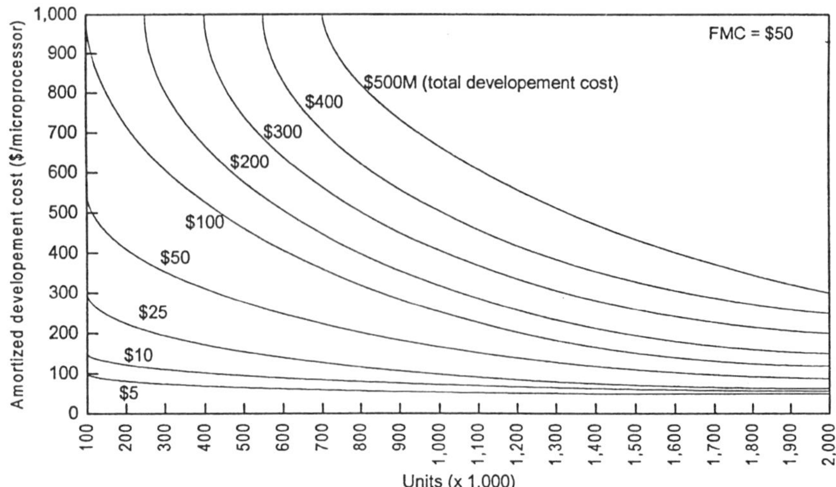


**Figure 3.19.** Amortized development cost versus the number of microprocessor chips sold (FMC—fixed manufacturing cost per microprocessor chip).
*Comment:* At 2 million units (microprocessor chips) sold, a $500 million total development cost amortizes at approximately $300 per unit.

from generation to generation, because being to the right of the $X$ axis causes Intel to operate under conditions of low-slope curves (now, it is clearer why Intel's products are stressed in this book).

Tredennick (1996) also discusses ways in which competition can be successful against Intel by: (1) increasing unit volumes (e.g., by being faster in penetrating into a new high-volume application) and (2) rendering the importance of the curve moot (e.g., by making a CPU that is a low-margin component in systems).

## 3.2. ADVANCED ISSUES

This section contains the author's selection of research activities that, in his opinion, have made important recent contributions to the field and are compatible with the overall profile of this book.

Jourdan et al. (1995) describe an effort to determine the minimal number of functional units (FUs) for the maximal speed-up of modern superscalar microprocessors. Analysis includes MC 88110, Sun UltraSPARC, DEC Alpha 21164 (IO—in-order superscalars), plus IBM 604 and MIPS R10000 (OoO—out-of-order superscalars). Basic characteristics of modern superscalar processors are given in Table 3.1.

A major conclusion of the study is that the number of FUs in modern superscalar microprocessor architectures needed to exploit the existing ILP is

**TABLE 3.1.  Basic Characteristics of Modern Superscalar Processors**

| Processor | PPC 604 | MC 88110 | UltraSPARC | DEC 21164 | MIPS 10000 |
|---|---|---|---|---|---|
| DataShip | 94 | 92 | 95 | 95 | 95 |
| Degree | 4 | 2 | 4 | 4 | 4 |
| IntegerUnit | 2 | 2 | 2 | 2 | 2 |
| ShiftUnit | 2 | 1 | 2 | 2 | 2 |
| DivideUnit | 1 | 1 | 2 | 2 | 2 |
| MultiplyUnit | 1 | 1 | 2 | 2 | 2 |
| FPAddUnit | 1 | 1 | 2 | 1 | 1 |
| ConvertUnit | 1 | 1 | 2 | 1 | 1 |
| FPDvideUnit | 1 | 1 | 2 | 1 | 1 |
| FPMultiplyUnit | 1 | 1 | 2 | 1 | 1 |
| DattCacheUnit | 1 | 1 | 1 | 2 | 1 |

*Source:* Jourdan (1995).
*Comment:* The most crucial issue here is the number of cache ports. A real two-port cache means that the cache memory complexity is doubled, which is the case with DEC Alpha 21164. However, the MIPS 10000 single-port cache, under certain conditions, can behave as a two-port cache. This is because the cache is designed to include two distinct modules. If the compiler is smart enough, it will schedule data fetches in such a way that the two single-port modules are always used concurrently.

from 5 to 9, depending on the application. If the degree of superscaling is four or more, the number of data cache ports becomes the bottleneck and has to be increased accordingly.

Conditions of the analysis imply the applications corresponding to the SPEC92 benchmark suite, a lookahead window of degree 2 to 8, and OoO execution architecture in all considered cases (where not included in the original architecture, the OoO capability is simulated). Effects of cache miss have not been studied (a large enough cache is implied). For details, the interested reader is referred to the original paper [Jourdan et al. 1995].

Wilson et al. (1996) propose and analyze the approaches for increasing the cache port efficiency of superscalar processors. A major problem is that, on one hand, cache ports are necessary for better exploitation of ILP, and on the other hand, they are expensive. The load/store unit data path of a modern superscalar microprocessor is given in Figure 3.20.

A solution offered by this research is to increase the bandwidth of a single port, by using additional buffering in the processor and wider ports in caches. Authors have proposed four different enhancements.

Conditions of this research imply split primary caches (two-way set-associative external 8-kB caches), a unified secondary cache (two-way set-associative 2-MB cache), and a main memory. Four ongoing cache misses are enabled by four special-purpose registers called MSHR (miss status handling registers). Concrete numerical performance results imply operating system SimOS and benchmark suite SPEC95.

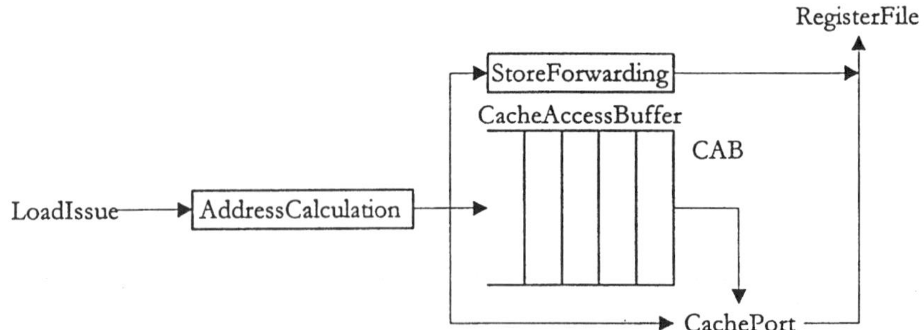


**Figure 3.20.** The load/store unit data path in modern SS microprocessors (CAB—cache address buffer). [*Source:* Wilson (1996).]
*Comment:* The structure supports four different enhancements:

1. Load all—if two or more loads from the same address are present in the buffer, the data that return from memory will satisfy all of them; this will not cause program errors, because the memory disambiguation unit already removed all loads that depend on not-yet-completed store instructions.

2. Load all wide—it is easy to widen the cache port (which means that "load all" can be applied to a larger data structure), so that the entire cache line (rather than a single word) can be returned from cache at one time; this is achieved by increasing only the number of interface sense amplifiers, which is not expensive.

3. Keep tags—if a special tag buffer is included (which holds tags of all outstanding data cache misses), newly arrived cache accesses that are sure to miss (because their addresses match some of the addresses in the tag buffer) can be removed from the cache access buffer, thus leaving room for more instructions that are potentially successful (i.e., likely to hit).

4. Line buffer—data returned from the cache can be buffered in some kind of L0 cache, which is fully associative, multiported, and based on the FIFO (first in–first out) replacement policy; if a line buffer contains the data requested by the load, the data will be supplied from the line buffer (good for data with a high level of temporal locality). Note that the four enhancements can be superimposed, to maximize performance.

Research performed at the Polytechnical University of Barcelona [Gonzalez and Gonzalez 1996] reports some very dramatic results. They have assumed the architecture of Figure 3.21, and they have tried to measure the real ILP in selected SPEC95 benchmarks. They have varied the parameters such as: (1) reorder buffer size (1024, 2048, 4096, 8192, 16384, and infinite), (2) memory ordering (in-order load, out-of-order load), (3) number of memory ports (1, 2, 4, no restrictions), and (4) register pressure (64, 128, 256, 512, infinite register count). The width of the fetch engine was chosen to be 64, which is much more than in contemporary microprocessors.

Surprisingly, an IPC (instructions per clock) of 9 to 50 was achieved in an ideal machine with the infinite number of resources and no memory dependencies, as indicated in Table 3.2. If memory dependencies are taken into

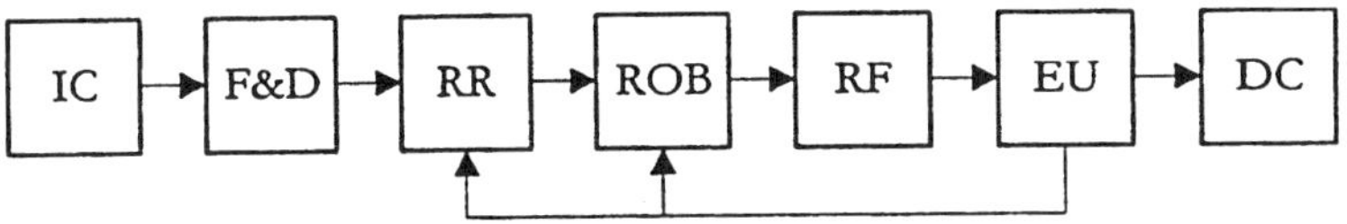

**Figure 3.21.** Block diagram of the simulated processor (IC—instruction cache; F&D —fetch and decode; RR—register renaming; ROB—reorder buffer; RF—register file; EU—execution unit; DC—data cache). [*Source:* Gonzalez and Gonzalez (1996).] *Comment:* Essentially, this is a traditional pipeline with two more stages included (register renaming and reorder buffer), as well as appropriate feedback links to these two new stages.

**TABLE 3.2.  The IPC for Various Reorder Buffer Sizes and No Memory Dependencies**[a]

| RB[b] Size<br>Program | 1024 | 2048 | 4096 | 8192 | 16384 | $\infty$ |
|---|---|---|---|---|---|---|
| Go | 30.63 | 41.00 | 47.69 | 49.53 | 50.13 | 50.13 |
| Compress | 31.59 | 43.19 | — | — | — | 43.19 |
| Vortex | 23.86 | 24.70 | 25.84 | 27.04 | 28.80 | 28.80 |
| Li | 17.97 | 18.11 | — | — | — | 18.11 |
| Fpppp | 16.04 | 18.63 | 21.93 | 25.53 | 28.40 | 28.40 |
| Applu | 29.85 | 31.43 | — | — | — | 31.43 |
| Wave5 | 18.56 | 18.56 | — | — | — | 18.56 |
| Swim | 16.71 | 16.71 | — | — | — | 16.71 |
| Turb3d | 9.02 | 9.03 | — | — | — | 9.03 |

*Source:* Gonzales and Gonzales (1996).

[a]A dash is used to indicate that the value did not change, that is, that the saturation point has been reached (saturation point is here defined as the value obtained by the infinite reorder buffer). Different benchmarks saturate at different reorder buffer sizes.

[b]RB—reorder buffer.

consideration, IPS drops down to the range of 4 to 19 for out-of-order load (Table 3.3), which is significantly more than was reported in a number of previous studies. The impact of the number of memory ports is given in Table 3.4, showing that 4-port memory should become mandatory as soon as permitted by technology. The impact of a smaller reorder buffer is given in Table 3.5. The register count limitation effects are summarized in Table 3.6, which advocated for 128 or even 256 flat (as opposed to windowed) registers.

The overall conclusion of the Gonzalez paper is that the available ILP is much longer compared to what current microprocessors can achieve. The major obstacles to considerably better microprocessor performance are memory dependencies.

One of the major problems of the superscalar microprocessors, in particular, and microprocessors, in general, is their flexibility. Execution units are fixed in their design; they execute a given set of operations, on a given set of

**TABLE 3.3. The IPC for In-Order and Out-of-Order Loads in the Presence of Memory Dependencies**[a]

|  | In-Order Load | Out-of-Order Load | No Dependencies |
|---|---|---|---|
| GO | 4.32 | 6.18 | 50.13 |
| Compress | 10.56 | 13.87 | 43.19 |
| Vortex | 5.17 | 5.76 | 28.80 |
| Li | 4.14 | 5.75 | 18.11 |
| Fpppp | 8.50 | 8.57 | 28.40 |
| Applu | 6.40 | 6.47 | 31.43 |
| Wave5 | 13.81 | 18.56 | 18.56 |
| Swim | 4.16 | 4.83 | 16.71 |
| Turb3d | 9.03 | 9.03 | 9.03 |

*Source:* Gonzalez and Gonzales (1996).

[a]The column without memory dependencies has been repeated from Table 3.2, so that the results can be compared. With most of the benchmarks, the drop in IPC is substantial.

**TABLE 3.4. IPC with One, Two, Four, or Infinitely Many Memory Ports**[a]

|  | 1 Port | 2 Ports | 4 Ports | No Dependencies |
|---|---|---|---|---|
| Go | 2.27 | 3.86 | 5.30 | 6.18 |
| Compress | 5.84 | 9.06 | 11.50 | 13.87 |
| Vortex | 2.12 | 3.63 | 4.94 | 5.76 |
| Li | 2.08 | 3.56 | 5.04 | 5.75 |
| Fpppp | 2.26 | 4.16 | 6.55 | 8.57 |
| Applu | 2.51 | 3.79 | 4.80 | 6.47 |
| Wave5 | 5.22 | 9.82 | 18.55 | 18.56 |
| Swim | 2.93 | 3.84 | 4.54 | 4.83 |
| Turb3d | 4.46 | 8.94 | 9.01 | 9.03 |

*Source:* Gonzalez and Gonzalez (1996).

[a]For comparison purposes, the out-of-order load column has been repeated from Table 3.3, as the case without any restrictions on the number of memory ports. No benchmark saturates at two memory ports, which is used in most contemporary microprocessors.

data types. Recent advances in field programmable gate array (FPGA) technology enable several reconfigurable execution units to be implemented in a superscalar microprocessor. Of course, full benefits of reconfigurability can be achieved only if appropriate compilers are developed.

A survey of reconfigurable computing can be found in Villasenor (1997). The first attempt at reconfigurable computing dates back to the late 1960s proposal by Gerald Estrin of UCLA, which was highly constrained by technology capabilities of those days. The latest attempts include a number of designs oriented to FPGAs with over 100,000 logic elements, including the effort of John Wawrzynek of UC Berkeley. An earlier research effort by Eduardo Sanchez and associates at the EPFL in Lausanne, Switzerland starts with the concept of reconfiguration at program or program segment bound-

**TABLE 3.5. The IPC as a Function of a Limited Size Reorder Buffer**[a]

|         | 32   | 64   | 128  | 256   | $\infty$ |
|---------|------|------|------|-------|----------|
| Go      | 4.29 | 5.43 | 6.02 | 6.17  | 6.18     |
| Compress| 4.11 | 5.37 | 6.92 | 10.12 | 13.87    |
| Vortex  | 4.78 | 5.65 | 5.74 | 5.74  | 5.76     |
| Li      | 4.39 | 5.42 | 5.75 | 5.75  | 5.75     |
| Fpppp   | 5.83 | 7.17 | 7.45 | 7.61  | 8.57     |
| Applu   | 4.37 | 5.25 | 5.51 | 5.57  | 6.47     |
| Wave5   | 4.52 | 5.90 | 6.77 | 7.58  | 18.56    |
| Swim    | 3.46 | 4.40 | 4.82 | 4.83  | 4.83     |
| Turb3d  | 7.11 | 8.11 | 8.39 | 8.39  | 9.03     |

*Source:* Gonzalez and Gonzalez (1996).

[a] Very few benchmarks saturate at 128 or 256 entry reorder buffer.

**TABLE 3.6. The IPC as a Function of a Limited Physical Register Count**[a]

|         | 64   | 128  | 256   | 512   | $\infty$ |
|---------|------|------|-------|-------|----------|
| Go      | 4.74 | 5.97 | 6.17  | 6.18  | 6.18     |
| Compress| 5.45 | 8.81 | 12.55 | 13.87 | 13.87    |
| Vortex  | 5.37 | 5.74 | 5.74  | 5.74  | 5.76     |
| Li      | 4.98 | 5.75 | 5.75  | 5.75  | 5.75     |
| Fpppp   | 6.46 | 7.43 | 7.61  | 7.75  | 8.57     |
| Applu   | 4.93 | 5.53 | 5.58  | 5.63  | 6.47     |
| Wave5   | 5.86 | 7.35 | 7.60  | 7.62  | 18.56    |
| Swim    | 4.20 | 4.83 | 4.83  | 4.83  | 4.83     |
| Turb3d  | 8.20 | 8.39 | 8.41  | 8.41  | 9.03     |

*Source:* Gonzalez and Gonzalez (1996).

[a] Register utilization is a function of the optimizing compiler characteristics.

aries [Iseli and Sanchez 1995] and progresses toward reconfiguration at the single instruction boundary level. Such a development is made possible by the latest technology trends, such as those by André Deffon and Thomas Knight at MIT, which imply FPGAs storing multiple configurations, and switching among different configurations in a single cycle (order of only tens of nanoseconds).

Important new trends in instruction-level paralellism related research are techniques that optimize system design and performance by turning to complexity reduction and dynamic speculation [Palacharla et al. 1997, Moshovos et al. 1997].

Hank et al. (1995) describe a research effort referred to as region-based compilation. The analysis is oriented to superscalar, superpipelined, and VLIW architectures. Traditionally, compilers have been built assuming functions as units of optimization. Consequently, function boundaries are not changed and numerous optimization opportunities get hidden. An example of an undesirable function-based code partition is given in Figure 3.22.

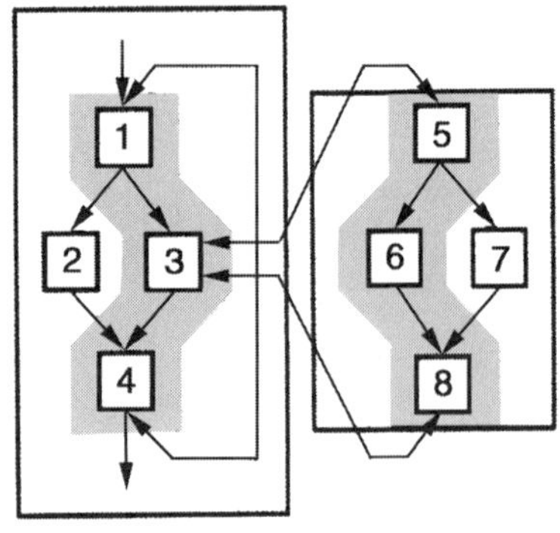



**Figure 3.22.** Example of an undesirable function-based code partition. Numbers refer to task units. [*Source:* Hank et al. (1995).]
*Comment:* If the two function codes are merged together, and the code partition into functions is redone by the compiler, chances are that much better code optimization conditions can be created.

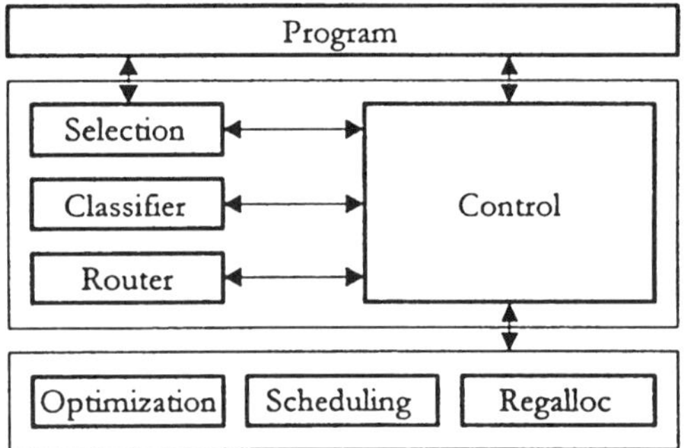



**Figure 3.23.** Example of a region-based compiler, inlining and repartition into regions. [*Source:* Hank et al. (1995).]
*Comment:* The essential elements that distinguish this compiler from traditional ones are the classifier unit and the router unit.

A solution offered by this research is that the compiler is allowed to repartition the code into more desirable units, in order to achieve more efficient code optimization. The block structure of a region-based compiler is given in Figure 3.23.

Conditions of this research imply that compilation units (CUs) are selected by the compiler, not by the software designer. An optimal size for the CU achieves better utilization of employed transformations. Profiler support is used to determine regions of code that enable better optimization. All experiments have been based on the multiflow technology for trace scheduling and register allocation.

Another approach of interest is to create architectures in which features are provided to facilitate compiler enhancements of ILP in all programs. One architecture type that follows this approach is explicitly parallel instruction computing (EPIC). The term was coined by Hewlett-Packard and Intel in their joint announcement of the IA-64 instruction set [Gwennap 1997b]. The

EPIC architectures require the compiler to express the ILP (of the program) directly to the hardware.

Techniques have been developed to represent control speculation, data dependence speculation, and predication. It has been shown that these techniques can provide a considerable performance improvement [e.g., August et al. 1998].

The rationale behind the EPIC architectures is as follows. Processors introduced before the year 1990 (for the most part) were able to execute at most one instruction per cycle. Processors introduced before the year 1995 (for the most part) were able to execute at most four instructions per cycle. Processors introduced after the year 2000 (for the most part) will be able to execute 16 or more instructions per cycle. All this implies an enormous pressure on compiler technology. One way to help is to migrate some of the load from the compiler domain into the architecture domain.

Many believe that control-flow misprediction penalties are the major source of performance loss in wide-issue superscalar processors. One approach to cope with the problem is to incorporate the selective eager execution (SEE) mechanism into the underlying processor architecture. One such effort is described in Klauser et al. (1998). The essence of the SEE approach is to execute both execution paths when a diffident branch occurs. A *diffident branch* is a branch that cannot be predicted with a high level of confidence. Obviously, executing both execution paths all the time is too complex, and (from the cost/performance point of view) not necessary if a branch can be predicted with a high level of confidence. The simulation results of Klauser et al. (1998) show that the SEE approach has a performance improvement potential of up to about 50% (maximum about 35% and average about 15% for selected benchmarks), for an eight-way superscalar architecture that has an 8-stage-deep pipeline. The wider the superscalar and the deeper the pipeline, the higher are the potentials of the SEE approach.

The success of wide-issue superscalar processors is highly dependent on the existence of adequate data bandwidth. One solution for the problem is to implement a larger number of ports to the data cache. Another solution is to use the so-called DDA (data decoupled architecture), as in Cho et al. (1999).

With support from compiler and/or hardware, early in the processor pipeline (before entering the reservation stations), the DDA approach partitions the memory stream into two independent streams and feeds each stream into a separate memory unit (access queue and cache). This has two advantages: (1) the cost and complexity of building a large cache with many ports is avoided, and (2) the splitting of the streams enables each stream to be treated with different specialized optimization techniques, which potentially results in a higher efficiency. In the case of Cho et al. (1999), one stream consists of the local variable accesses, and the other stream includes the rest. The first stream is fed into the specialized local variable cache (LVC), and optimizations such as fast data forwarding or access combining can be utilized.

Studies show that a considerable ILP is present among instructions that are dynamically far away from each other. With this in mind, Vajapeyam et al. (1999) propose a hardware mechanism called *dynamic vectorization* (DV), which quickly builds a large logical instruction window. The DV converts repetitive dynamic instruction sequences into a vector form, thus enabling concurrent processing of instructions in the current program loop and those from far beyond the current program loop. The DV mechanism helps in cases when the static control flow is too complex, and compile time vectorization cannot be very successful. Evaluations using SPECint92 have shown that a relatively large portion of dynamic instructions can be captured in a vector form, making the speed-up of two or more. Traditional instruction windows have a length of a few tens to a couple of hundred instructions. With the help of DV, an instruction window with a size of up to a thousand instructions or more could be fully utilized.

The essence of DV is as follows. It detects the repetitive control flow at run time and captures the corresponding dynamic loop body in a vector form. Consequently, multiple loop iterations are issued from the single copy of the loop body in the instruction window. This eliminates the need to refetch the loop body for each loop iteration, which frees the fetch stage to fetch post loop code (instead of the loop code, as in traditional solutions). The end result is that a much larger instruction window can be built.

From the discussion presented so far, one could conclude that all new microprocessors follow the described trend of maximal exploitation of the ILP. However, that is not the case. The IBM S/390 single-chip mainframe represents an interesting new experiment with a processor that is not super-scalar [Slegel et al. 1999]. Instead of focusing on execution of more than one instruction per cycle, it focuses on the minimization of the number of cycles needed to execute instructions of the ESA/390 architecture; this architecture stresses complex instructions needing tens, hundreds, or even thousands of clock cycles to execute.

The author and his associates were not very active in the field of ILP, except for side activities on related projects. For details, see Helbig and Milutinović (1989) and Milutinović (1996b).


## PROBLEMS

**3.1.** What type of code includes more ILP: (**a**) scientific code, (**b**) artificial-intelligence-oriented code, or (**c**) data processing code? Explain why.

**3.2.** Design a control unit (CU) for an *N*-issue superscalar microprocessor. Derive the formula that gives the CU transistor count as a function of *N*.

**3.3.** Design a control unit (CU) for an $N$-deep superpipelined microprocessor. Derive the formula that gives the CU transistor count as a function of $N$.

**3.4.** Some years ago, many believed that MLP = 4 (machine-level parallelism equal to 4) is the maximum that makes sense to implement. These days researchers talk about MLP = {16, 32}. Check the open literature and try to find out about the asymptotic value of MLP. Explain different opinions.

**3.5.** Explain differences between superpipelined, superscalar, and VLIW code, using a simple code example. Which one of the three approaches is potentially the best one, if the available compiler technology is not very sophisticated?

**3.6.** What type of unit is more necessary in modern microprocessors: IU or FPU? Explain why and give an application example for which the opposite applies.

**3.7.** Comment on emulating a CISC with a RISC. Discuss the expected number of RISC instructions per CISC instruction, for various RISC architecture types.

**3.8.** At the time of reading this book, what is more critical: (**a**) the transistor count to implement all needed functional units or (**b**) extracting the ILP from a typical application that is necessary to keep busy all the functional units that the technology can support?

**3.9.** What is a more natural solution for a GaAs technology microprocessor: (**a**) superscalar architecture or (**b**) superpipelined architecture? Explain why.

**3.10.** Design a simulator of a simple microprocessor that can be used to experiment with various superscalar and superpipelined architectures, using realistic application code. Is it possible to modify Limes (see Appendix B of this book), in order to achieve such a goal?

# CHAPTER 4

# Prediction Strategies

This chapter includes two parts: one on branch prediction strategies and one on data prediction strategies. Chronologically, branch prediction comes first, and data prediction represents a newer development.

## 4.1. BRANCH PREDICTION STRATEGIES

This part includes two sections. The first one covers basic issues (background). The second one discusses advanced issues.

### 4.1.1. Basic Issues

A straightforward method to cope with the negative effects of branching (as far as execution time) is multiple path execution. Machines like the IBM 370, model 168, and IBM 3033 were among the first to adopt this type of strategy. The strategy boils down to doubling of the resources used, at each branch encountered, prior to the first branch outcome resolved. After that, computation results from the wrong branch path are flushed, and the doubling is applied to the next branch in order.

The rest of the presentation is oriented to branch prediction algorithms based on the correlation principle. The correlation principle means that the future outcome of a branch is usually predictable, based on past outcomes of the same or related branches. Branch prediction strategies of this type are classified as hardware, software, or hybrid.

Hardware branch prediction strategy (BPS) implies dynamic prediction based on branch target buffer (BTB) and/or branch prediction buffer (BPB). The latter is sometimes referred to as the *branch history table* (BHT). Each branch has two essential parameters: the target address and the branch condition. The first one is taken care of by the BTB, while the second one is taken care of by the BPB. Of course, as will be seen later, in addition to BTB and BPB, appropriate additional resources are also necessary. All hardware BPS schemes assume that the outcome of a branch is stable over a period of time.

Software BPS implies static prediction based on preannotating techniques (to point to the parallelism to be exploited) and prearranging techniques (to detect and eliminate dependencies, in order to increase the level of ILP). All software BPS schemes assume that each high-level language (HLL) program context is characterized with typical outcomes of branches.

Hybrid BPS implies that the hardware and compiler are cooperating together. The most popular approaches are based on predicated (conditional) and speculative (renaming) instructions. Note that some authors classify predicated and speculative instructions in different ways.

### *4.1.1.1. Hardware BPS*

In the remainder of the text, the terms hardware BPS and dynamic BPS are used interchangeably. Dynamic decisions in this context are typically done in hardware. Actually, each BPS can also be implemented (dynamically, at run-time) in software; however, in this particular environment, there is no time for the slower software implementation.

Typical BTB includes one entry for each recently executed branch instruction. That entry contains several fields: (1) a branch instruction address—full address or only low-order bits, (2) one bit telling about the outcome of the branch instruction, and (3) an address of the branch target.

The fact that only low-order bits can be used means that several branches may map into the same BTB entry. However, if the number of low-order bits is high enough, according to the locality principle, it is very likely that the information in the entry will refer to the right branch instruction.

The fact that only one bit is used to tell about the outcome of the branch instruction can be utilized for two purposes: (1) to eliminate all entries referring to a recently executed branch instruction, for cases when the branch was not taken, and (2) to eliminate the one-bit entry telling about the outcome. Consequently, most BTBs include entries only for taken branches.

Typical BPB does not include the branch target address. This means less complexity, but a slowdown, since the branch target address has to be looked up elsewhere. Also, BTB must include entries for both taken and not-taken branches, which means more complexity if a joint BTB/BPB is to be designed, because such a unified resource must have entries for both taken and not-taken branches; consequently, some machines include both a BTB and a BPB.

The PowerPC 620 is one of the machines including both a BTB and a BPB. Actually, PowerPC 620 includes a variation of a BTB, which has one or more target instructions, instead of or in addition to the target address. Because of the added complexity, a BTB access takes longer, which may complicate the critical path design. However, this solution enables the effect referred to as branch folding. Branch folding means that unconditional branches always take zero cycles to execute (if BTB hit, the instruction is already in the CPU) and conditional branches may also take zero cycles to execute (assuming that condition evaluation is fast enough).

The CRISP machine includes another variation that enables branch folding for indirect branches. Their target address varies at run time; this complicates the conventional design, which implies one fixed target address per entry. The CRISP solution by Ditzel and McLellan starts from a statistical analysis stating that procedure returns contribute with about 90% to the total number of indirect branches. It includes a stack of return addresses, and return addresses are pushed or popped at call or return. The top of the stack is treated as the target address field of the BTB. Consequently, only the top-of-the-stack target address is used in conjunction with each specific branch. If the depth of the stack is larger or equal to the nesting level of procedures, the target address prediction will be absolutely correct.

The DLX machine by Hennessy and Patterson (1996) uses the same solution. It includes a return buffer for a nesting depth up to 16. This number was obtained from statistical analysis of SPEC89, aimed at the optimal performance/complexity ratio.

The simplest predictor is referred to as the 2-bit predictor, introduced by Jim Smith. It is shown in Figure 4.1. The 2-bit predictor yields a considerably better performance than the 1-bit predictor. This is because the 2-bit predictor mispredicts only at one of the two branches forming a loop, while the 1-bit predictor mispredicts at both branches. A 3-bit predictor yields a performance that is only slightly better, at a 50% greater cost. Consequently, only 2-bit predictors make sense, both in the case of the simplest predictor of Figure 4.1, and the more complex predictors to be discussed later.

Prediction accuracy of a BTB depends on its size, as indicated in Figure 4.2. This figure shows that before the saturation point is met, the logic design strategy of the BTB may help, since the set-associative approach works somewhat better than the direct-mapped approach. However, after the saturation point is reached, logic design strategy is of no consequence. This figure also shows that the saturation point is reached at about 2-kB entries in the BTB. Several other studies claim that a 4-kB entry BTB is about equally as good as an infinite BTB. Note that it is not sufficient that the branch instruction is located in the BTB; what is also important is that the prediction must be correct. In other words, in real designs, the cost of a misprediction is typically the same as the cost in the case of a BTB miss!

Another solution is presented in Figure 4.3, where the prediction related information is included into the instruction cache, in addition to the standard information, which is the addressing information and the code. Prediction related information includes two fields: (1) the successor index field and (2) the branch entry index field.

The successor index field contains two subfields: (1) the address of the next cache entry predicted to be fetched and (2) the address of the first instruction (in that entry) that is predicted to be executed. The length of the successor index field depends on cache size (the number of instructions that

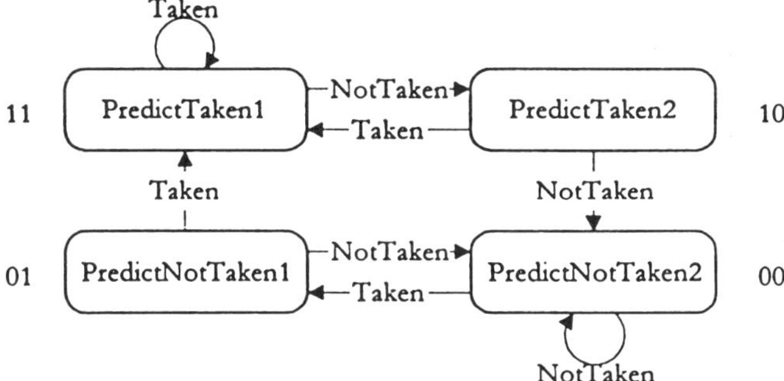


**Figure 4.1.** States of the 2-bit predictor; avoiding the misprediction on the first iteration of the repeated loop (nodes—states of the scheme; arcs—state changes due to branches). [*Source:* Hennessy and Patterson (1996).]

*Comment:* State 11 means branch very likely. State 10 means branch likely. State 01 means branch unlikely. State 00 means branch very unlikely. Note that two (rather than three) mispredictions move the state machine from 11 to 00, and vice versa. The best initial state is "branch likely," because a randomly chosen branch instruction is more likely to branch than not, but not very likely to branch. A condition-controlled loop is typically executed several times. The last execution of the loop condition statement must result in a misprediction (that misprediction cannot be avoided, unless a special loop count estimation algorithm is incorporated, like the one in Chang and Banerjee (1995). If a 1-bit predictor is used, the predictor bit gets inverted on the misprediction, and the very next execution of the loop condition statement will also result in a misprediction (in spite of the fact that loops are very unlikely to execute only once). However, if a 2-bit predictor is used, an "inertia" is incorporated into the system, and the predictor needs two mispredictions before it switches from 11 to 00. If the average loop count is $N$, the 2-bit predictor, compared to the 1-bit predictor, has a misprediction that is at least $(2/N - 1/N)\%$ better (a number that justifies the exclusive use of 2-bit predictors). On the other hand, a 3-bit predictor brings a negligible performance improvement and a 50% complexity increase, compared to the 2-bit predictor (another fact that fully justifies the exclusive use of the 2-bit predictor).

fit into the cache). The relative size of the two subfields depends on the number of instructions per cache entry. For example, a 1-MB direct-mapped cache for a machine with 64-bit instructions, and 8 instructions per cache entry, requires a 17-bit successor index field ($N_1 = 14$ bits to address the cache entry and $N_2 = 3$ bits to address an instruction within the entry). This is so because a 1-MB cache holds 128K 8-byte (or 64-bit) instructions, and these instructions are organized in 16K cache entries (8 instructions per cache entry). To address one of the 16K cache entries, one needs 14 bits. To address one instruction within an entry, one needs 3 bits.

The branch entry index field specifies the location (within the current cache entry) of the branch instruction that is predicted to be taken. Consequently, instructions beyond the branch point are predicted not to be executed.

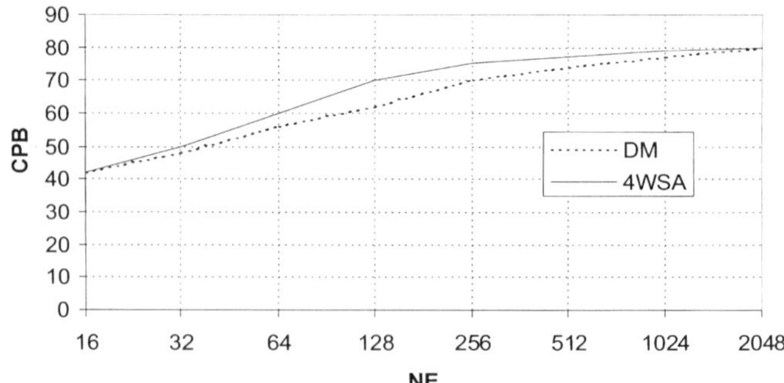

**Figure 4.2.** Average branch target buffer prediction accuracy; a BTB with 2-kB entries is about the same in performance as an infinite BTB (CPB—percentage of correctly predicted branches; 4WSA—4-way set-associative; DM—direct mapped; NE—number of entries). [*Source:* Johnson (1991).]
*Comment:* It is important to notice that the set-associative approach is better than the direct-mapped approach only if the BTB is too small, or not large enough.

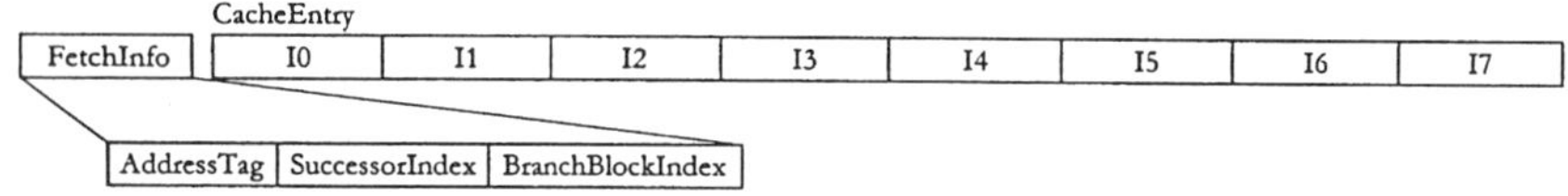

**Figure 4.3.** Instruction cache entry for branch prediction (I—cache entry). [*Source:* Johnson (1991).]
*Comment:* Efficiency of the approach depends on the number of instructions in each cache entry. Other approaches to the incorporation of prediction into cache memory are possible, too. Their elaboration is left as an exercise for the students.

This organization represents a way to incorporate the prediction related information into the cache. The successor index field specifies where to start the prefetching from (instructions before the one pointed to by the successor index field are not needed). The branch entry index field specifies the branch instruction to which the prediction information refers (an instruction beyond the one pointed to by the branch entry index field is likely not to be executed).

An improvement of 2-bit predictors is referred to as the *two-level predictor* (a more accurate name would be the *two-level 2-bit predictor*, since the 2-bit predictor is an element of a two-level predictor). The scheme was introduced by Yeh and Patt. Two-level predictors use the information on the behavior of other branches (at other addresses), in order to do prediction about the currently executing branch instruction (at the current address). Practically always (although theoretically not always), two-level predictors show better performance compared to 2-bit predictors. A two-level 2-bit (2, 2) branch predictor is shown in Figure 4.4.

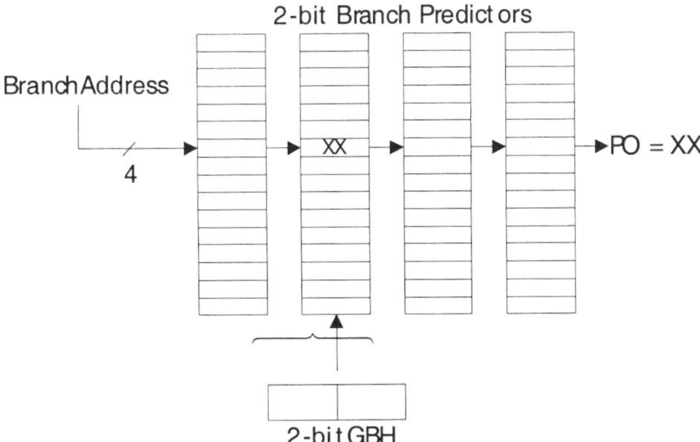


**Figure 4.4.** A $(2, 2)$ branch predictor using a 2-bit global history to select one of the four 2-bit predictors (GBH—global branch history; PO—prediction outcome). [*Source:* Hennessy and Patterson (1996).]
*Comment:* A global branch history register of the size 9 or 10 bits, and the size of one single (vertical) vector branch predictor equal to 2048 or 4096 entries, seems to be a good price/performance compromise.

The example $(2, 2)$ branch predictor from Figure 4.4 includes four vectors of 2-bit predictors. The first one is for the case when the most recent two branches (on any two addresses) were $(0, 0)$, the second one is for the case when the most recent two branches (on any two addresses) were $(0, 1)$, and so on. In this context, $(0, 0)$ means that the most recent two branches (on any addresses) were not taken, and so forth. Information about the outcome of the most recent branches (on any addresses) is kept in a register called *global branch history* (GBH). Therefore, for a given branch, prediction depends on both the address of the branch (horizontal entry into the matrix of 2-bit predictors) and the contents of the GBH register (vertical entry into the matrix of 2-bit predictors).

The example described above belongs to the category of global predictors, because there is only one GBH, and its contents refer to the most recent branches at any address. It will be seen later that one can also talk about per address predictors, where the number of GBH registers is equal to the number of entries in each vertical vector of 2-bit predictors.

The basic rationale behind global predictors is explained using the code example from Algorithm 4.1 and the explanation from Table 4.1. Algorithm 4.1 includes a section of HLL (high-level language) code and its MLL (machine-level language) equivalent. From the explanation in Table 4.1 one can see that if a previous branch (B1) is taken, there is a high probability that the next branch (B2) will also be taken. Branches B1 and B2 are related through the semantics of the code, and that is what global predictors rely on.

```
HLL:

    if (d==0)

        d=1;

    if (d==1)

MLL (d assigned to r_1):

    bnez    r_1, l_1        ; branch B1 (d ≠ 0)

    addi    r_1, r_0, #1    ; d=0, so d ← 1 (note: [r_0 = 0])

1:  subi    r_3, r_1, #1

    bnez    r_3, l_2        ; branch B2 (d ≠ 1)

...


l_2:
```

**Algorithm 4.1.** Example code (HLL—high-level language; MLL—medium-level language). This example implies that register r0 is hardwired to zero, which is typical for a number of RISC microprocessors. [*Source:* Hennessy and Patterson (1996).]

**TABLE 4.1. Example Explanation—Illustration of the Advantage of a Two-Level Predictor with 1-Bit History**[a]

| $D$init | $D = 0$?[c] | $B_1$ | $D$ before $B_2$ | $D = 1$? | $B_2$ |
|---|---|---|---|---|---|
| 0 | Yes | Not taken | 1 | Yes | Not taken |
| 1 | No | Taken | 1 | Yes | Not taken |
| 2 | No | Taken | 2 | No | Taken |

[a] It is important to underline that a number of other program segments would result in the same pattern of taken/not-taken relationship, which is a consequence of the programming discipline and compiler design. In other words, parts of the code that are semantically unrelated can be strongly related as far as prediction-related issues.

[b] If $B_1$ is *not_taken*, then $B_2$ will also be *not_taken*, which can be utilized to achieve better prediction.

[c] $D$init—initial $D$.

[d] $D$before $B_2$—value of $D$ before $B_2$.

In general, one can talk about an $(M, N)$ predictor; it uses the behavior of the last $M$ branches to select one of the $2^M$ predictor vectors of length $L$, each one consisting of $N$-bit predictors. One study claims that it is not unreasonable that $M$ goes all the way up to $M = 9$. Previous discussion argues that it does not make sense to go beyond $N = 2$.

Note that each 2-bit predictor corresponds to a number of different branches, because only a subset of address bits is used to access different 2-bit predictors in one vertical vector of 2-bit predictors. Consequently, the prediction may not correspond to the branch currently being executed, but to another one with the same set of low-order address bits. However, most of the time, the prediction will correspond to the branch currently being executed, because of the locality principle. In some cases, the prediction will not correspond to the branch currently being executed, but the scheme will still work, because the same programming discipline results in a similar pattern of branchings/nonbranchings.

Figure 4.5 represents another viewpoint of looking at the above-mentioned types of branch predictors. Different viewpoints enable the reader to create a better understanding of the issue.

The basic scheme is now referred to as 2bC, as it is named in a number of research papers. It is seen that only a subset of $n$ branch address bits is used to access the single vector of $2^n$ 2-bit registers ($n = \log L$). The cost (complexity) of the scheme is given by

$$C(2bC) = 2 \cdot 2^n \text{ bits}$$

The global scheme is now referred to as GAs, as it is named in a number of research papers. Again, only a subset of lower $n$ address bits is used to access one of the vertical $2^m$ vertical branch prediction vectors ($m = M$). The one to be accessed is determined by the contents of the branch history shift register (BHSR). The acronyms BHSR and GBH refer to the same thing. One horizontal vector of 2-bit predictors is referred to as PHT (pattern history table). The cost of the scheme is given by

$$C(GAs) = m + 2^{m+n+1} \text{ bits}$$

The per address scheme is referred to here as PAs, as it is named in a number of research papers. Everything is the same, except that each PHT is associated with a different BHSR. Consequently, the cost of the scheme is given by

$$C(PAs) = m \cdot 2^n + 2^{m+n+1} \text{ bits}$$

assuming the same number of local branch history registers and rows in PHT.

Figure 4.6 covers four different schemes of much smaller cost and (most of the time) of only slightly lower performance. The simplification is based either on an interesting run-time property or on an efficient compile-time effort.

In the case of gshare [introduced by McFarling (1995)], the lowest $n = m$ bits of the branch address are exclusive ORed with the contents of the single global $m$-bit BHSR. The value obtained points to one of $2^m$ 2-bit predictors. The exclusive OR operation enables each vertical vector from previous schemes to be substituted with a single 2-bit predictor. Consequently, the complexity of the scheme drops down considerably. Fortunately, empirical studies state that performance of the scheme drops down only slightly, due to

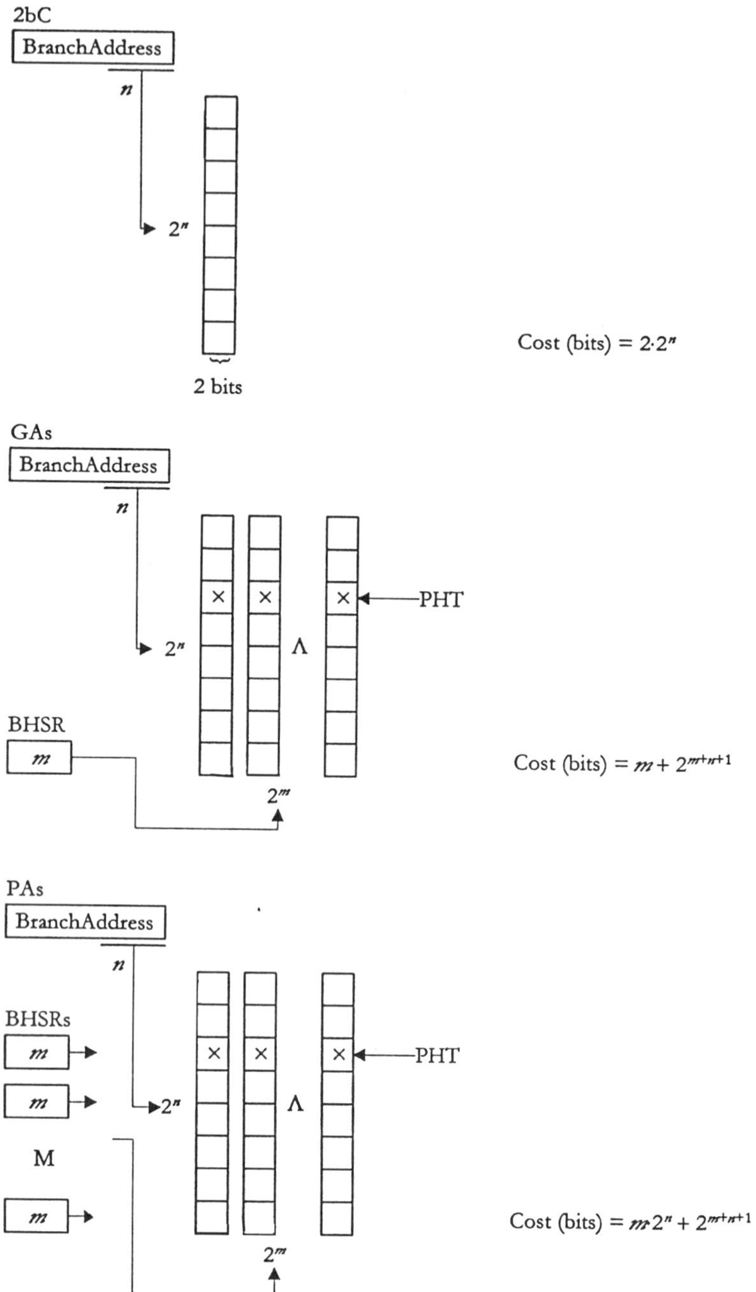



**Figure 4.5.** Schemes 2bC, GAs, and PAs (PHT—pattern history table; BHSR—branch history shift register). [*Source:* Evers et al. (1996).]
*Comment:* Common characteristic of all three schemes is that the stress is on increased performance, rather than decreased complexity.

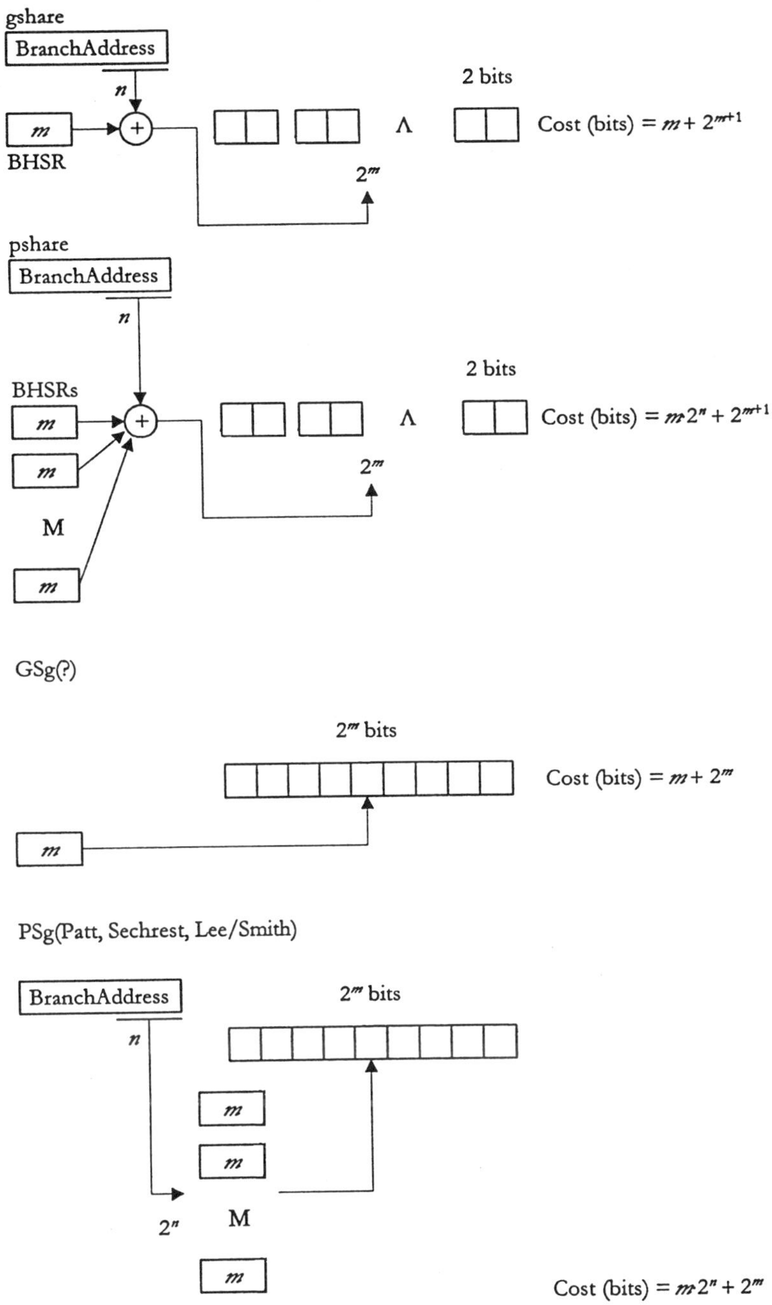

gshare

BranchAddress

$n$

$m$

BHSR

2 bits

$\wedge$

$2^m$

Cost (bits) $= m + 2^{m+1}$

pshare

BranchAddress

$n$

BHSRs

$m$

$m$

M

$m$

2 bits

$\wedge$

$2^m$

Cost (bits) $= m2^n + 2^{m+1}$

GSg(?)

$2^m$ bits

Cost (bits) $= m + 2^m$

$m$

PSg(Patt, Sechrest, Lee/Smith)

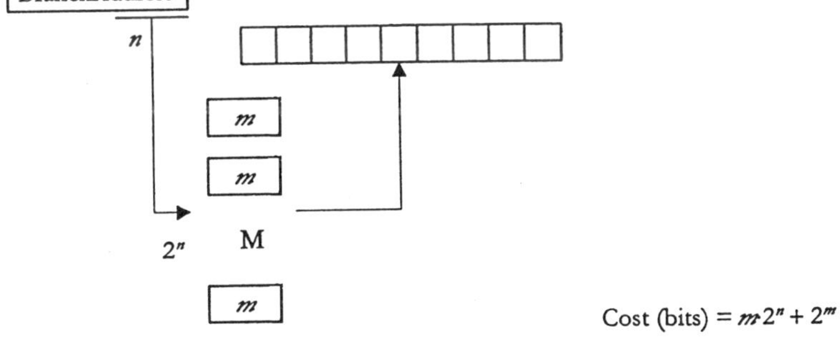


**Figure 4.6.** Schemes gshare, pshare, GSg, and PSg (BHSR—branch history shift register). [*Source:* Evers et al. (1996).]
*Comment:* The common characteristic of all three schemes is that the stress is on decreased complexity, rather than increased performance.

the code locality and programming culture issues discussed above. The cost of this scheme is given by

$$C(\text{gshare}) = m + 2^{m+1} \text{ bits}$$

In the case of pshare, the lowest $n = m$ bits of the branch address are exclusive ORed with the per address BHSR corresponding to the same address as the branch instruction being currently executed. Basically, everything is the same, except that the global treatment is substituted by the per address treatment. The cost of this scheme is given by

$$C(\text{pshare}) = m \cdot 2^m + 2^{m+1} \text{ bits}$$

In the case GSg, predictors are based on compile-time prediction (rather than run-time prediction). Consequently, predictors are one-bit wide (rather than 2 bits wide). Scheme GSg is a global scheme. This means that the address of the current branch is not relevant. What is relevant is the contents of the $m$-bit BHSR, which points to one of the $2^m$ one-bit predictors. This scheme only has a theoretical value. It does not have any practical value, in spite of its extremely low-cost function:

$$C(\text{GSg}) = m + 2^m \text{ bits}$$

Finally, in the case of PSg (introduced, in various forms, independently by Lee and Smith, by Sechrest [Sechrest et al. 1996], and Patt), the lowest $n$ bits of the branch address are used to select one of the $2^n$ $m$-bit BHSRs of the per address type. The selected $m$-bit BHSR is used to point to one of the one-bit predictors, inside the array of one-bit predictors; the length of this array is $2^m$. This scheme has some practical value, and is used in low-cost and hybrid schemes. Its cost is given by

$$C(\text{PSg}) = m \cdot 2^m + 2^m \text{ bits}$$

Note that the warmup time of the schemes based on one-bit predictors is equal to zero; this characteristic has been utilized in some hybrid branch predictors to be discussed later.

### 4.1.1.2. Software BPS

In the remaining text, the terms *software BPS* and *static BPS* will be used interchangeably. Static decisions in this context are typically done at compile time. Actually, each static decision can also be implemented (dynamically, at run time) inside the operating system; however, in this particular environment, there is no time for the slower operating system execution.

Static branch prediction schemes (in the narrow sense) use the information that is gathered before program execution, either by static code inspection or by appropriate profiling.

The simplest schemes in this category assume that conditional branches are either always taken (as was the case with the Stanford MIPS-X experimental RISC machine), or always not taken (as was the case with the Motorola MC88000 commercial RISC machine).

A more sophisticated solution has been used in the PowerPC architecture; the compiler can use two types of branches—one that provides better execution time if the branch is taken and one that provides better execution time if the branch is not taken. In this context, the compiler can consult a profiler, before the appropriate conditional branch instruction is scheduled.

Static branch prediction schemes (in the wide sense) also include those based on preannotating, software scheduling, local and global code motion, trace scheduling, loop unrolling, software pipelining, and other more sophisticated techniques. All these schemes should only conditionally be treated as static branch prediction schemes. Often, their primary goal is to eliminate branches or to minimize penalties of wrong decision, rather than to predict them; however, the final consequence may be the same.

Preannotating implies the insertion of special instructions, to help in obtaining the predefined goal, either at coding time, or at profiling time, or at compile time.

Software scheduling implies rearrangement of code at compile time, with hints from the programmer or the profiler. The motion is either local (within the basic block boundaries) or global (across the basic block boundaries). A basic block starts at one of three points: (1) at the very beginning of the code, (2) at the instruction following a branch, and (3) at the labeled instruction—which is a possible target of a branch. The basic block ends at one of the following three points: (1) at the very end of the code, (2) at the branch instruction—the branch instruction itself is treated as being a part of the basic block it terminates, and (3) at the instruction immediately before the labeled instruction. Each code rearrangement must be accompanied by appropriate code compensations, so that the semantic structure of the code is not violated, no matter which way the execution proceeds.

Trace scheduling is the principal technique used in conjunction with VLIW (very long instruction word) architectures. Here, global code motion is enhanced with techniques to detect parallelism across conditional branches, assuming a special type of architecture (VLIW).

Loop unrolling is a technique used on any type of architecture, to increase the amount of sequentially executable code, specifically, to increase the size of basic blocks. Although the technique can be used in conjunction with any type of architecture, it gives the best results in conjunction with VLIW architecture and trace scheduling.

Software pipelining (symbolic loop unrolling) is a technique to pipeline the operation from different loop iterations. Each iteration of a software-pipelined loop includes instructions from different iterations of the original loop.

Note that many of the software techniques can be implemented in hardware, directly or indirectly. For example, the Tomasulo algorithm is a hardware equivalent of the software pipelining algorithm.

Software BPS will not be further elaborated on here. For more information on these issues, the interested reader is referred to specialized literature.

### 4.1.1.3. Hybrid BPS

Hybrid BPS includes techniques with elements of both hardware BPS and software BPS. Since the hardware/software boundary is often ambiguous, the classification to hardware, software, and hybrid schemes has to be considered conditionally.

In this book, predicated instructions and speculative instructions are treated as the basic two approaches to hybrid BPS. However, in other sources, these two approaches are classified differently, and/or other possible approaches are treated as a part of the hybrid BPS group.

*4.1.1.3.1. Predicated Instructions.* A predicated instruction includes a condition that is evaluated during its execution. If the condition is evaluated true, normal execution proceeds. If the condition is evaluated false, noop execution proceeds (a cycle is wasted). Actually, this is an if–then construct with a minimal body (one instruction body). Efficient utilization of predicated instructions requires the compiler and the hardware to cooperate, and that is why the predicating is classified here into the hybrid approaches.

The most common predicated instruction in modern microprocessors is predicated register-to-register move. However, predicated instructions are not a new invention. A form of predicated instruction can be found even in the first microprocessor of the x86 series—in the Intel 8086. The Intel 8086 instruction set includes the conditional REP prefix, which can be treated as a primitive form of predicating.

Predicated instructions help to eliminate branches in some contexts: (1) where one has if–then with minimal body, or (2) where the code can be rearranged to create if–then with minimal body. These contexts happen in numerical codes (e.g., when the absolute value is to be computed) and in the symbolic code (e.g., when the repetitive search is to be applied).

Usefulness of predicated instructions is limited in a number of cases: (1) when the movement of a predicated instruction across a branch creates a code slowdown—this may occur because a canceled instruction does take execution cycles; (2) when the moment of condition evaluation comes too late—the sooner the condition is evaluated, the better the speed of the code;

(3) when the clock count of predicated instructions is too large—this happens easily in some architectures; and (4) when exception handling may become a problem, due to the presence of a predicated instruction.

Still, as indicated above, it is rather a rule than an exception that modern microprocessor architectures do include predicated instructions. The above-mentioned predicated register-to-register move is included in architectures such as DEC Alpha, SGI MIPS, IBM PowerPC, and Sun SPARC. The PA (Precision Architecture) approach of HP is that any register-to-register instruction can be predicated (not only the move instruction).

### 4.1.1.3.2. *Speculative Instructions.*

A speculative instruction is executed before the processor knows if it should execute or not, that is, before it is known if the prior branch instruction is taken. The control unit and optimizing compiler in microprocessors that implement speculative execution support the following scenario: first, branch is predicted; second, the next instruction—either the target address instruction or the next address instruction, depending on the outcome of the prediction—is made speculative and is executed; third, some run-time scheduling is done, in order to optimize the code, that is, to increase the efficiency of speculation; and fourth, if prediction was a miss, the recovery action is invoked and completed.

Some of the actions defined above are the hardware's responsibility; others are the compiler's (or, in principle, even the operating systems) responsibility; still others are either a combined responsibility or can be treated one way (hardware) or the other way (compiler). That is why this book classifies speculating in the hybrid group.

There are two basic approaches to speculation: (1) the compiler schedules a speculative instruction and hardware helps recover it, if it turns out that speculation was wrong (here, speculation is done at compile time), and (2) the compiler does a straightforward code generation and the branch prediction hardware is responsible for speculation (here, speculation is done at run time).

Exception handling is less critical with speculative instructions, compared to predicated instructions (remember, the previous section specifies that exception handling is one of the limiting factors of predicated execution).

In principle, there are two types of exceptions: (1) program errors and the like, when exception causes the program to terminate, and (2) page faults and the like, when exception causes the program to resume. No matter which type of exception is involved, in principle, there are three techniques that can be used to handle the problem, if speculation is used in a microprocessor architecture. However, only the first technique (to be explained next) is directly applicable to both types of exceptions. The other two techniques are directly applicable only to the second type of exception.

The three techniques used to handle exceptions in architectures with speculation, when a speculative instruction causes an exception, are as follows:

1. The hardware and/or operating system are expanded with constructs to handle the exception; these constructs are invoked each time an exception happens.

2. Each register (in the set of general-purpose registers) is expanded with a set of special status bits called "poison bits"—these bits are set whenever a speculative instruction writes a register; also, these bits are reset when the instruction no longer has the speculative status; at the time of exception, appropriate poison bits are tested, and a fault is generated if some other instruction, beyond the speculative instruction, selects a poisoned register for read.

3. The architecture is expanded with a mechanism called "booster," which is essentially a type of hardware renaming mechanism, and is responsible for moving instructions past the branches, labeling each instruction during the period while it is speculative, and guarding the results of the labeled instructions inside a special-purpose renaming buffer (in this way, results of speculative instructions are kept in a special-purpose register file, and poison bits are not needed as a part of the general register file).

Speculation can be implemented in hardware or in software.

Hardware-based speculation is complex to design (which is not the problem for high-production-volume microprocessors) and transistor count consuming (which is not the problem in conditions when one has to find efficient use for the growing number of transistors on a single VLSI chip). On the other hand, hardware-based speculation offers the speed advantage (which is so important for the microprocessor market) and the reusability advantage (which is so important for fast design of new generations of microprocessors). Consequently, hardware-based speculation has been the solution of choice for modern microprocessors.

It is interesting to note that hardware-based speculation has its roots in the product development of the 1960s. Hardware-based speculation was used in a number of CDC and IBM commercial products of those days. Of course, today's speculation has not much to do with how the developers of the 1960s saw it. The concept surfaced in the 1960s, but today's solutions include almost nothing from solutions of those days.

The Thornton scoreboarding algorithm (developed for CDC 6600) is an early approach to dynamic scoreboarding. It does not include speculation, but it can be modified to include speculation.

The Tomasulo renaming algorithm (developed for IBM 360/91) is also an early approach to dynamic scoreboarding. It includes a primitive form of speculation, and it can be modified to include very sophisticated forms of speculation.

This book assumes that the reader is fully acquainted with the details of the Thornton–Tomasulo algorithm. Any reader who is not is referred to either the original papers [Thornton 1964, Tomasulo 1967] or to a well-known textbook [Hennessy and Patterson 1996].

### 4.1.2. Advanced Issues

This part of the chapter contains the author's selection of research activities that, in his opinion, (1) have made an important contribution to the field in recent time and (2) are compatible with the overall profile of this book.

Numerous research papers describe efforts to combine two or more branch prediction strategies, in order to increase the probability of prediction. Such predictors, referred to as *hybrid predictors*, include two or more of the predictors described in the sections on branch predictors, plus a selection mechanism.

McFarling has introduced a dynamic (run-time) selection mechanism for a two-component hybrid BPS (referred to as *branch selection*); at run time, for each branch, it uses a 2-bit up–down saturating counter, to determine which of the two component predictors is more appropriate to use in the case of that specific branch; this selection mechanism can easily be expanded for the case of a multicomponent hybrid BPS.

Chang, Hao, Yeh, and Patt have introduced a static (compile-time) selection mechanism for multicomponent hybrid BPS (referred to as *branch classification*); at compile time, for each individual branch instruction, the branch predictor is determined that suits that particular branch best. Consequently, each component branch predictor does prediction for the branches for which it is best suited.

In a follow-up research effort, using a profiler, their compiler classifies branches into three categories: (1) mostly taken, (2) mostly not taken, and (3) others. Authors propose the use of their compile-time selection mechanism (branch classification) for the first two groups of branches, and the McFarling run-time selection mechanism (branch selection) for the third group of branches.

Multicomponent branch predictors cost more (very few resources in each individual component predictor can be efficiently shared among several component predictors); however, they perform better. The question is, how much better they are under realistic conditions, when periodic context switches impact the behavior of individual predictors. The most efficient predictors need lots of time to warm up. On the other hand, predictors with zero or short warmup times are not nearly as efficient at steady state, as is the case with the most efficient but slow-warming predictors.

Obviously, it is a question of what component predictors to select, how to combine them for the best possible overall performance, and how to obtain reliable information on the overall performance, under conditions of periodic context switches. Several papers are dedicated to these issues.

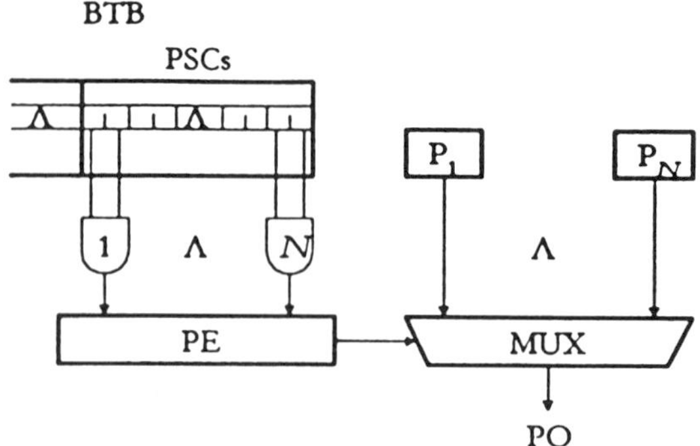



**Figure 4.7.** Predictor selection mechanism (BTB—branch target buffer; PSC—prediction selection counter; $P_1$—predictor 1; $P_N$ Predictor $N$; PE—priority encoding; PO—prediction outcome). [*Source:* Evers (1996).]
*Comment:* The priority encoding mechanism can be made flexible, so that different priority encoding algorithms can be evaluated (a good exercise for students).

A paper by Evers, Chang, and Patt [Evers et al. 1996] introduces the multihybrid BPS, referred to here as MH-BPS. For the same implementation cost, it provides better performance compared to some two-component hybrid BPS schemes.

As indicated in Figure 4.7, it is based on an array of 2-bit up–down predictor selection counters (PSCs) similar to those from the McFarling branch selection mechanism, except that McFarling includes two 2-bit selection counters per branch and Evers et al. include $N$ 2-bit selection counters per branch; each BTB entry is extended with one PSC.

The initial value of all PSC entries in Figure 4.7 is 3, and a priority logic is used if several predictors are equal (see the order of component predictors in Table 4.2). If, among the predictors for which the contents of the PSC was 3, at least one was correct, the PSCs for all incorrect predictors were decremented. If none of the predictors with PSC = 3 was correct, the PSCs of all correct predictors are incremented. This algorithm guarantees that at least one PSC is equal to 3. The complexity of the described multihybrid selection mechanism is 2CL, where L is the number of entries in the BTB and C is the number of component predictors.

Table 4.2 defines two things: (1) the priority order for the included component predictors—2bC has the highest priority and AlwaysTaken has the lowest priority and (2) the component cost for each component predictor included in a given version of the MH-BPS. Different versions of MH-BPS differ in the overall bit count. The simplest version of MH-BPS from Table 4.2 includes 11 kB. The most complex version of MH-BPS includes 116 kB. If 1 bit takes four transistors, the overall transistor count is in the range of about 350 kTr to about 3.5 MTr, which is more than the entire Intel Pentium.

**TABLE 4.2. Multihybrid Configurations and Suboptimal Priority Ordering 95.22 / 95.65**

| | HybPredSiz[a] (kB) | | | | |
| | ~ 11 | ~ 18 | ~ 33 | ~ 64 | ~ 116 |
| | | | CompCost[b] (kB) | | |
|---|---|---|---|---|---|
| SelectMech[c] | 2 | 2.5 | 3 | 3 | 3 |
| 2bC | 0.5 | 0.5 | 0.5 | 0.5 | 0.5 |
| GAs | — | 2 | 2 | 4 | 8 |
| Gshare | 4 | 8 | 16 | 32 | 64 |
| Pshare | 4 | 5.25 | 7.5 | 20 | 36.25 |
| loop | — | — | 4 | 4 | 4 |
| AlwaysTaken | 0 | 0 | 0 | 0 | 0 |

[a]HybPredSiz—hybrid predictor size.

[b]CompCost—component cost.

[c]SelectMech—selection mechanism.

Research by the Evers group [Evers et al. 1996] demonstrates that the optimal priority encoding algorithm provides the hit ratio, which is only slightly larger than the hit ratio provided by the priority encoding algorithm selected in that paper [Evers et al. 1996]: 95.22% versus 95.65%.

The quoted numbers set the stage for a small digression discussing the overall complexity of the MH-BPS. As already indicated, a complexity of 116 kB means more than the transistor count of most of the single-chip microprocessors from the 1980s. However, the transistor count of VLSI chips keeps growing, and their use has to be defined. If the performance improvement provided by MH-BPS is higher than the performance improvement for other possibilities, then the MH-BPS may become a standard element of many future microprocessors on chips with more than 10 MTr. Therefore, the question boils down to the exact performance benefit of the MH-BPS. The answer can be found in Evers et al. (1996). For a predictor size of approximately 64 kB, the MH-BPS achieves a prediction accuracy of 96.22%, compared to 95.26% for the best two-component BPS of the same cost. At first glance, this difference does not look spectacular (only 0.96%, or less than 1%). However, what matters is the difference in misprediction percentage, which drops from 4.74% to 3.78%, or over 25%. It is the misprediction that is costly and decreases the number of cycles in the execution time. Therefore, at second glance, the difference does look spectacular and places this research among the most exciting ones.

Note that the above data are obtained for SPECint92 (which is, according to many, not a sufficiently large application suite for this type of research and includes user code only) and for systems in which the contents of the prediction tables are flushed after periodical context switches (which is, according to many, not the optimal way to treat prediction tables after periodical context switches).

Table 4.2 tells us that the multihybrid predictor of 11 kB includes, in addition to the selection mechanism of about 2 kB, the following component predictors: 2bC, Gshare, Pshare, and AlwaysTaken. The multihybrid predictor of 116 kB includes a somewhat more costly selection mechanism of about 3 kB, plus the following component predictors: 2bC, GAs, Gshare, Pshare, loop, and AlwaysTaken ["loop" is a simple scheme that predicts the number of iterations for each loop branch—for details, see Chang and Banerjee (1995)].

The authors claim that the selection of component predictors was guided by the following rationales: (1) large dynamic predictors have better accuracy at steady state but longer warmup time after context switch; (2) smaller dynamic predictors have worse accuracy at steady state but shorter warmup time after context switch, which means better accuracy during the initial period after context switch; (3) static predictors have zero warmup time, which means the best accuracy for a very short period immediately after context switch; and (4) a price/performance analysis has eliminated the predictors not included in Table 4.2, due to their marginal price/performance.

All predictors taken into consideration in this study are summarized in Table 4.3, from a point of view different from that in the previous presentation of the same facts (different viewpoints and perspectives lead to a much deeper understanding of the issues).

Table 4.3 includes alternative descriptions of the branch prediction algorithms covered in this book. Receiving information from different sources is an important prerequisite for better understanding of essential issues.

The study in Gloy et al. (1996) is based on the IBS traces that include both system and user code (as indicated before, SPECint92 includes only user code) and a larger number of static branches (which means a more realistic environment). It also analyzes the systems in which the prediction tables are not flushed after the periodic context switches (which many believe is a better way to go, since different contexts are coded by the programmers using the same software design methodologies and consequently produce code with similar run-time prediction-related characteristics).

The study by Gloy, Young, Chen, and Smith from Harvard University implies the BPS model shown in Figure 4.8 (another way of representing the same model) and includes the component predictors shown in Figure 4.9. The impact of zero warmup time and short warmup time predictors is smaller if no flushing is involved. This fact had an impact on the selection of component predictors shown in Figure 4.9.

Major conclusions of the study are twofold: (1) better prediction accuracy results are obtained if prediction tables are not flushed after periodic context switches, and (2) results for traces that include both user code and system code differ from the results based only on user code.

The study by Sechrest et al. (1996) is based on extremely long custom-made traces (both user and system code), and it claims that even the longest

**TABLE 4.3. Single-Scheme Predictors—Algorithms and Complexities**

| Predictor | Algorithm | Cost (bits) |
|---|---|---|
| 2bC | A 2-bit counter predictor consisting of a 2000-entry array of 2-bit counters | $2^{12}$ |
| GAs($m, n$) | A global variation of the two-level adaptive branch predictor consisting of a single $m$-bit global branch history and $2^n$ pattern history tables | $m + 2^{m+n+1}$ |
| PSg($m$) | A modified version of the per address variation of the two-level adaptive branch predictor consisting of 2000 $m$-bit branch history registers and a single pattern history table (each PHT entry uses one statically determined hint bit instead of a 2bC); the version of PSg used in this study is the PSg(algo) | $2^{11}m + 2^{m}$ |
| gshare($m$) | A modified version of the global variation of the two-level adaptive branch predictor consisting of a single $m$-bit global branch history and a single pattern history table | $m + 2^{m+1}$ |
| pshare($m$) | A modified version of the per address variation of the two-level adaptive branch predictor consisting of 2000 $m$-bit branch history registers and one pattern history table; as in the gshare scheme, the branch history is XORed with the branch address to select the appropriate PHT entry | $2^{11}m + 2^{m+1}$ |
| loop($m$) | An AVG predictor where the prediction of a loop's exit is based on the iteration count of the previous run of this loop; a 2000-entry array of two $m$-bit counters is used to keep the iteration counts of loops; in this study, $m = 8$ | $2^{12}m$ |
| Always Taken[a] | | 0 |
| Always Not Taken[b] | | 0 |

*Source:* Evers et al. (1996).

[a]AlwaysTaken—branch always taken (MIPS 10000).

[b]AlwaysNotTaken—branch always not taken (Motorola 88110).

standard application suites give traces that are not long enough to measure with sufficient precision the real accuracy of various branch predictors. The study assumes the BPS model shown in Figure 4.10 (still another way to represent the same basic model), and the authors have used the benchmarks presented in Table 4.4.

Sechrest, Lee, and Mudge claim that control of aliasing (when many branches map into the same entry of the branch history table, due to the lower number of address bits used) and interbranch correlation (when various branches impact each other) are crucial for the prediction success of a scheme.

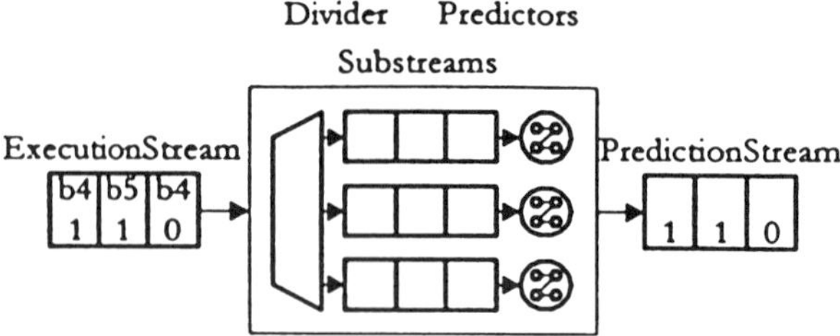


**Figure 4.8.** Another model of BPS. [*Source:* Gloy et al. (1996).]
*Comment:* This is an alternative method of modeling the branch prediction system covered in this book. Receiving information from different sources is an important prerequisite for better understanding of essential issues.

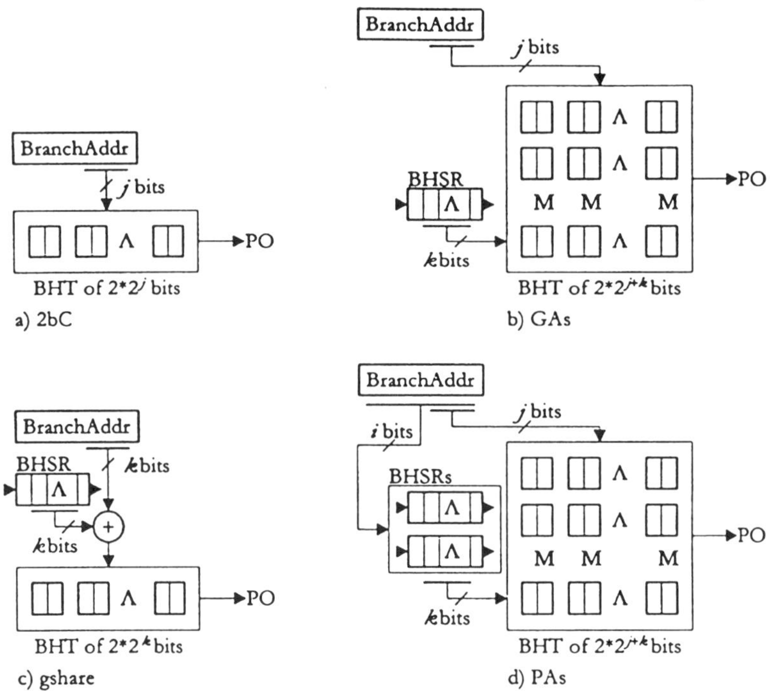


**Figure 4.9.** Explanation of four BPS approaches (BranchAddr—branch address; BHSR—branch history shift register; BHT—branch history table; PO—prediction outcome). [*Source:* Gloy et al (1996).]
*Comment:* This figure singles out the four schemes used in the research of Gloy et al. (1996). Note that those authors and Evers et al. (1996) have chosen a different set of schemes in their research.

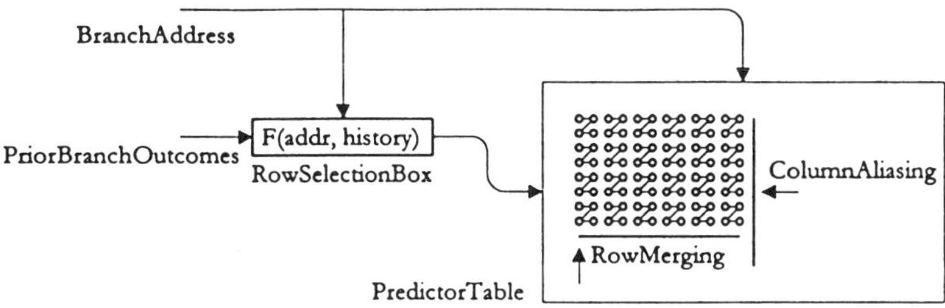


**Figure 4.10.** Yet another model of BPS. [*Source:* Sechrest et al. (1996).]
*Comment:* This figure includes still another alternative method of modeling the branch prediction system covered in this book. Receiving information from more sources is an important prerequisite for a better understanding of essential issues. The reader should make an effort to understand the real reasons for using different symbolics.

An important new trend in branch prediction research implies techniques that reduce negative branch history interference and improve target prediction for indirect branches [Sprangle et al. 1997, Chang et al. 1997].

In object-oriented (OO) code, the relative amount of indirect branches is higher; consequently, as the impact of OO programming increases, it becomes more and more important to predict indirect branches as accurately as possible. Driesen and Holzle (1998) investigate a number of different two-level predictors tuned to indirect branching. They start with predictors that use full-precision addresses and unlimited-size tables, and they gradually come to limited-precision addresses and limited-size tables of acceptable performance and complexity. For indirect branches, their two-level predictor achieves a misprediction rate of about 10% (with a 1K-entry table); their hybrid predictor (for the same table size) achieves the misprediction rate of about 9% (in real microprocessors of the mid- to late 1990s, this rate varies from 20% to 25%).

Evers et al. (1998) have tried to determine which characteristics of branch behavior make predictors perform well. They quantify the reasons for predictability and show that not all predictability is captured by two-level adaptive branch predictors, which means that there is still ample room for new advances in the field. They also show that only a very few previous branches are needed for a correlation-based predictor to be accurate, and that these branches are typically very close to the branch being predicted (in most cases, two or three); this means that new predictors can be deviced, which are not only better performancewise but also less complex.

Along similar lines is the research of Juan et al. (1998), who propose a third level of adaptibility for branch prediction. Traditional two-level predictors combine a part of the branch address and a fixed amount of global history. However, optimal history length (from the performance point of

**TABLE 4.4. Benchmarks—SPEC versus IBS**

| B[a] | DI[b] | DCB[c] (%[TI])[d] | SCB[e] | N[f] |
|---|---|---|---|---|
| compress | 83,947,354 | 11,739,532 (14.0%) | 236 | 13 |
| eqntott | 1,395,165,044 | 342,595,193 (24.6%) | 494 | 5 |
| espresso | 521,130,798 | 76,466,489 (14.7%) | 1784 | 110 |
| gcc | 142,359,130 | 21,579,307 (15.2%) | 9531 | 2020 |
| xlisp | 1,307,000,716 | 147,425,333 (11.3%) | 489 | 48 |
| sc | 889,057,008 | 150,381,340 (16.9%) | 1269 | 157 |
| groff | 104,943,750 | 11,901,481 (11.3%) | 6333 | 459 |
| gs | 118,090,975 | 16,308,247 (13.8%) | 12852 | 1160 |
| mpeg_play | 99,430,055 | 9,566,290 (9.6%) | 5598 | 532 |
| nroff | 130,249,374 | 22,574,884 (17.3%) | 5249 | 228 |
| real_gcc | 107,374,368 | 14,309,867 (13.3%) | 17361 | 3214 |
| sdet | 42,051,812 | 5,514,439 (13.1%) | 5310 | 508 |
| verilog | 47,055,243 | 6,212,381 (13.2%) | 4636 | 850 |
| video_play | 52,508,059 | 5,759,231 (11.0%) | 4606 | 757 |

*Source:* Sechrest et al. (1996).

[a] Benchmarks.

[b] Dynamic instructions.

[c] Dynamic conditional branches (percentage of total instructions).

[d] TI—total instructions.

[e] Static conditional branches.

[f] Number of static branches constituting 90% of total DCB.

*Comment:* Note the minor difference in the number of dynamic instructions, between the benchmarks in the upper part of the table (SPEC) and the benchmarks in the lower part of the table (IBS). Also, note that the number of static branches constituting 90% of total dynamic conditional branches is extremely small in the SPEC case and much larger (on average) in the IBS case. This means that in IBS a much larger percentage of the branch population has an impact on the overall happenings in the system. Sechrest and co-workers (1996) believe that the latter size difference is crucial for correct understanding of the essential issues (impact of aliasing, impact of warmup, etc.). In other words, what matters (when it comes to proper performance evaluation) is the size of the branch population, not the size of the code.

view) depends on code type, input data characteristics, and frequency of context switches. This means that fixed predictors will never be as efficient as those that dynamically determine the optimal history length and adapt to it. Those authors propose the DHLF (dynamic history length fitting) method that is applicable to any predictor type based on global branch history; it uses a BHR size that is equal to the maximal history length of interest. If the DHLF method is combined with Gshare, one obtains the DHLF-Gshare predictor, which is able to XOR any number of history bits with the PC bits of the branch instruction to be predicted. The major question is how to determine (dynamically) the optimal number of history bits. This is done by testing periodically all possible history lengths and choosing the one that provides the minimal number of mispredictions. The effects of history length on the misprediction rate are shown in Figure 4.11, for two different SPECint95 benchmarks (gcc and go).

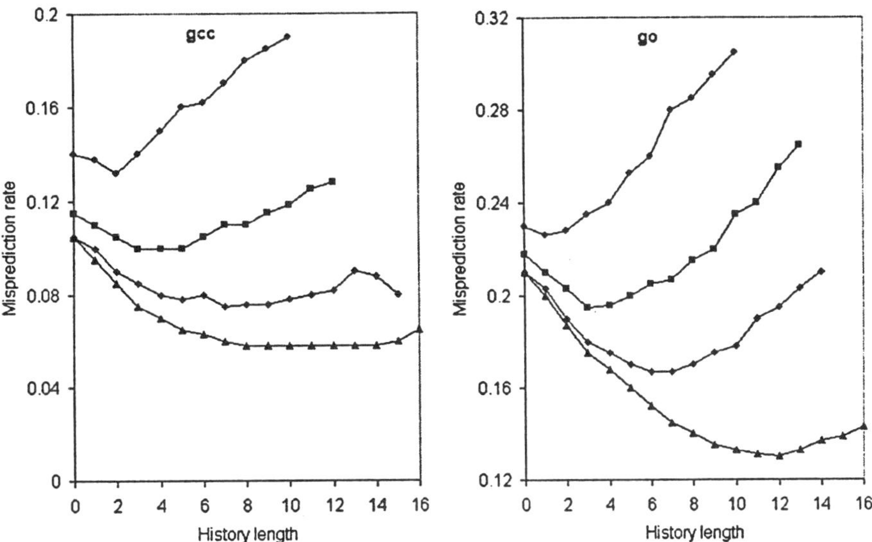


**Figure 4.11.** Effect of history length on the misprediction rate for selected SPECint95 benchmarks using a gshare predictor.

## 4.2. DATA PREDICTION STRATEGIES

This part of the chapter includes two sections, the first discusses basic issues and the second proceeds to advanced issues. Note that data prediction strategies cannot be treated through analogies with branch prediction strategies. Important conceptual differences have to be kept in mind constantly while reading this part.

### 4.2.1. Basic Issues

Data dependencies limit the available ILP. In classical architectures, an instruction cannot be executed until after the results are produced by all previous instructions on which the current one is data-dependent. Techniques to overcome the negative effects of data dependencies can be classified into three different groups:

1. Elimination techniques
2. Concealment techniques
3. Prediction techniques

Elimination techniques are based on dependence collapsing by fusing [e.g., Montoye et al. 1990] or on appropriate global code optimization approaches. Dependence collapsing by fusing implies combining the data-dependent instructions into a single one, using a fused functional unit (logically speaking, a form of encapsulated "if–then–else," where the dependency issue is migrated into a lower execution layer).

Concealment techniques are based either on run-time efforts [e.g., Smith and Sohi 1995] or on compile-time efforts [e.g., Hwu et al. 1995]. The essence of these techniques is parallel execution of nonconsecutive data-independent instructions. While waiting for needed data to get ready, other instructions that are data-independent can be executed in parallel, through run-time or compile-time "shuffling" of instructions.

Prediction techniques are typically of the data value prediction type [e.g., Lipasti and Shen 1996] or the data address prediction type [e.g., Gonzalez and Gonzalez 1997]. Data value prediction generates the "complete" solution and consequently achieves better performance. The essence of data prediction is in speculating the result of a data-producing instruction (based on past behavior of previous instances of the same instruction or the related instructions). The speculated (predicted) value is used as input for speculative execution, which is later (after the real result becomes available) either validated (with *no loss* of additional cycles) or reexecuted (with a *loss* of additional cycles). The rest of the text is dedicated to data value prediction techniques.

### 4.2.1.1. Essence of Data Value Prediction

First, the data value prediction problem implies predicting one out of $2^W$ values, where $W$ is the wordlength of the processor. Remember, in the case of the branch prediction problem, it is necessary to predict one out of only two values (branching or not branching). Consequently, the data prediction problem is a more serious one.

Second, the data prediction techniques can be applied only to value-producing instructions. Consequently, the question is how frequent these instructions are in a typical code. As one can see from Table 4.5, for typical

**TABLE 4.5. Percentage of Value-Producing Instructions (Dynamical Statistics)**[a]

| Program | % |
| --- | --- |
| go | 79 |
| compress | 74 |
| m88ksim | 72 |
| li | 69 |
| gcc | 57 |

[a]Selected benchmark programs belong to the category of general-purpose code. Statistics of this sort are application-dependent. One can expect different average values on scientific code or on artificial intelligence code.

programs found in typical benchmark suites, this percentage varies from about 50% to about 80%, on average. Note that only dynamical statistics (code in execution) are relevant here, not static statistics (code in memory).

The section on basic issues presents details of some important data value prediction techniques, from two landmarking references in the field [Sodani and Sohi 1997, Wang and Franklin 1997]:

1. Dynamical instruction reuse [Sodani and Sohi 1997]
2. Last outcome predictor [Lipasti and Shen 1997]
3. Stride based predictor [Wang and Franklin 1997]
4. Two-level predictor [Wang and Franklin 1997]
5. Hybrid predictors [Wang and Franklin 1997]

Once again, it is very dangerous to apply analogies with branch prediction when reasoning about data value prediction. In some aspects it is possible; in other aspects, it can lead to fatal misunderstandings.

### 4.2.1.2. Algorithms for Data Value Prediction

For each selected algorithm, one will be able to find here the basic schematic and the explanation with minimal redundancy. For more details, the interested reader is referred to the original references.

*4.2.1.2.1. The Dynamical Instruction Reuse Predictor.* In essence, no prediction is done here. Only the operand values are detected, and the result is reused, if available. This is potentially useful for long latency operations, because cycles for the execution part of the current instruction are saved.

A block scheme of the dynamic instruction reuse predictor is given in Figure 4.12. If a 1024-entry FABuf is used, the execution part can be skipped in about 33% of data-producing instructions (dynamical statistics). Effects of that skipping are architecture dependent.

*4.2.1.2.2. The Last Outcome Predictor.* The essence here is in predicting the value, which is equal to the last one produced (e.g., when the same static value-producing instruction was executed last time). The block scheme of the last outcome predictor is given in Figure 4.13.

Interestingly, this scheme, which is extremely simple, gives surprisingly good results. For the PowerPC architecture and for SPEC92 applications, the prediction accuracy is 49%. If the last four different values are stored, and if the scheme were able *always* to pick the correct value (if one of the four was correct), then the prediction accuracy would be 61%.

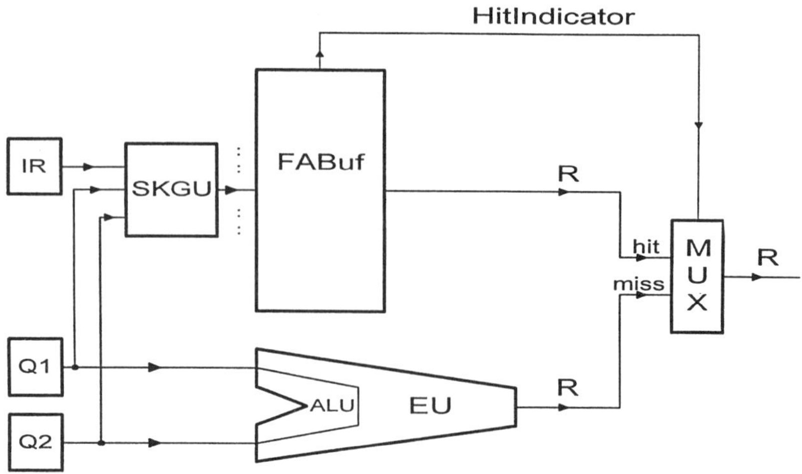


**Figure 4.12.** Block scheme of the dynamic instruction reuse predictor [ALU—arithmetic–logic unit; SKGU—search key generator unit; EU—execution unit; FABuf—fully associative buffer; HitIndicator—indicator of the hit in the fully associative buffer; Qi—operand #1 ($i = 1, 2$); R—result].
*Comment:* Performance of the scheme depends on the size of the FABuf. Unfortunately, the larger the size of the FABuf, the larger the complexity of the search mechanism.

*4.2.1.2.3. Design Issues of Importance.* The major problem when designing a data value predictor is how much history to use for speculation or prediction. If too little history is used, the prediction accuracy is poor. If too much history is used, design overhead gets high and the prediction process takes more time to complete.

The most crucial design parameter is data value locality. The main question is how many of the last $n$ produced values are reused. Table 4.6 gives the percentage of eligible instructions reusing one of the last 16 produced values, for the MIPS R2000 architecture and the SPEC92 applications.

The probability of one of the last 16 being reused is pretty high. However, the storage needed to keep all that history is prohibitively high. Consequently, the next question is how many of the last 16 produced values are unique.

The answer to that question is given in Figure 4.14. For most programs, statistically speaking, the number of unique values is relatively low. For example, in the case of m88ksim, about 65% of value-producing instructions (i.e., about 65% of eligible instructions) create only four different values in the last 16 value-producing activities.

In conclusion, statistical data support the hypothesis that the storage that covers the last 16 value productions can considerably be reduced. One

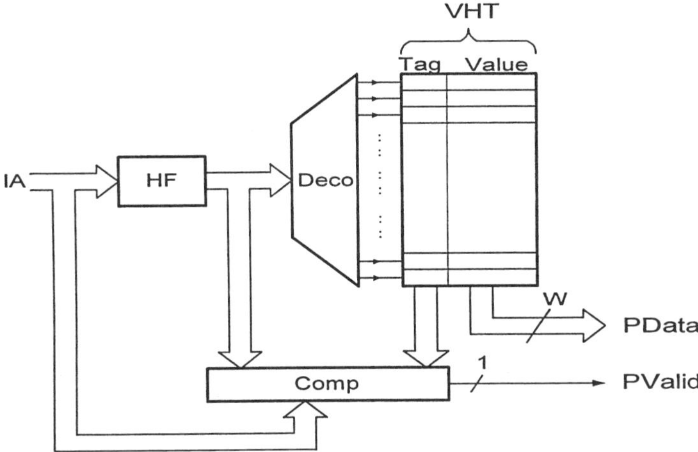


**Figure 4.13.** Block scheme of the last outcome predictor [Comp—comparator; Deco —decoder; HF—hash function; IA—instruction address; PData—predicted data (*W* bits); PValid—prediction valid indicator (1 bit); just the VHT hit; Tag—tag storing the identity of the currently mapped instruction; VHT—value history table; Value—value produced during the last execution of the currently mapped instruction]. *Comment:* Note that only a subset of instructions are data-producing. Consequently, after the hash function is applied (truncation of leading bits) several instructions map into the same VHT entry. However, for locality reasons, it is most likely that the last produced value will be stored in the value entry of the VHT.

**TABLE 4.6. Percentage of Eligible Instructions Reusing One of the Last 16 Values**

| Program | % |
|---|---|
| go | 73 |
| compress | 56 |
| m88ksim | 91 |
| li | 81 |
| gcc | 75 |

reasonable compromise is to have only four storage registers for each data value producing instruction.

*4.2.1.2.4. The Stride-Based Predictor.* If results vary by a constant stride, then it is easy to predict the result of the next instance of the same static instruction. If a relatively large number of instructions follow this pattern, benefits from a stride-based predictor could be relatively high.

The stride-based approach was used successfully for data prefetching. It works well for data prediction, too, because a relatively large percentage of instructions include (1) loop-controlling variables and/or (2) array stepping variables.

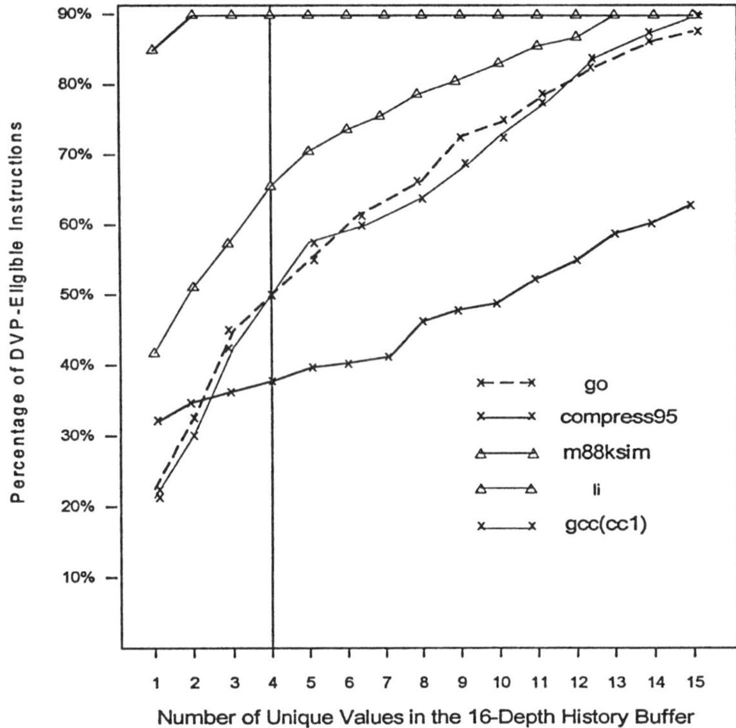


**Figure 4.14.** Number of unique values in the last-16-buffer of eligible instructions (cumulative distribution) ($X$—number of unique values in the last-16-buffer of eligible instructions; $Y$—percentage of register result producing instructions with $X$ or less unique values).
*Comment:* Case $X = 1$ is taken care of by the last outcome predictor; 22–85% of instructions have all their last 16 results the same. Case $X = 4$ needs a more accurate predictor; 38–91% of instructions have four or fewer unique values in the last-16-buffer.

A block scheme of the stride-based predictor is given in Figure 4.15. Stride value predictors work well only while in the steady state, which necessitates the incorporation of a special bit called STATE (into the VHT). The state transition diagram of the stride value predictor is given in Figure 4.16.

With the help of Figures 4.15 and 4.16, it is relatively straightforward to understand the operation of the stride-based predictor. The entire algorithm with necessary explanations is given in Algorithm 4.2.

It is obvious that the last outcome predictor and the stride predictor cover different data-producing behaviors. Consequently, the clear next step is to combine the two approaches. This issue and related problems are covered in a later section.

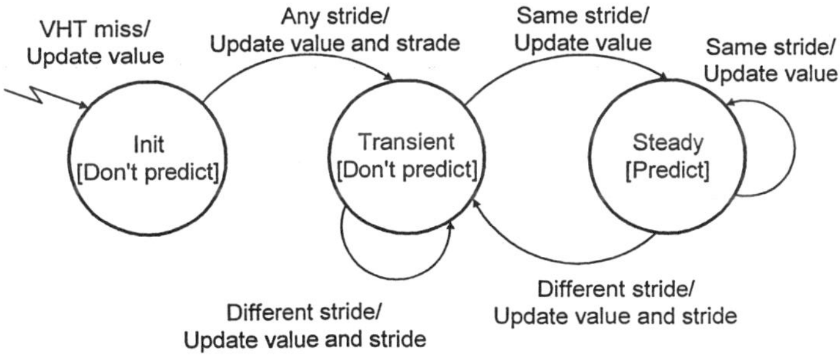

**Legend:**

TAG = {subset/transform (AddressBits)}

STATE = {Init, Transient, Steady}

VALUE = {LastValueEncountered}

STRIDE = {Dj - Dk}; j=k+1, k=1,2,...



**Figure 4.15.** Block scheme of the stride value predictor [TAG = {subset/transform (AddressBits)}; STATE = {Init, Transient, Steady}; VALUE = {LastValueEncountered}; STRIDE = {Dj − Dk}; j = k + 1, k = 1, 2, … ].



**Figure 4.16.** State transition diagram of the stride predictor.

## The Stride Detection Phase

- At the first execution instance—NO prediction is made;
  when the instruction produces its result—an entry
  is allocated in the VHT:
    - (i)  The result D1 is stored into the VALUE field
           of the allocated entry
    - (ii) The STATE field of the entry is set to {Init}

- On the next instance of the same instruction,
  if STATE={Init} then NO prediction is made;
  however, it is assumed that the stride value is calculatable
  after the second instance of an instruction,
  and the following computation is done:

  S1 = D2 - Value(VHT)

  S1 - Stride

  D2 - Value generated by the second instance of the instruction
  Value(VHT) - Value generated by the first instance (DI)

  After the computation is completed, the following updates
  are done:

    - (i)    VALUE = {D2}

    - (ii)   STRIDE = {S1}

    - (iii)  STATE = {Transient}

- If STATE = {Transient} then still NO prediction is made

  on the next instance of the same instruction;
  however, it is assumed that the time has come to declare that

  > A STABLE STRIDE DOES EXIST
  > IF THE NEXT RESULT CREATES THE STRIDE
  > WHICH IS THE SAME AS THE ONE IN THE VHT.

  The following calculation is done:

  > S2 = D3 - Value(VHT)

  The following updates are done:
    - (i)    VALUE = {D3}
    - (ii)   STRIDE = {S2}
    - (iii)  STATE =

      > {Steady} if S2 = S1
      > {Transient} if S2 < > S1—no change of the STATE field

- While STATE = {Steady} predictions are done by adding [VALUE] +
  [STRIDE].

- When Sj < > Sk, j = k + 1, then STATE = {Transient} and prediction
  STOPs.

- After STATE = {Steady}, then prediction RESUMEs.

**Algorithm 4.2.**  Algorithm of the stride predictor.

*4.2.1.2.5. The Two-Level Predictor.* As already mentioned, a substantial percentage of dynamic instructions have four or fewer unique values in their most recent history. The question is which one to select. The first step is to encode properly the four relevant outcomes. The second step is to incorporate a selection mechanism. The third step is to form a two-level predictor, similar to what is done in branch prediction.

A block scheme of a two-level value predictor is given in Figure 4.17. The core of the first level is a VHT (value history table). The core of the second level is a PHT (pattern history table). Various fields in the VHT and PHT are listed in Algorithm 4.3.

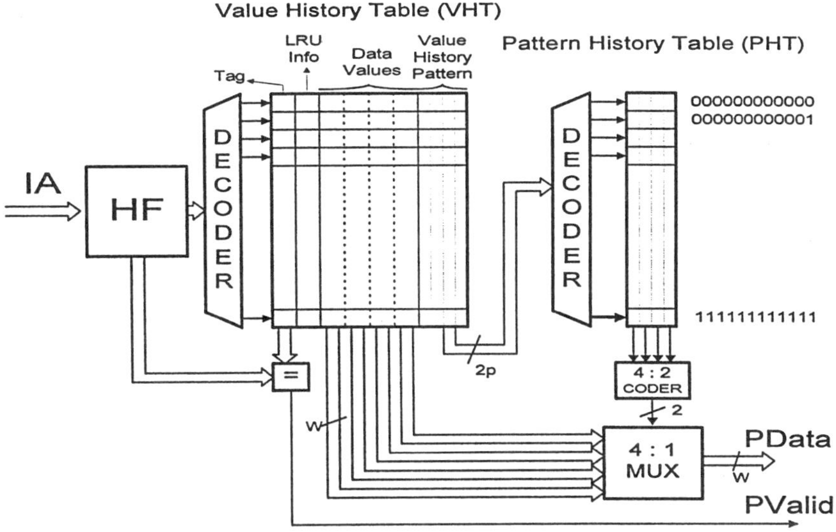


**Figure 4.17.** Block scheme of a two-level value predictor.

```
The VHT (of the first level) has four fields:

TAG  - Business as usual
DATA - Four subfields for up to four recent unique values;
               the four values are associated with the
               encoding {00,01,10,11}
LRU  - Keeping track of the order in which the 4 data values
               were seen; when the fifth value appears, the
               least recently seen value goes out
VHP  - Value History Pattern;
               the last P outcomes are kept for each
               instruction in VHT
```

**Algorithm 4.3.** Description of the VHT and the PHT.

Similar to the branch predictor, when a value prediction is to be made for an instruction, the following steps are to be completed:

1. The appropriate VHT entry is selected.
2. The TAG field is checked, to see if the entry corresponds to the current instruction (you may remember that a number of different instructions may map to the same entry).
3. If yes, the VHP value is used to select the PHT entry.
4. The maximum of the four counter values is selected, and the corresponding value is declared as a "predicted value." If there is a tie, either the value related to the last outcome is selected, or one of the values is selected at random.

Note that the prediction is made only if the maximum is above the prespecified threshold. If it is below the threshold, no prediction is made (because it is assumed that the prediction quality would be low).

Updating of the VHT entry is done in two steps: (1) contents are shifted by 2 bits and (2) the new outcome is entered into the bits left vacant. Updating of the PHT entry is also done in two steps: (1) the counter selected (corresponding to the correct outcome) is incremented by 3 (or less, if there are three results in saturation) and (2) all other counters are decremented by 1 (unless they are already at zero).

Note that the updating parameters (3 for incrementation and 1 for decrementation) are obtained empirically (after experimenting with various values).

*4.2.1.2.6. The Hybrid Predictor.* No single scheme provides good prediction for each and every application. As in the case of branch prediction, the solution is in creating a hybrid predictor. In theory, there are a number of possible ways to create a hybrid predictor, with two or more components.

One possible way to create a hybrid predictor is to combine a two-level predictor and a stride-based predictor. In that case, the VHT has to be expanded with two additional fields: STATE and STRIDE. A block scheme of such a hybrid predictor is given in Figure 4.18.

In the case of the scheme from Figure 4.18, the prediction algorithm includes the following steps:

1. The appropriate VHT entry is selected and the TAG field is checked.
2. In parallel, the VHP field (of the "two-level" part) and the STATE field (of the "stride-based" part) are read out.
3. If the selected PHT entry has the maximum count value above the threshold, then the two-level predictor is responsible for prediction; otherwise, the stride-based predictor makes the prediction, unless the value of STATE is different from {ready}, in which case no prediction is made (that is a sign of low prediction quality).

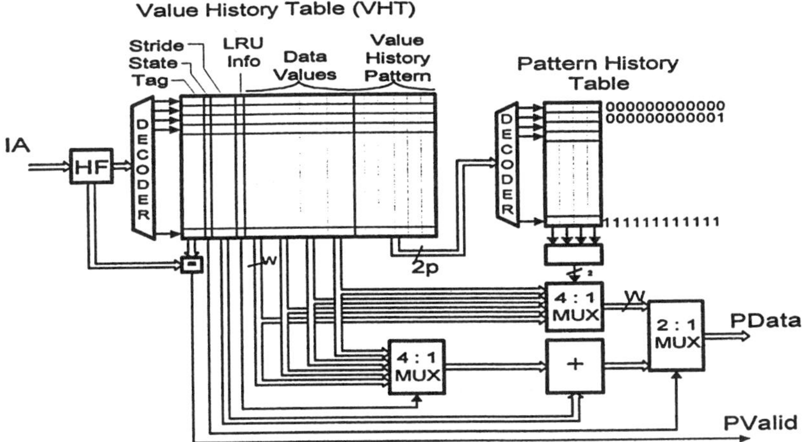


**Figure 4.18.** Block scheme of a hybrid (two-level and stride-based) predictor.

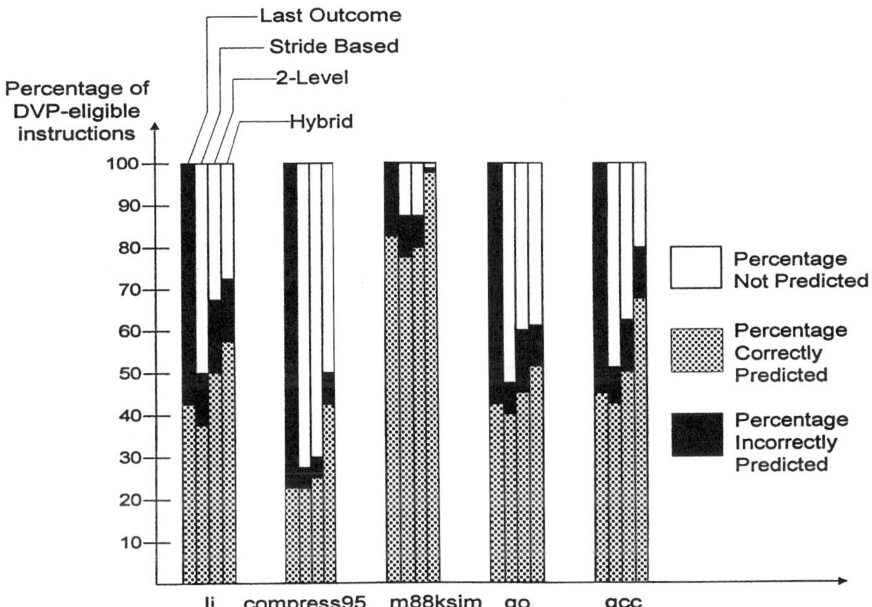


**Figure 4.19.** Results of a simulation-based comparison of different predictors.

Complexity of the hybrid value predictor is fairly large, so the question is how much performance improvement was enabled by all this additional complexity.

Figure 4.19 shows the results of a simulation study to compare performance of different value predictor types. This study assumed the MIPS-I instruction set architecture and the integer part of the SPEC95 application suite. The study implied the VHT with 4K direct-mapped entries and a PHT also with 4K entries (counters saturate at 6, while the threshold was equal to 6). The simulation was run for 100 million instructions or until completion of the program, whichever comes first.

Figure 4.19 gives results for the following metrics (relative to the total number of eligible instructions): (1) percentage of instructions correctly predicted, (2) percentage of instructions mispredicted, and (3) percentage of instructions not predicted. Obviously, these three items sum up to 100%. The major conclusion is that the hybrid predictor can offer a prediction rate of up to 98% (m88ksim). Unfortunately, the average is much lower, while the overall complexity is significant. These facts cause some people to view value prediction with limited optimism.

## 4.2.2. Advanced Issues

After a detailed presentation of basic issues, results of the follow-up research will be presented briefly. Only essential features will be emphasized, and pointers to original sources will be given.

The trace-based predictor [Wang and Franklin 1997] is founded on the following rationale. If there are two updates to a register in a trace, only the second update is live at the end of the trace, and only it has to be predicted. Less prediction means less bandwidth (predictor supplies less data) and less memory (predictor stores less data).

A trace-based predictor can be value-oriented or stride-oriented. Also, a two-level or a hybrid trace-based predictor can be formed.

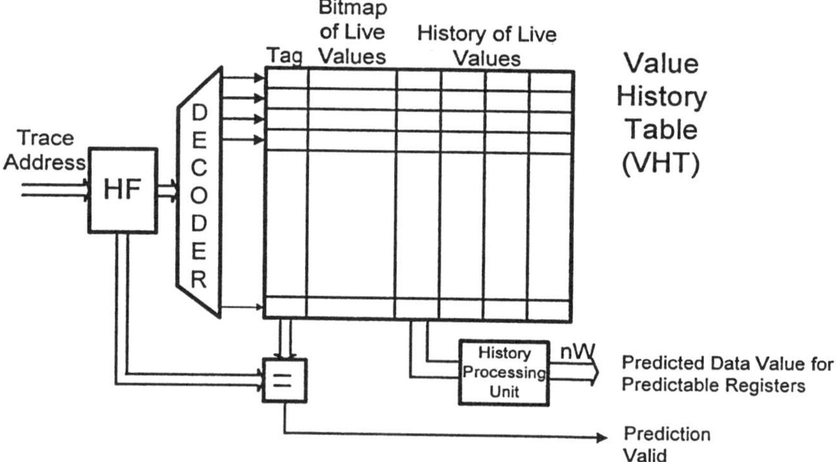


**Figure 4.20.** Value-oriented block scheme of a trace-based predictor.

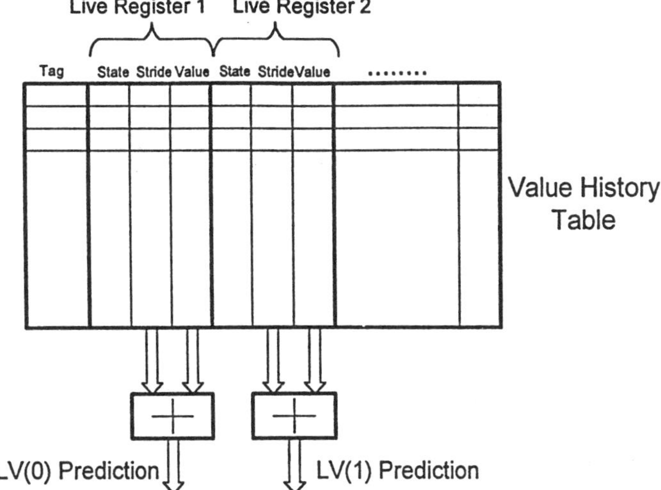


**Figure 4.21.** Stride-oriented block scheme of a trace-based predictor.

A block scheme of a value-oriented trace-based predictor is given in Figure 4.20. The history fields keep track of multiple register values. A bitmap is used to store the mapping from the live register values to instructions that produce the values.

A block scheme of a stride-oriented trace-based predictor is given in Figure 4.21. Note that different live registers are associated with different sections of the predictor. The combined stride and last value (of a live register) represent the prediction of the next value.

Performance data for various forms of trace-based predictors are given in Figure 4.22. Performance improvements are not always very dramatical, so the complexity reduction seems to be the major benefit.

The major question of data prediction is understanding its limitations in realistic machines. Results from one study [Gabbay 1998] show that instruction fetch bandwidth and issue rate have a very significant impact on the efficiency of value prediction. A hardware solution that speeds up the value prediction, by exploiting low-level parallelization opportunities, has been proposed.

Another promising approach is *selective value prediction* [Calder et al. 1999]. So far, research efforts did not take into consideration the impact of limited predictor capacities and realistic misprediction penalties. That same study [Calder et al. 1999] filters out certain instruction types (values produced by those instructions are not taken into consideration) and gives priority to other instruction types (those within the longest data dependence path in the processor's active instruction window). For example, filtering away all instructions except load instructions is potentially useful, because load instructions are responsible for most program latencies.

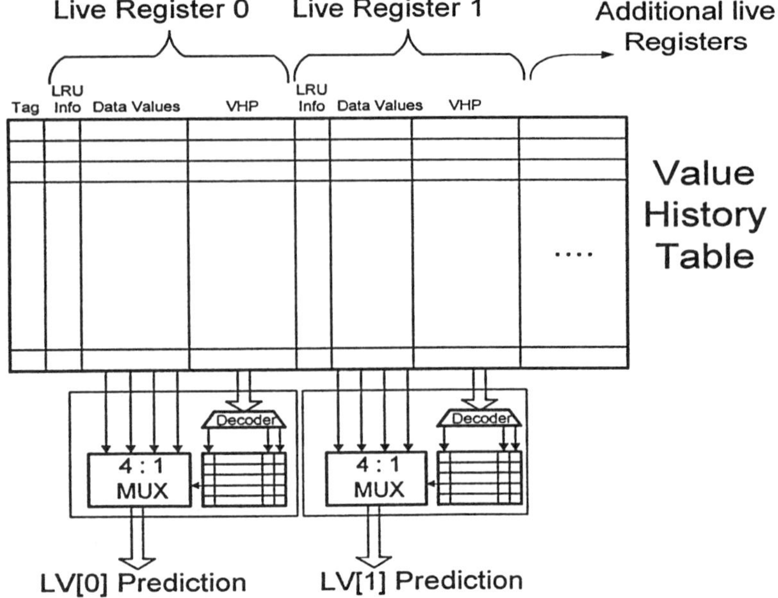


**Figure 4.22.** Simulation results for trace-based predicators.

For more advanced topics, see the special issue of *Microprocessors and Microsystems* [Tabak 1998], as well as more recent papers presented at major computer architecture conferences [e.g., Yoaz et al. 1999, Bekerman et al. 1999, Calder et al. 1999, Tullsen and Seng 1999].

The author and his associates were not very active in the field of prediction strategies, except for side activities on related projects. For more details, see Helbig and Milutinović (1989) and Milutinović (19xx).

## PROBLEMS

**4.1.** Construct a block diagram of branch predictors for all microprocessors mentioned in this book. If some relevant data are missing from this book or the open literature, assume reasonable values, and continue with the work. Calculate the approximate transistor count for each scheme.

**4.2.** Explain the strengths and weaknesses of global and per address predictors. Compare both performance and complexity.

**4.3.** Introduce an alternative predictor selection mechanism, and compare its advantages and/or drawbacks relative to the mechanism from

Figure 4.7. What is the potential performance improvement, and what is the added transistor count?

**4.4.** Calculate the exact transistor count for each predictor type mentioned in Table 4.2. Include the control circuitry into the calculation as well.

**4.5.** Design a detailed control unit for the multihybrid predictor. What is the transistor count of the control unit?

**4.6.** Figure out the percentage of value-producing instructions in a program of your choice. What type of code results in a higher percentage of value-producing instructions?

**4.7.** Construct two pieces of code and compare their suitability for the dynamic reuse data predictor. Explain which feature causes one of the two pieces of code to work better.

**4.8.** Design a detailed control unit for the last outcome data predictor. What is the transistor count?

**4.9.** Design a detailed control unit for the two-level data predictor. What is the transistor count?

**4.10.** Design a detailed control unit for the hybrid data predictor. What is the transistor count?

# CHAPTER 5

# The Input/Output Bottleneck

The so-called input/output bottleneck is defined as a discrepancy between the larger speed of processing elements in the system and the smaller speed of input/output elements of the system. According to the overall structure of the book, after the basic issues are discussed, selected advanced issues will be covered.

## 5.1. BASIC ISSUES

The overall system performance is frequently limited by I/O devices. The slowdown results for several reasons. Two of them are the most important: (1) monitoring of the I/O process consumes processor cycles and (2) if I/O supplies input data, the processing must wait until data are ready.

### 5.1.1. Types of I/O Devices

The I/O devices are divided into three major groups: (1) data presentation devices at the user interface, for processor–user communications; (2) data transport devices at the network interface, for processor–processor communications; and (3) data storage devices at the storage interface, for processor–storage communications. Of course, devices with dual or even triple roles are not uncommon in present-day machines.

Table 5.1 defines data rates for traditional presentation devices. Table 5.2 defines data rates for traditional transport devices. Table 5.3 defines data rates for traditional data storage devices. The I/O bottleneck has existed for years as a "medium" to "high" bottleneck, which is threatening to become a "major" bottleneck. However, these threats have not materialized, primarily because of technological advances. Consequently, the numbers in the three figures have to be considered conditionally, and as a lower bound.

Data presentation devices can successfully be made autonomous in their operation and typically require minimal processor interaction. This is less the case for data transport devices. Most of the processor workload comes from data storage devices. Consequently, most of the attention in the text to follow is dedicated to data storage devices (for both uniprocessor and multiprocessor environments).

**TABLE 5.1. Data Rates for Some Traditional Data Presentation Devices**[a]

| Devices | Data Rates[b] |
|---|---|
| Sensors | 1 B/s – 1 kB/s |
| Keyboard entry | 10 B/s |
| Communications line | 30 B/s – 200 kB/s |
| CRT display | 2 kB/s |
| Line printer | 1–5 kB/s |
| Tape cartridge | 0.5–2 MB/s |

*Source:* Flynn (1995).

[a]*Key:* CRT—cathode ray tube; kB/s—kilobytes per second; MB/s—megabytes per second.

[b]The numbers in this table change over time. The change is especially dramatic in the case of communications lines.

**TABLE 5.2. Data Rates for Some Traditional Data Transport Devices**[a]

| Devices | Data Rates | Maximal Delay |
|---|---|---|
| Graphics | 1 MB/s | 1–5 s |
| Voice | 64 kB/s | 50–300 ms |
| Video | 100 MB/s | ~ 20 m |

*Source:* Flynn (1995).

[a]The speed of the video depends on the type of encoding used, and it is expected to increase in coming years.

**TABLE 5.3. Data Rates for Some Traditional Data Storage Devices**[a]

| Devices | Access Time | Data Rate | Capacity |
|---|---|---|---|
| Disk | 20 ms | 4.5 MB/s | 1 GB |
| Tape | $O(0.1\ s)$ | 3–6 MB/s | 0.6–2.4 GB (per cartridge) |

*Source:* Flynn (1995).

[a]This is perhaps the field where technological changes of I/O devices are most dramatic. The speed gap between the central processing units and the data storage devices is still widening.

Technological progress of data storage devices is especially fast. For example, currently, an EIDE (enhanced integrated drive electronics) disk has an average access time of 11.5 ms, supports a 16.6-MB/s data transfer rate, and has a capacity of 4 GB. Since technology data change so quickly in time, the interested reader should check the WWW presentations of major disk technology vendors, if state-of-the-art information is needed (e.g., `http://www.wdc.com/products/drivers/drive_specs/AC34000.html`).

Also, at the present time, tape technology is characterized by an average access time of $O(0.1\ \text{s})$, an effective data transfer rate of 1.2 MB/s, and a capacity of 20 GB. For state-of-the-art information, see the WWW presentations of major tape vendors (e.g., `http://www.interface_data.com/idsp2140.html`).

The CD-ROM technology has become very popular. At the present time, it is characterized by an average access time of 150 ms, a data transfer rate of 1800–2400 kB/s (speed 12× and 16×), and capacity of 650 MB (standardized). For state-of-the-art information, see the WWW presentations of major vendors (e.g., `http://www.teac.com/dsp/cd/cd_516.html`).

The new DVD-ROM (DVD—digital versatile disk) standard implies an access time of 4.7 ns, a data rate of 9.4 MB/s, and a capacity of 17 GB. For details, see the WWW presentations of major manufacturers (e.g., `http://www.toshiba.com/taisdpd/dvdrom.htm`).

### 5.1.2. Types of I/O Organization

The I/O devices in general and data storage devices in particular can be organized in three different ways: (1) program-controlled I/O, (2) interrupt-driven I/O, and (3) DMA-managed I/O.

Internal microprocessor design is crucial for efficient support of I/O. The P/M (processor/memory) bus represents the major interface between the processor and its I/O. Consequently, design of the P/M bus represents a major challenge during the microprocessor design process.

Issues of importance when designing an efficient I/O interface are to find (1) the physical location, (2) the path to the physical location, and (3) the data, with no or only minimal processor involvement.

Major types of I/O coprocessors for uniprocessor machines are (1) multiplexer channel, based on many low-speed devices; (2) selector channel, based on one high-speed device with the capability to perform data assembly and disassembly efficiently; and (3) block channel, based on a combination of the first two approaches.

Major types of I/O coprocessors for multiprocessor machines, where it is important that devices be accessible to all processors in the system, no matter where they are physically located, are (1) a single low-speed asynchronous bus, (2) multiple high-speed synchronous buses, and (3) a combination of the above, with some intelligence added into the communications process.

### 5.1.3. Storage System Design for Uniprocessors

Access time is a major problem with storage devices. This problem can be made less pronounced if appropriate storage buffers are added. If these buffers exploit some type of locality principle, they are called *disk caches*. Disk caches are typically performed in a semiconductor technology, which means a shorter access time. They are useful for two reasons: (1) because the spatial locality is present in the disk data access pattern (access happens in

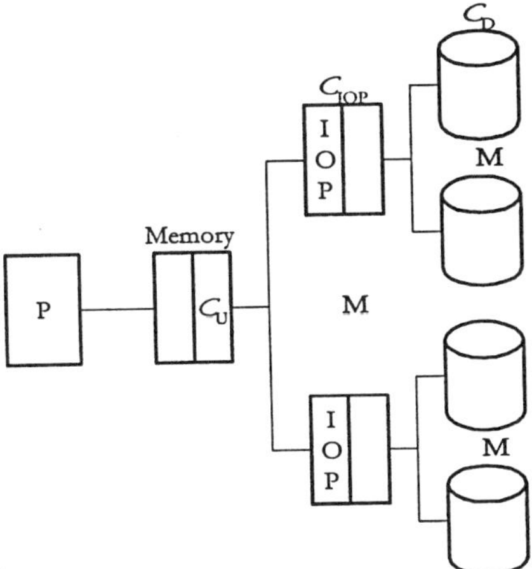


**Figure 5.1.** Three possible locations for disk cache buffers (P—processor, IOP—input/output processor). [*Source:* Flynn (1995).]
*Comment:* In disk caches, the spatial locality component is much more dominant than the temporal locality component. In processor caches, both types of locality are present, spatial more with complex data structures, and temporal more with single variables, like loop control, process synchronization, and semaphore variables.

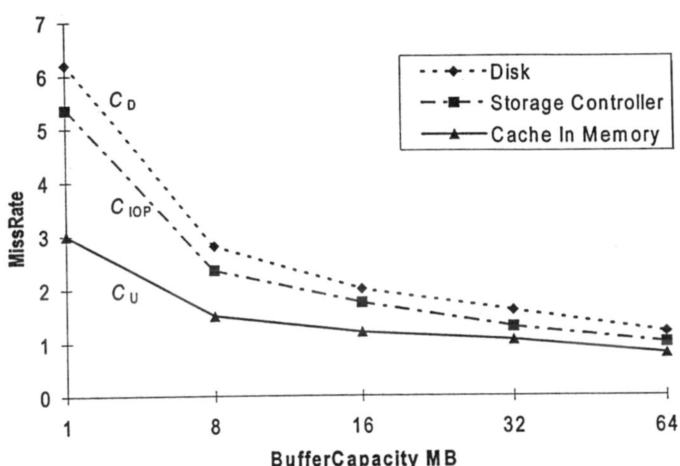


**Figure 5.2.** Miss ratios for three different locations of disk cache buffers ($C_D$—disk; $C_{IOP}$—storage controller; $C_U$—cache in memory). [*Source:* Flynn (1995).]
*Comment:* If disk access prediction algorithms are used, each of the three different locations implies a different algorithm type, which enlarges the performance differences of the three different approaches (predictors in the memory have access to more information related to prediction, compared to those located in the I/O processor, and especially to those located in the disk itself).

System structure:

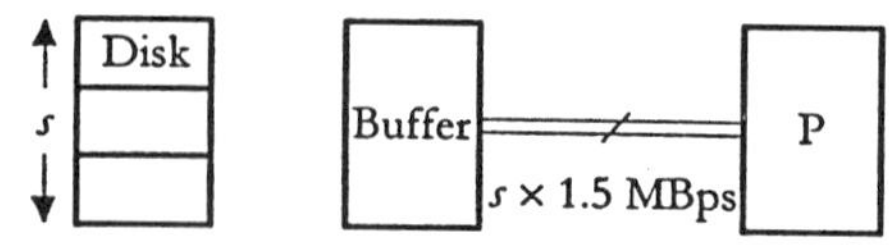

Access structure:

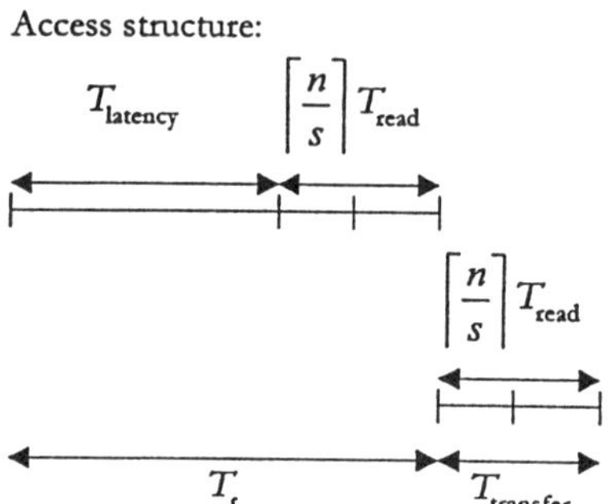

**Figure 5.3.** Structure of a disk array (P—processor; s—number of disks acting as a single unit). [*Source:* Flynn (1995).]
*Comment:* This figure implies that latency time is greater than the transfer time; that relationship is technology dependent, and may change over time.

sequences, as far as data addresses), which can be utilized for prefetch purposes; and (2) because the disk timing access pattern includes the no-activity periods (access happens in bursts, as far as data transfer), which can be utilized for prediction purposes.

Figure 5.1 defines three possible locations for disk cache buffers: (1) in the disk subsystem, (2) in the storage controller, and (3) in the main memory. Figure 5.2 gives typical miss ratios for three different locations of disk cache buffers.

Disk access implies two different activities: read and write. Different methods have been used to improve the read and the write access.

Disk arrays represent a method to improve disk read. Bytes of each block are distributed across all disks. Consequently, the time to read a file and the time to do the buffer-to-processor transfer improve. In many cases of interest, this speed-up is almost linear. This means, that if $N$ disks are used, the speed-up will be equal to about $N$.

System structure and access structure typical of disk arrays are given in Figure 5.3. Numerics that demonstrate the speed-up derived from a disk array organization are given in Algorithm 5.1.

Disk logs represent a method to improve disk write. Data are first collected in a log buffer, until their size becomes equal to the size of a disk

$$T_{transfer} = n \cdot \frac{T_{read}}{s}$$

For $(1, s)$ configurations, $n = E(f)$ and

$$T_{service} = T_{latency} + \frac{E(f)}{s} T_{read}$$

$$T_{transfer} = \frac{E(f)}{s} T_{read}$$

For (1.16),

$$T_{service} = 17.5 + \frac{3.4}{16}(2.6)$$

$$= 17.5 + 0.55 = 18.1 \text{ ms}$$

$$T_{transfer} = 0.55 \text{ ms}$$

For (1,8), we would have

$$T_{service} = 17.5 + \frac{3.4}{8}(2.6)$$

$$= 18.6 \text{ ms}$$

$$T_{transfer} = 1.1 \text{ ms}$$

**Algorithm 5.1.** Numerics of a disk array ($n$—number of blocks in a file; $s$—number of disks acting as a single unit). Variations of performance-related computations represent potentially good test questions. [*Source:* Flynn (1995).]

access unit. Consequently, disk access is characterized by a minimal ratio of overhead time to transfer time.

### 5.1.4. Storage System Design for Multiprocessor and Multicomputer Systems

Major contributors to the overall traffic for I/O in multiprocessor and multicomputer systems are (1) virtual memory and virtual I/O traffic, to support the requirement that each and every I/O device is accessible to all nodes in the system; and (2) metacomputing and metaprocessing, to support the requirement that each and every I/O device be accessible to all heterogeneous architectures representing different nodes in the system.

Table 5.4 defines the I/O requirements of the so-called Grand Challenge applications [Patt 1994]. These applications have been defined as the major research-driving forces in the fields of I/O technology and heterogeneous computing.

**TABLE 5.4. I / O Requirements of Grand Challenge Applications**[a,b]

| Application | I/O Requirements | Storage |
|---|---|---|
| Environmental and earth sciences | | |
| Eulerian air-quality modeling | Current 1 GB/model, 100 GB/application; projected 1 TB/application | S |
| | 10 TB at 100 model runs/application. | A |
| Four-dimensional | 100 MB–1 GB/run | S |
| data assimilation | 3-TB database; expected to increase by orders of magnitude with the earth observing system—1 TB/day. | A |
| Computational physics | | |
| Particle algorithms | 1–10 GB/file; 10–100 files/run | S |
| in cosmology and astrophysics | 20–200 MB/s | IOB |
| Radio synthesis imaging | 1–10 GB. | S |
| | HiPPI bandwidths minimum | IOB |
| | 1 TB | A |
| Computational biology | | |
| Computational quantum | 150 MB (time-dependent code) | S |
| materials | 3 GB (Lanczos code) | |
| | 40–100 MB/s | IOB |
| Computational fluid and plasma dynamics | | |
| High-performance | 4 GBs of data/4 h | S |
| aircraft simulation | 40 MB–2 GB/s disk, 50–100 MB/s disk to 3 in. storage (comparable to HiPPI/Ultra) | IOB |
| Computational fluid | 1 TB | A |
| and combustion dynamics | 0.5 GB/s to disk, | IOB |
| | 45 MB/s to disk for visualization | |

*Source:* Patt (1994).

[a] Key: S—secondary; A—archival; IOB—I/O bandwidth.

[b] All Grand Challenge applications belong to the domain of scientific computing. With the recent innovations in Internet technology, applications from the business computing domain become even more challenging.

Different companies employ different disk interface models in order to solve the I/O bottleneck in multiprocessor and multicomputer systems. The starting point for reading on these issues is Patt (1994). It covers the traditional approaches, and most of them can be generalized using the architecture from Figure 5.4.

The Intel Touchstone Delta model implies a 2D array ($16 \times 32$) of processing element nodes with 16 I/O nodes on the sides of the array. The Intel Paragon model uses inexpensive I/O nodes that can be placed anywhere within a mesh.

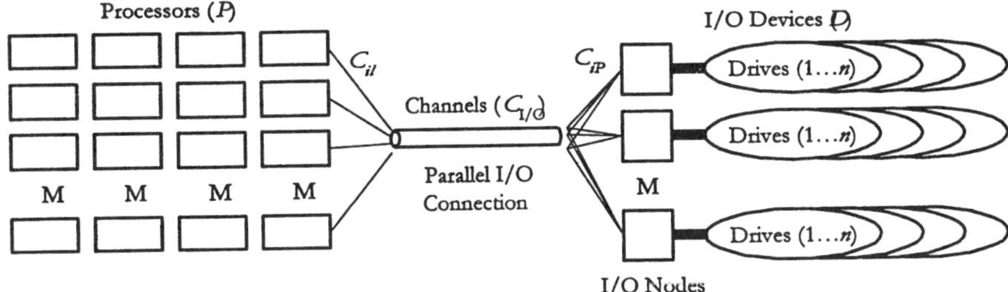


**Figure 5.4.** Parallel I/O subsystem architecture ($n$—number of drives). [*Source:* Patt (1994).]

*Comment:* The field of I/O for parallel processing is progressing more through the technological advances, rather than through the advances in architecture.

The Thinking Machines CM-5 model is based on fewer I/O processors, each one characterized with a relatively high I/O bandwidth.

The best sources of information on the state of the art in the field are conferences that include sessions on I/O and manuals on the latest products for the Grand Challenge applications.

At the present time, the Encore Infinity SP model, based on an internal architecture of the reflective memory type, according to many, is believed to be the best I/O pump available (the reflective memory model and the author's involvement are discussed in a later section).

Network interface technologies are considered crucial, especially for high-demand applications such as Grand Challenge or multimedia. Table 5.5 sheds some light on the capacities and characteristics of some traditional approaches.

Once again, issues covered so far are only those that, according to this author, represent the major problems to be solved by designers of future

**TABLE 5.5. Network Capacities and Characteristics**[a]

| Type | Bandwidth | Distance | Technology |
| --- | --- | --- | --- |
| Fiber channel | 100–1000 MB/s | LAN[b], WAN[c] | Fiberoptics |
| HiPPI | 800 MB/s or 1.6 GB/s | $\leq 25\,m$ | Copper cables (32 or 64 lines) |
| Serial-HiPPI | 800 MB/s or 1.6 GB/s | $\leq 10$ km | Fiberoptics channel |
| SCI | 8 GB/s | LAN | Copper cables |
| Sonet/ATM | 55–4.8 GB/s | LAN, WAN | Fiberoptics |
| N-ISDN | 64 kB/s, 1.5 MB/s | WAN | Copper cables |
| B-ISDN | $\leq 622$ MB/s | WAN | Copper cables |

[a] Quoted numbers are subject to change due to technological advances; for state-of-the-art numbers, the reader is referred to related WWW pages. Also, new standards are emerging (e.g., IEEE FireWire).

[b] Local area networks (up to several meters).

[c] Wide area networks (up to several kilometers).

microprocessors and multimicroprocessors on a single chip. The latest developments, only on these issues, are covered in the next section.

## 5.2. ADVANCED ISSUES

This section contains the author's selection of research activities that, in his opinion, have made an important contribution to the field in recent time and are compatible with the overall profile and mission of this book.

Hu and Yang (1996) describe an effort to optimize the I/O write performance. The result of their research implies a small log disk used as a secondary disk cache to build a disk hierarchy. A small RAM buffer collects write requests and passes them to the log disk when it is idle. In this way, one obtains performance close to the same-size RAM for the cost of a disk. Conditions of their analysis imply that the temporal component of data is relatively high. The higher the temporal locality, the higher the performance of this approach.

Maquelin et al. (1996) describe an effort to improve the efficiency of handling of incoming messages in message passing environments. The result of their research implies a hardware extension that limits the generation of interrupts to cases where polling fails to provide a sufficiently quick response. Conditions of their research imply that the message arrival frequency is the criterion for selection of interrupt versus polling. This solution is promising in environments typical of the future distributed shared memory multimicroprocessors on a chip.

An important new trend in I/O research implies issues in minimization of negative effects of deadlocks, as well as minimization of negative effects of multiple failures in redundant array of independent disks (RAID) architectures [Pinkston and Warnakulasuriya 1997, Alvarez et al. 1997].

In the case of disk array architectures, declustered organizations are used to achieve fast reconstruction of a failed disk content. A crucial design issue is data layout. The most common data layout organization is the well-known stripes organization. The six desirable properties of declustered organizations are [Holland and Gibson 1992]:

1. *Single-Failure Correcting.* No two units of the same stripe are mapped to the same disk, to make recovery possible from a single disk crash.

2. *Distributed Parity.* All disks have the same number of check units mapped to them, to balance the accesses to check units during writes or when a failure has occurred.

3. *Distributed Reconstruction.* There is a constant number of stripes such that, for every pair of disks, that constant number of stripes has units mapped to both disks (to ensure that access to surviving disks during online reconstruction are spread evenly).

4. *Large Write Optimization.* Each stripe contains a contiguous interval of the client's data, to process a write of all $(k - 1)$ data units without prereading the prior contents of any disk.

5. *Maximal Parallelism.* Whenever a client requests a read of $n$ contiguous data units, all $n$ disks are accessed in parallel.

6. *Efficient Mapping.* The functions that map client addresses to array locations are efficiently computable, with low time and space requirements.

A layout is ideal if all six properties are present. An architecture that, under certain conditions, can be treated as ideal, has been described [Alvarez et al. 1998].

Finally, a few words about hardware accelerators. Conditionally, they also belong into the I/O category. In the large plethora of various products, only three will briefly be presented here: (1) the audio accelerator EMU10K1 by the Joint E-mu/Creative Technology Center [Savell 1999], (2) the graphics accelerator by Compaq and Mitsubishi Electric Research Laboratory [McCormack et al. 1999], and (3) the multimedia accelerator by AMD [Oberman et al. 1999]. As indicated in the introductory part of this book, accelerators are an important component of future microprocessing. The special-purpose chip accelerators of today are the on-chip accelerators of tomorrow.

The audio accelerator EMU10K1 [Savell 1999] was originally designed for the consumer computer market (videogames). It includes both a high-quality music synthesizer and a powerful audio effects processor. The chip is implemented on about 2.5 MTr and includes a 64-channel wave table synthesizer with a per channel sample rate converter, digital filter, envelope generator, low-frequency oscillator, and the routing/mixing logic. Digital audio data are stored in system memory. The core of the system is a powerful audio effects processor. It efficiently supports the applications that produce music using the abstract language called MIDI. One of the characteristics of this chip is that MIDI commands can originate from both external sources and internal memory. The audio effects processor is especially efficient for generation of the 3D audio. The core of the special effects processor is the multiply-accumulate unit, which essentially places this accelerator into the category of digital signal processors.

The graphics accelerator Neon [McCormack et al. 1999] is tuned to WindowsNT 2D rendering and OpenGL 3D rendering. The core of the chip is a vector of eight identical processor memory units. Each of the eight units includes 4–16 MB of SDRAM (100-MHz synchronous dynamic RAM), a memory controller, and the pixel processor. The CPU of the main system can send commands to the Neon chip via a 64-bit PCI bus using programmed I/O, or it can use DMA to move commands from main memory of the main system directly to the Neon chip. Commands coming from outside are first parsed, then passed through the fragment generator to determine object boundaries, and finally forwarded to a crossbar facing the eight processor memory units. After the processing is completed, the Neon memory controller initiates the forwarding of final data to the video controller, which is responsible for screen refreshment (up to $1600 \times 1200$ pixels at 76 Hz).

The multimedia accelerator AMD 3DNow! is a part of the AMD K6-2 effort, which introduces 21 new instructions focusing on major bottlenecks in multimedia and floating-point-intensive applications [Oberman et al. 1999]. This enables faster frame rates on high-resolution scenes, more accurate physical modeling of real-world phenomena, and so on. The 3DNow! architecture operates on multiple operands in parallel (SIMD) and represents an extension of the x86 MMX architecture, which included the initial 57 instructions for multimedia applications. In addition to speed-up that comes from new instructions, there is also a speed-up that comes from organizational innovations, like introduction of new register structures and prefetching. In other words, the traditional SIMD architecture, enhanced with a more sophisticated (application-tuned) register memory and a properly designed prefetching (with elements of pipelining), results in a much better price/performance ratio.

The author and his associates were not very active in the field of I/O bottlenecks, except for side activities on related projects [Ekmecic et al. 1996, Raskovic et al. 1995].


## PROBLEMS

**5.1.** Consult the open literature, and create a table with data rates of modern presentation devices, for the year in which you are reading this book. What is the approximate unit cost for each device in your table?

**5.2.** Consult the open literature, and create a table with data rates of modern transport devices, for the year in which you are reading this book. What is the approximate unit cost for each device in your table?

**5.3.** Consult the open literature, and create a table with data rates of modern storage devices, for the year in which you are reading this book. What is the approximate unit cost for each device in your table?

**5.4.** Create a table with URLs for major manufacturers of presentation, transport, and storage devices. What are the trends that you have noticed, after a careful examination of each offer?

**5.5.** Disk access happens in bursts. In between the data transfer bursts, there are relatively long silent periods, which can be used for prediction of what data may be needed (from the disk) next. That data can be prefetched into the disk cache. Try to create a prediction mechanism for prefetch (from the disk) into the disk cache. Develop a schematic that implements the proposed solution.

**5.6.** Create a block diagram of the disk log approach for improving the disk write. Develop a detailed schematic.

**5.7.** Consult the reference [Ekmecic et al. 1996] and discuss the suitability of each presented concept for efficient large-scale I/O. Repeat the same for various applications of interest (e.g., for applications from Table 5.4).

**5.8.** Multimedia applications are especially demanding when it comes to I/O. Consult the open literature and make a survey of I/O for multimedia applications.

**5.9.** A two-node reflective memory system can be used as an I/O "pump" with fault-tolerance. All I/O data can be reflected at both nodes, for the case that one node fails. Consult the available literature on reflective memory and create a block diagram of such a solution. Discuss pros and cons.

**5.10.** Create a detailed block diagram of the architecture presented in Alvarez et al. (1998). Explain how it satisfies each one of the six requirements for an "ideal" declustered architecture.

# CHAPTER 6

# Multithreaded Processing

This chapter includes two main sections. The first one covers basic issues (background) while the second one discusses advanced issues (state-of-the-art).

## 6.1. BASIC ISSUES

This chapter gives an introduction to multithreaded processing—mainly the elements that are, in the author's opinion, of importance for future microprocessors on a chip.

There are two major types of multithreaded machines: (1) coarse-grained, based on task-level multithreading; and (2) fine-grained, based on instruction-level multithreading.

Task-level multithreading implies that switching to a new thread is done on the context switch. Instruction-level multithreading implies that switching to a new thread is done on every cycle.

The principal components of a multithreaded machine are (1) multiple activity specifiers, such as multiple program counters and multiple stack pointers; (2) multiple register contexts; (3) thread synchronization mechanisms, such as memory-access tags and two-way joins; and (4) mechanisms for fast switching between threads.

A common question concerns the difference between a thread and a process. An important difference between threads and processes is that each process has its own virtual address space while multiple threads run in the same address space; consequently, thread switching is faster than process switching. According to another definition, threads are supported mostly at the architecture level, and processes are supported mostly at the operating system level. For example, major multithreading constructs such as start, suspension, and continuation, of a thread, are usually supported on the ISA (instruction-level architecture) level. Major processing constructs like start, suspension, and continuation, of a process, are usually supported at the OS (operating system) level.

Projects oriented to multithreading include, but are not limited to, Tera (Smith at Tera Computers), Monsoon and T (Arvind at MIT in cooperation with Motorola), Super Actor Machine (Gao at McGill University), EM-4 (Sakai, Yamaguci, and Kodama at ETL in Japan), MASA (Halstead and Fujita at Multilisp), J-Machine (Dally at MIT), and Alewife (Agarwal at MIT). Many of these machines include elements of other concepts, such as dataflow, message passing multicomputing, and distributed shared memory multiprocessing. Consequently, all these efforts can be classified in a number of different ways. Again, for details, see the original papers or Iannucci et al. (1994). For an interesting discussion about relationships between dataflow and multithreading, see Silc et al. (1998).

### 6.1.1. Coarse-Grained Multithreading

The first coarse-grained MIMD (multiple instruction, multiple data) machine based on multithreading was the heterogeneous element processor (HEP), built at Denelcor in 1978. Structure of the HEP system is given in Figure 6.1. It includes up to 16 processing element modules (PEMs), a number of data memory modules (DMMs), an input/output controller (IOC), and a multi-stage interconnection network (ICN) based on a number of packet switching units (PSUs), which can be bypassed if necessary, using a set of local access paths (LAPs).

The internal structure of a PEM is shown in Figure 6.2. Each PEM can run up to eight user threads and up to eight system threads. Details can be found in the original papers or in Iannucci et al. (1994).

Coarse-grained multithreading is important as a method of combining existing microprocessors into larger and more powerful systems. However, fine-grained multithreading (to be discussed in the next section) is important for the existing instruction-level parallelism, in order to increase the speed of single-chip machines.

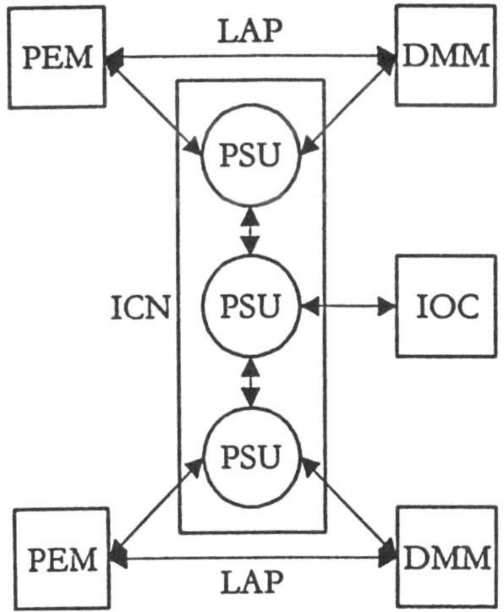


**Figure 6.1.** Structure of the HEP multiprocessor system. [*Source:* Iannucci et al. (1994).] *Comment:* The HEP is based on packet switching. More recent heterogeneous processing projects tend to explore more shared memory [Ekmecic et al. 1995, 1996].

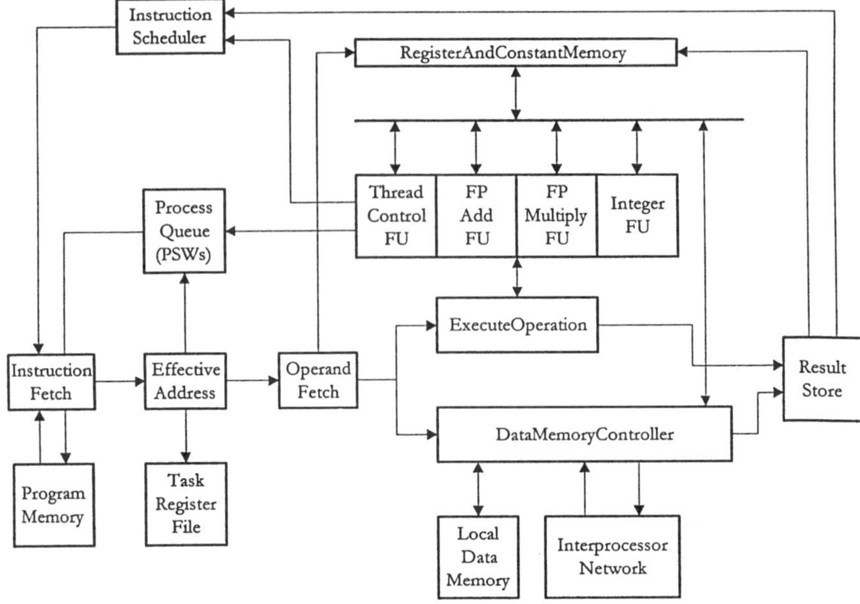

**Figure 6.2.** Structure of the HEP processing element module (FU—functional unit; FP—floating-point). [*Source:* Iannucci et al. (1994).]
*Comment:* This structure can be conditionally treated as a superscalar with a special thread control functional unit.

### 6.1.2. Fine-Grained Multithreading

As far as this author is concerned, among various approaches to fine-grained multithreading, the approach referred to as *simultaneous multithreading* (SMT) seems to be the most promising candidate for incorporation into next-generation microprocessors on a single chip.

In traditional fine-grained multithreading, only one thread issues instructions in each cycle. In SMT, in each cycle, several independent threads issue instructions simultaneously to multiple functional units of a superscalar. This approach provides higher potentials for utilization of resources in a wide-issue superscalar architecture. One study [Tullsen et al. 1995] indicates that, on an eight-issue superscalar processor, the performance can be up to about 4 times better compared to the same superscalar processor without multithreading, and up to about 2 times better compared to traditional fine-grained multithreading on the same superscalar architecture.

Figure 6.3 describes the essence of SMT. The term *horizontal waste* refers to unused slots within one cycle of a superscalar machine. Horizontal waste is a consequence of the fact that one thread often does not include enough instruction-level parallelism. The term *vertical waste* refers to cases in which

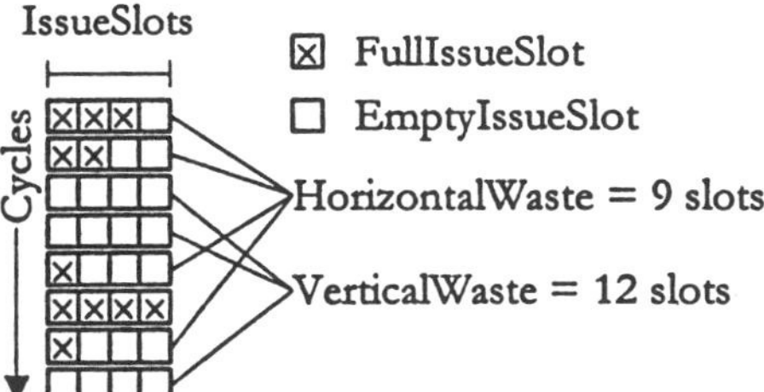



**Figure 6.3.** Empty issue slots: horizontal waste and vertical waste. [*Source:* Tullsen et al. 1995).]

*Comment:* Potential efficiency of filling the empty slots (see Table 6.2) and minimal architectural differences in comparison with traditional superscalars [Eickmeyer et al. 1996] make this approach a prime candidate for incorporation into next-generation microprocessor architectures.

an entire cycle is wasted, because various hazards have to be avoided at run time. In Figure 6.3, which describes traditional fine-grained multithreading, the horizontal waste is equal to 9 slots, and the vertical waste is equal to 12 slots; consequently, the total waste is equal to 21 slots. If SMT were used, other threads could fill in the unused slots, which, in turn, could bring the total waste down to zero.

In conclusion, superscaling is not efficient for vertical waste, traditional fine-grained multithreading is not efficient for horizontal waste, but SMT is efficient in both cases.

Table 6.1 lists the major sources of wasted issue slots and possible latency hiding and/or latency reducing techniques that can be used to solve the problems. It is essential to understand that all listed latency hiding and latency reducing techniques have to be used properly before one turns to SMT for further improvement of run-time performance. Consequently, one must be careful about the interpretation to benefits of SMT—it may appear as being more efficient than is realistically, unless all techniques from Table 6.1 have been used properly. The research by Tullsen et al. (1995) recognizes the problem and provides a realistic analysis that sheds important light on SMT.

A performance comparison of SMT and various other multithreaded multiprocessor approaches is given in Table 6.2. The results are fairly optimistic in favor of SMT, despite the fact that the study is based on a somewhat idealized case of SMT. The study concludes that SMT paves the way to 8-issue and 16-issue superscalars, while the techniques used in the microprocessors of the mid- to late 1990s, applied to superscalar machines, do not enable designers to go beyond the 4-issue superscalars.

**TABLE 6.1. Causes of Wasted Issue Slots and Related Prevention Techniques[a,b]**

| Source of Wasted Issue Slots | Possible Latency—Hiding or Latency—Reducing Techniques |
| --- | --- |
| Instruction TLB miss, data TLB miss | Decrease the TLB miss rates (e.g., increase the TLB sizes); hardware instruction prefetching; hardware or software data prefetching; faster servicing of TLB misses |
| I-cache miss | Larger, more associative, or faster instruction cache hierarchy; hardware instruction prefetching |
| D-cache miss | Larger, more associative, or faster data cache hierarchy; hardware or software prefetching; improved instruction scheduling; more sophisticated dynamic execution |
| Branch misprediction | Improved branch prediction scheme; lower branch misprediction penalty |
| Control hazard | Speculative execution; more aggressive if conversion |
| Load delays (first-level cache hits) | Shorter load latency; improved instruction scheduling; dynamic scheduling |
| Short integer delay | Improved instruction scheduling |
| Long integer, short FP, long FP delays | (Multiply is the only long integer operation, divide is the only long FP), shorter latencies; improved instruction scheduling |
| Memory conflict | (Accesses to the same memory location in a single cycle), improved instruction scheduling |

*Source:* Tullsen et al. (1995).

[a]*Key:* TLB—translation look-aside buffer; FP—floating-point operation.

[b]It is crucial that appropriate prevention techniques are completely utilized before the simultaneous multithreading is used.

## 6.2. ADVANCED ISSUES

This section contains the author's selection of research activities that, in his opinion, have made important contributions to the field recently and are compatible with the overall profile of this book.

Eickmeyer et al. (1996) describe an effort at IBM to use an off-the-shelf microprocessor architecture in the form of multithreading. The effort is motivated by the fact that memory accesses are starting to dominate the execution time of uniprocessor machines.

The major conclusion of this study is that multithreading is an important avenue to more efficient multiprocessor/multicomputer environments. For details, see Eickmeyer et al. (1996).

Conditions of the analysis apply [Eickmeyer et al. 1996] in object-oriented programming for online transactions processing. This does not mean that the approach is not efficiently applicable to other environments of interest.

**TABLE 6.2. Comparison of Various (Multithreading) Multiprocessors and an SMT Processor**[a,b]

| Purpose of Test | Common Elements | Specific Configuration | T |
|---|---|---|---|
| Unlimited FUs:<br>  equal total issue bandwidth,<br>  equal number of register sets<br>  (processors or threads) | Test A: FUs = 32S<br>IssueBw = 8, RegSets = 8<br>Test B: FUs = 16<br>IssueBw = 4, RegSets = 4<br>Test C: FUs = 16<br>IssueBw = 8, RegSets = 4 | M: 8 thread, 8-issue<br>MP: 8 1-issue<br>SM: 4 thread, 4-issue<br>MP: 4 1-issue<br>SM: 4 thread, 8-issue<br>MP: 4 2-issue | 6.64<br>5.13<br>3.40<br>2.77<br>4.15<br>3.44 |
| Unlimited FUs:<br>  test A, but limit SM to 10 FUs | Test D:<br>IssueBw = 8, RegSets = 8 | SM: 8 thread, 8-issue, 10 FU<br>MP: 8 1-issue procs, 32 FU | 6.36<br>5.13 |
| Unequal issue BW:<br>  MP has up to 4 times<br>  the total issue bandwidth | Test E: FUs = 32<br>RegSets = 8<br>Test F: FUs = 16<br>RegSets = 4 | SM: 8 thread, 8-issue<br>MP: 8 4-issue<br>SM: 4 thread, 8-issue<br>MP: 4 4-issue | 6.64<br>6.35<br>4.15<br>3.72 |
| FU utilization:<br>  equal FUs, equal issue bw,<br>  unequal reg sets | Test G: FUs = 8<br>IssueBw = 8 | SM: 8 thread, 8-issue<br>MP: 2 4-issue | 5.30<br>1.94 |

*Source:* Tullsen et al (1995).

[a]*Key:* T—throughput (instructions/cycle); FU—functional unit.

[b]Note that the number of instructions per cycle scales up almost linearly with the increase of the issue width.

Tullsen et al. (1996) describe a follow-up effort on the SMT project, aimed at better characterization of the SMT design environment, so that more realistic performance figures can be obtained. The major goal of the project is to pave the way for implementation that extends the conventional wide-issue superscalar architecture using the principles of SMT.

The major principles of their study are (1) minimizing the changes to the conventional superscalar architectures, which makes SMT more appealing for incorporation into future versions of existing superscalar machines; (2) making any single thread to be only slightly suboptimal, which means that performance of a single thread can be reduced by at most 2%; and (3) achieving maximal improvement over existing superscalar machines when multiple threads are executed. In other words, Tullsen et al. (1996) insists that performance of multiple threads should not be targeted at the cost of slowing a single thread for more than about 2%.

Conditions of their analysis imply a modified multiflow compiler working with eight threads, and a specific superscalar architecture along the lines of SGI 10000. This study takes into account the fact that the increased number of multiple activity specifiers (register files, etc.) either slows down the microprocessor clock or causes some operations that are of the one-cycle type in traditional superscalars to become of the two-cycle type in SMT. In spite of the relatively pessimistic timing conditions of the study, SMT still demonstrates a relatively optimistic throughput of 5.4, versus the traditional superscalar throughput of 2.5, when all architectural details are the same, except that SMT runs multiple threads and traditional superscalar runs a single thread.

An important new trend in high-performance microprocessors implies the combination of simultaneous multithreading on one side and multithreaded vector architectures and/or multiscalar processor architectures on the other side [Espasa and Valero 1997, Jacobson et al. 1997].

An issue of special importance, which helps a wider acceptance of a new concept, is to show performance in a variety of scenarios of importance. Lo et al. (1998) show the performance analysis results for SMT in the case of database workload. Results promise that SMT has a future in this important application field.

Multithreading and cache designs are highly correlated. However, the problem is not widely studied in the open literature. The paper by Kwak et al. (1999) represents one such effort and proposes a novel multithreaded virtual processor model, which enables an easier evaluation of the interaction between multithreading and cache performance.

In general, multithreading is especially useful if programs are, prior to execution, partitioned into threads. If that is not the case, the benefits of multithreading are limited. With this in mind, Chappell et al. (1999) propose the SSMT (simultaneous subordinate microthreading) approach to enhance the performance of the primary thread, which proves to be useful for overall performance.

### 6.2.1. How to Use One Billion Transistors on a Chip

Before moving to multiprocessor issues, it is interesting to see what researchers think about the best way to put together the architecture of a uniprocessor on a chip, if the available transistor count is one billion transistors. The following discussion is based on the special issue of *IEEE Computer* dedicated to one-billion-transistor (1-BTr) architectures [Burger and Goodman 1997].

Lipasti and Shen (1997) believe that the best solution for a 1-BTr (billion-transistor) single-chip microprocessor is the *superspeculative microarchitecture*—an architecture that employs a broad spectrum of speculative techniques to maximize the overall performance. Figure 6.4 shows the structure of Superflow—an architecture for which Lipasti and Shen (1997) predict potential performance of 9 instructions per cycle and realizable performance of 7.3 instructions per cycle for SPECint95.

The structure of Figure 6.4 assumes the following on-chip resources: (1) core CPU with 32 instructions issued per cycle, (2) value prediction table of 32 kB, (3) classification table of 8000 entries (2 bits/entry), (4) dependence

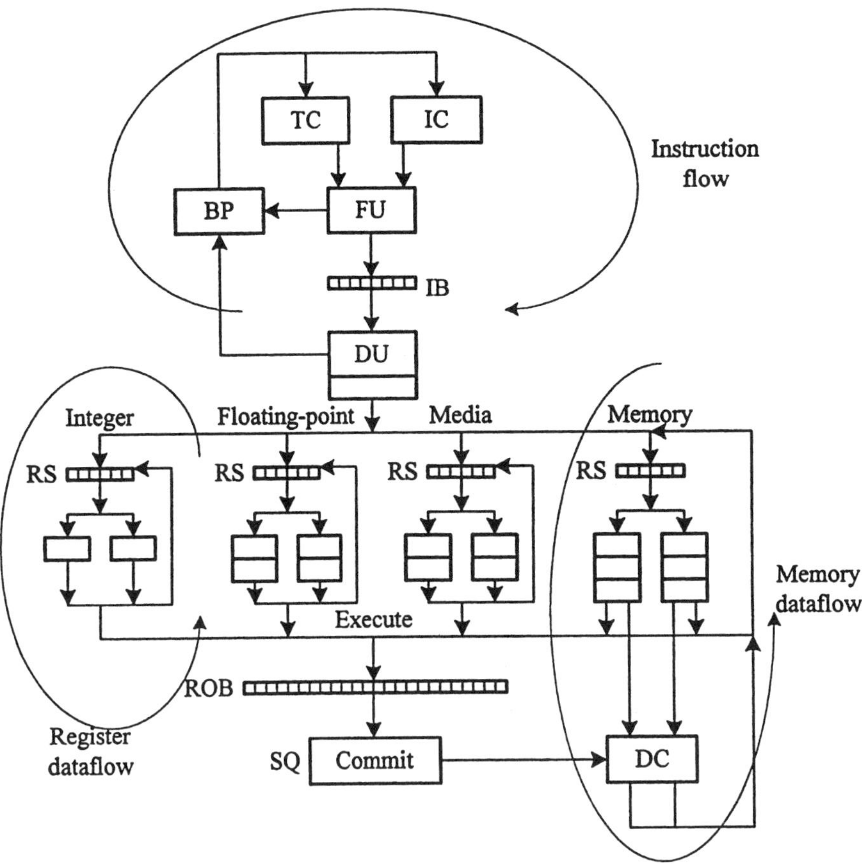



**Figure 6.4.** Architecture of Superflow (BP—branch predictor; DC—data cache; DU —decode unit; FU—fetch unit; IB—instruction buffer; IC—instruction cache; ROB —reorder buffer; RS—reservation station; SQ—store queue; TC—trace cache). *Comment:* The reorder buffer is assumed to include 128 entries, the fetch width is assumed to be 32, and a delay of 10 cycles for L1 miss is assumed, as well as a queue store of 128 entries. Estimated performance is 1.3 instructions per second for SPECint95.

predictor of 8000 entries (7 bits/entry), (5) alias prediction table of 8000 entries (7 bits/entry), (6) pattern history tables (two tables) of 64 entries each (2 bits/entry), (7) trace cache of 64 kB, (8) L1 instruction cache of 64 kB, (9) L1 data cache (4 ports) of 64 kB, and (10) L2 cache of 16 MB.

The transistor count contribution of all resources specified above is expected to be as follows: (1) 128 MTr, (2) 1.6 MTr, (3) 0.1 MTr, (4) 22 MTr, (5) 11 MTr, (6) 1.6 MTr, (7) 3.1 MTr, (8) 3.1 MTr, (9) 12.6 MTr, (10) 805.3 MTr, and 1.6 MTr for miscellaneous small resources.

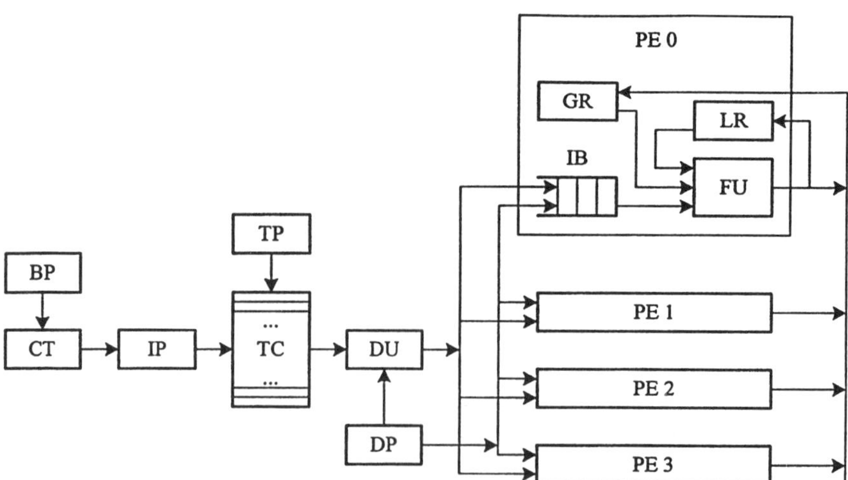

**Figure 6.5.** Architecture of a trace processor (BP—branch prediction; CT—construction of trace; DP—data prediction; DU—dispatch unit; FU—functional unit; GR—global registers; IB—instruction buffer; IP—instruction preprocessing; LR—local registers; PE—processing element; TC—trace cache; TP—trace prediction).

*Comment:* Aggregate performance of the 4-PE system is estimated at 16 instructions per cycle.

Smith and Vajapeyam (1997) argue that the best solution for a 1-BTr on-chip uniprocessor is the *trace processor*. Instruction fetch hardware unwinds programs into traces. Traces are placed into a trace cache. A special trace fetch unit reads traces from the trace cache and forwards them to the appropriate processing elements. With appropriate control mechanisms and processing elements, a peak throughput of one trace per clock cycle can be achieved. Traces are hardware-generated (not compiler-generated as in Sohi's *multiscalar processor* [Sohi et al. 1995]). Multiple superscalar pipelines are used.

The architecture of a trace processor is shown in Figure 6.5. It includes four processing elements. With the help of the branch predictor, traces are constructed and preprocessed, to capture data dependencies and to determine resource requirements. The trace cache includes a next-trace predictor and is linked to the dispatch unit. Processor elements include instruction buffer, fundamental units, local and global registers, and can be fed either from the dispatch unit or the data predictor.

Smith and Vajapeyam (1997) did not go into detailed transistor count estimation for different resources of the 1-BTr trace processor. It is believed that the best allocation of transistors per resources is determined after a simulation study is completed (which evaluates various configurations, assuming six transistors per memory cell).

Patt et al. (1997) believe that the solution is a very-high-performance uniprocessor on a chip, with chips interconnected to form an SMP. Since 1 BTr is still a limited number, some potentially useful resources will have to be excluded from the chip. The primary candidates for exclusion from the chip are the resources that are not sensitive to large interchip delays. Consequently, those resources to be included are the ones that are highly sensitive to large interchip delays. According to Patt et al. (1997), resources to be included in their entireness are those that support aggressive speculation, which implies strong dynamic branch prediction and very wide superscalar processing. These resources include (1) a large trace cache, (2) a large number of reservation stations, (3) a large number of pipelined functional units, (4) a large enough data cache, and (5) sufficient conflict resolution and data forwarding logic.

Patt et al. (1997) argue that issue widths of up to 16 or 32 instructions per cycle are quite reasonable, as well as reservation stations to accommodate 2048 instructions, or 24–48 highly optimized pipelined functional units. Professor Yale Patt believes that designers will not run out of transistors before they run out of their use for supporting a single instruction stream. Of course, in order to exploit all these features efficiently, and in order to enable such a uniprocessor to run at its peak speed, advances are needed also in areas like compilation technology and CAD tools.

Figure 6.6 shows a possible structure of a 1-BTr uniprocessor, as suggested by Patt et al. (1997). The structure from Figure 6.6 assumes that 640 MTr are used for the second-level cache, 240 MTr for the trace cache, 60 MTr for the execution core, 48 MTr for the branch predictor, and 32 MTr for data caches.

Kozyrakis et al. (1997) argue that the best way to achieve a 1-BTr uniprocessor on a single chip is to place the entire main memory on the same chip, together with a regular CPU such as R10000 or similar. This assumes that DRAM technology is used. It is argued that such a uniprocessor will be perfectly scalable—one only adds more main memory—a resource that is always needed and never large enough. Such an approach is generally referred to as IRAM, or *intelligent RAM*.

Architecture of an IRAM-based 1-BTr CPU is given in Figure 6.7. It is simple. It includes a RISC processor (e.g., one of the year 2000) plus main memory. An important argument of Kozyrakis et al. (1997) is that DRAM can hold up to 50 times more data than the same area devoted to the caches. Another advantage of IRAM is that it can be combined with a number of different CPU architecture types for better tuning to a given application.

The author and his associates were not very active in the field of multithreading, except for side activities on related projects. For details, see Helbig and Milutinović (1989) and Milutinović et al. (1988b).

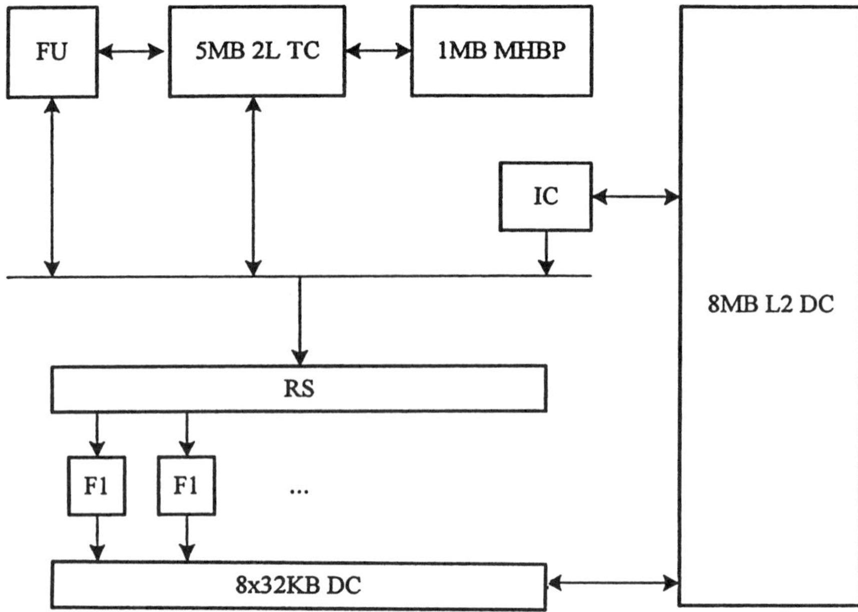

**Figure 6.6.** One possible structure of a 1-BTr one-chip microprocessor [DC—data cache; IC—instruction cache; Fi—functional unit ($i = 1, 2, \ldots$); FU—fill unit; MHBP —multihybrid branch predictor; RS—reservation station; TC—trace cache].
*Comment:* Estimated performance is from about 11 to about 14 instructions per second for SPECint95.

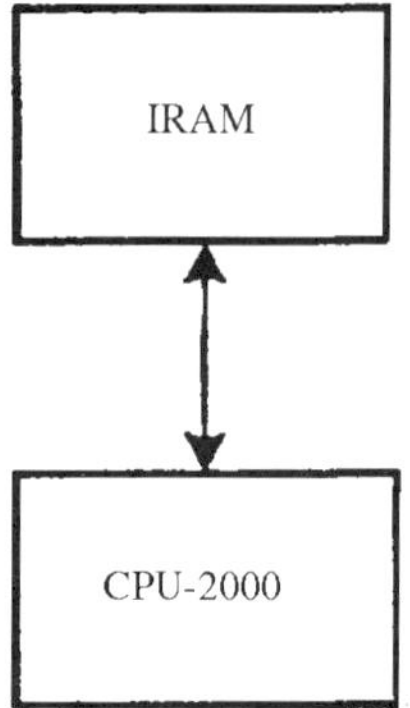

**Figure 6.7.** The IRAM-based 1-BTr microprocessor on a single chip (CPU-2000—a CPU typical of the generation 2000; IRAM—intelligent RAM).
*Comment:* The peak performance of the UC Berkeley V-IRAM microprocessor is expected to be 16 GFLOPS (or 128 GOPS at 64 bits per operation).

## PROBLEMS

**6.1.** Create a piece of symbolic graphics that illustrates essential differences between a thread and a process. What are the major issues to emphasize?

**6.2.** Create a table with major characteristics (and URLs) of coarse-grained multithreading projects mentioned in this book. Add more examples not included in this book.

**6.3.** Create a piece of symbolic graphics that illustrates essential differences between coarse-grained multithreading and dataflow architectures. What are the major issues to emphasize?

**6.4.** How would you expand the HEP architecture to include an efficient support for multimedia? Provide a block diagram, like the one in Figure 6.2, with appropriate explanations.

**6.5.** Think about the incorporation of the trace cache into the HEP architecture, and how the concept of trace cache would have to be modified, in order to fit better into the HEP-type architecture. Provide a block diagram.

**6.6.** Create a table with major characteristics (and URLs) for a set of fine-grained multithreading projects. Concentrate on both performance and complexity issues.

**6.7.** Create your own short piece of code, and calculate both the horizontal and the vertical wastes, for a structure that is similar to the one in Figure 6.3. Which type of waste is larger, and why?

**6.8.** Provide a block diagram of a control unit for the SMT machine described in Tullsen et al. (1996). Give a detailed design.

**6.9.** Explain in detail all latency hiding and latency reducing techniques from Table 6.1. Rate the effectiveness of each technique. Make your educated guess estimates and rank them with grades 10 (best) to 1 (worst).

**6.10.** Create a piece of symbolic graphics that illustrates essential differences between SMT and superscalar architectures. What are the major issues here?

# CHAPTER 7

# Caching in Shared Memory Multiprocessors

Shared memory multiprocessing (SMP) is one of the most widely used concepts in the implementation of modern multiprocessor workstations. The SMP architecture is essentially an MIMD architecture. It includes a shared memory, an interconnection network, and a set of independent processors with caches (only L1 or both L1 and L2). All processors share the same logical address space and are implemented as a single physical unit. Many successful implementations are based on off-the-shelf memory chips, standard buses, and off-the-shelf microprocessors. This makes the concept relatively easy to implement.

## 7.1. BASIC ISSUES

The programming model of SMP machines is relatively simple and straightforward. It resembles the programming model of single-instruction single-data (SISD) machines. Therefore, the existing code can easily be restructured for reuse, and the new code can easily be developed using well-known and widely accepted programming techniques.

Unfortunately, the SMP systems are not scalable beyond some relatively low number of processors. This number changes over time and depends on the speed of the components involved (processors, bus, memory). For the current technology of components involved, the price/performance ratio starts dropping after the number of processors reaches 16; it starts dropping sharply after the number of processors passes 32. Consequently, SMP systems with more than 16 nodes are rarely considered.

One of the major problems in implementing the SMP is cache consistency maintenance. This problem arises if two or more processors bring the same shared data item into their local private caches. While this data item is being read from local private caches of the processors–sharers, consistency of the system is maintained. However, if one of the processors–sharers executes a write and changes the value of the shared data item, all subsequent reads (of that same data item) may result in a program error. Cache consistency is maintained using appropriate cache consistency maintenance protocols. They can be implemented in hardware, in software, or by using hybrid techniques. Here we cover only the basic notions of hardware protocols. For advanced aspects of hardware protocols, the interested reader is referred to Tomašević and Milutinović (1993). Software protocols will not be elaborated on in this book, due to their marginal use in modern computer systems today. For detailed information on software protocols, interested readers are referred to Tartalja (1996). Hybrid protocols try to combine the best of the two extreme approaches (fully hardware approach vs. fully software approach). They represent a promising new avenue for ongoing research.

There are two major approaches to hardware-based cache consistency maintenance: (1) snoopy protocols and (2) directory protocols. Both approaches are elaborated on in the sections to follow, in the context of cache consistency maintenance.

As will be seen later (in the section on distributed shared memory systems), one can also talk about the memory consistency maintenance, in systems with one logical address space, which is implemented using a number of different physical memory modules. Such systems, referred to as *distributed shared memory* (DSM), typically consist of a number of interconnected clusters—one cluster representing an SMP system. In SMP systems, snoopy protocols are the most efficient solutions for maintenance of cache consistency. In DSM systems, snoopy protocols continue to be the most efficient approach to cache consistency maintenance (at the SMP level), while the directory protocols represent the most efficient solutions for maintenance of memory consistency (at the DSM level). In order to justify these statements, details of snoopy and directory protocols are presented next.

### 7.1.1. Snoopy Protocols

In snoopy protocols, information related to cache consistency maintenance is fully distributed, because the cache consistency maintenance mechanism is built into the controllers of local private caches. Major elements of the cache consistency maintenance mechanism are broadcast and snooping. In other words, each cache controller uses the shared bus to broadcast all consistency maintenance related information, and all cache controllers constantly monitor (snoop) the bus; each individual cache controller fetches from the bus any information that is relevant for that particular processing node.

In general, snoopy protocols are ideally suited for bus-based multiprocessors; this is because broadcast is the only operation supported by typical buses, and snoopy protocols only need the broadcast operation. Snoopy protocols have a low implementation cost; however, they are characterized by a limited scalability, for reasons discussed earlier.

There are two basic classes of snoopy protocols: (1) write-invalidate (WI) and (2) write-update (WU). The criterion for classification is the type of

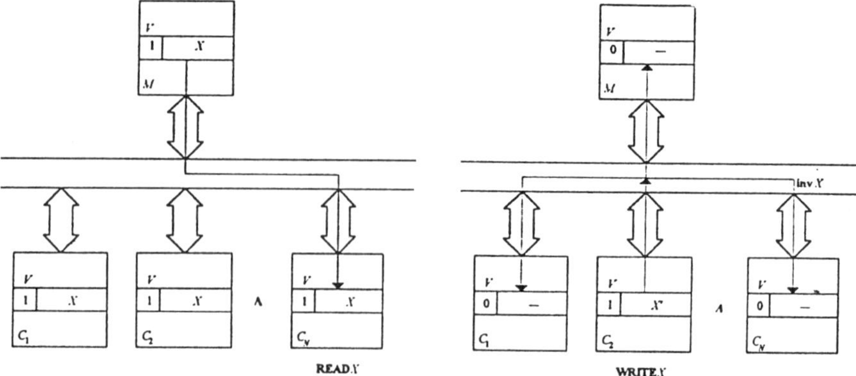


**Figure 7.1.** Explanation of a write-invalidate protocol (M—memory; $C_i$—cache memory of processor $i$; V—validity bit). [*Source:* Tomešević and Milutinović (1993).]
*Comment:* Write-invalidate protocols generate much less traffic on the interconnection network during consistency maintenance activities. One bus cycle is sufficient to invalidate the entire block. During that cycle, active lines are address bus and invalidation control (one line). Data bus is not used and could be utilized for other purposes. However, write-invalidate protocols generate more bus traffic later, if and when an invalidated data item is needed again.

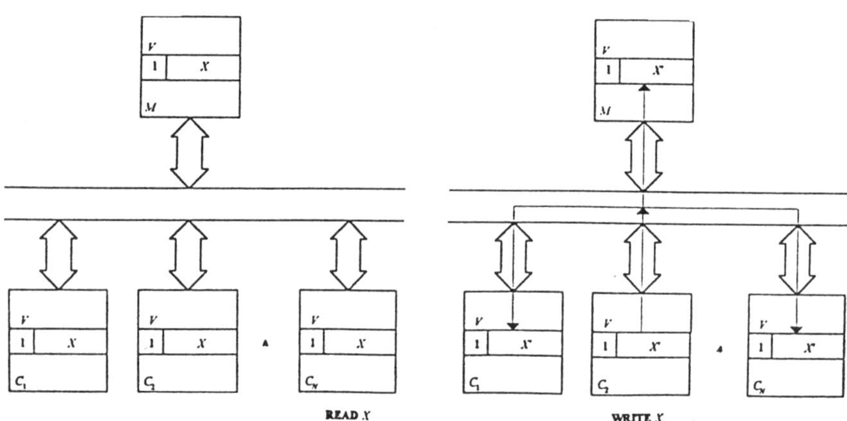


**Figure 7.2.** Explanation of a write-update protocol. [*Source:* Tomešević and Milutinović (1993).]
*Comment:* Write-update protocols generate much more traffic on the interconnection network during the consistency maintenance activities. The number of bus cycles needed is equal to the number of words in the block. In each bus cycle, both address and data lines are busy and cannot be used for other purposes. However, write-update protocols generate no bus traffic later, if and when the updated data are used again (by individual processors from their local caches).

action utilized: to avoid inconsistency or to force consistency. The two protocol classes are explained next. They are described using the examples in Figures 7.1 and 7.2.

### 7.1.1.1. *Write-Invalidate Protocols*

Write-invalidate protocols allow the processors to do a simultaneous read of all cached copies. However, only one processor has permission to write during a given time interval. Other processors are given permission to write at other time intervals.

Write to a shared copy is preceded by an invalidation signal. After the broadcast of the invalidation signal is completed, all other shared copies are invalidated (validity bit V set to zero), and the originator of the invalidation signal has permission to write to its shared copy (which is not shared any further since other copies are invalid now). Note that invalidation comes first, and writing comes second. If the order of invalidation and writing is reversed, it is possible that some other processor reads a stale copy, during the time interval after the shared copy was written by one processor and the validity bits were set to zero in the other processors.

The memory copy (if any) has to be either invalidated (if the write-invalidate protocol is of the writeback type) or updated (if the write-invalidate protocol is of the writethrough type). The term *writeback* indicates that memory is updated when the block with the newly written value is to be replaced. The term *writethrough* denotes that memory is updated immediately after the new written value is deposited into the local private cache.

The example in Figure 7.1 shows the read case (never a problem) and the write case (combined with the writeback approach). The new reader (cache number $N$) simply reads in the shared value and uses it. The new writer (cache number 2) first invalidates the other caches and the memory, and then it writes its own cache (invalidation signal is typically denoted as inv$X$).

Write-invalidate protocols are economical to implement. They generate relatively low bus traffic—only the invalidation signal in the example of Figure 7.1. However, they are characterized with one important disadvantage. If the newly written copy is to be shared in the future, all other processors needing that copy will have to run through a read miss cycle, which slows down the code being executed. This problem is cured in the write-update protocols discussed next.

### 7.1.1.2. *Write-Update Protocols*

Write-update protocols allow multiple processors with write permission and support the so-called distributed write approach. This means that the updated word is immediately broadcast to other sharers.

Main memory can be updated at the same time or not. In other words, in principle, both writeback and writethrough approaches can be used. In practice, writeback is typically used for private data, and writethrough is typically used for shared data.

The example in Figure 7.2 shows the read case (never a problem) and the write case (combined with the writethrough approach). The new reader (cache number $N$) simply reads in the shared value and uses it. The new writer (cache number 2) writes its own cache, but it also forces the newly written value into the other caches as well as the memory.

Write-update protocols are not so economical to implement, since additional bus lines are needed. They generate more bus traffic control signals as well as data, and addresses are transferred across the bus in the example of Figure 7.2. However, as indicated earlier, other sharers of the same new value will have that value readily available when needed and will not have to run through the potentially expensive read miss cycle. On the other hand, the updating may bring a data item that will never be used by a remote processor and may force the purge of a data item that may be needed later.

### 7.1.1.3. MOESI Protocol

The MOESI protocol is a consequence of an effort to combine the properties of the many existing snoopy protocols, and to enable different processors based on different protocols to be used as building blocks for a powerful SMP system. This protocol has been devised through the effort to develop the cache consistency maintenance superset protocol for the IEEE Futurebus standard.

It is well known that cache memory can be of the writethrough or writeback type. In the first case, memory is updated immediately on each cache update. In the second case, memory is updated, when the cache block (line) is replaced.

Several studies have concluded that the writeback approach provides greater reduction in bus traffic and better performance [Sweazey and Smith 1986]. Therefore, the rest of this presentation assumes the writeback approach.

For the writeback cache, each block (line) in the cache may be assigned to one of the five states. These five states can be specified with 3 bits: V (validity), E (exclusiveness), and O (ownership).

Shared data is either *valid* or *invalid*. Note that the term "valid" relates to the consistency maintenance protocol and not to the semantics of the code.

Valid data residing in caches can be *exclusive* or *shared* (*nonexclusive*). Exclusive data are data that exist only in one cache and must match the copy in main memory. Shared data are data that exist in more than one cache, or whose number is unknown (and may be even one or zero).

Valid data residing in caches can also be classified as *owned* or *unowned*. In this context, *ownership* means responsibility for the accuracy or validity of data marked as valid (in general, case main memory is not responsible, as data in main memory may not be valid). In the general case, data are owned uniquely either by one of the cache memories or by the main memory. The owner is responsible for the following activities: (1) updating of the main memory, (2) passing of ownership to another entity—another cache or main memory, and (3) owning cache substitutes for main memory in transfers. The main purpose of the ownership category is to distinguish a copy that has been modified from the copy in main memory.

Here we assume that main memory is not responsible for tracing the state of its data. This is the responsibility of data caches. Therefore, the *state* of each data item is stored in caches.

In theory, with 3 state bits, one can create eight states. However, some states make no sense (exclusiveness or ownership of an invalid data item). Consequently, only five states make sense:

1. Exclusive owned
2. Sharable owned
3. Exclusive unowned
4. Sharable unowned
5. Invalid

As mentioned earlier that ownership helps distinguish a modified copy from the copy in main memory. Consequently, the term *modified* can be used instead of the term *owned*, so the five states defined above can be labeled as follows:

1. Exclusive modified, or modified
2. Sharable modified, or owned
3. Exclusive unmodified, or exclusive
4. Sharable unmodified, or sharable
5. Invalid

These five states define the acronym MOESI. Figure 7.3 illustrates the way in which V, E, and O bits combine to create five states of the MOESI model. Note that states M, O, E, and S are actually state pairs (only I is not a state pair, but a single state). Also, note the difference between the O bit and the O state.

The MOESI model represents an extrapolation and concatenation of models found in the preceding research. Before the MOESI model was established, researchers used different terms for the state pairs M, O, E, and S. For example, M is known as *modified, exclusive modified*, or *exclusive owned*. O is known as *owned, sharable modified, sharable owned*, or *sharable responsible*. E is known as *exclusive, exclusive unmodified*, or *exclusive unowned*. S is known as *sharable, sharable unmodified, sharable unowned*, or *shared*. Note that S state does not imply that main memory is valid.

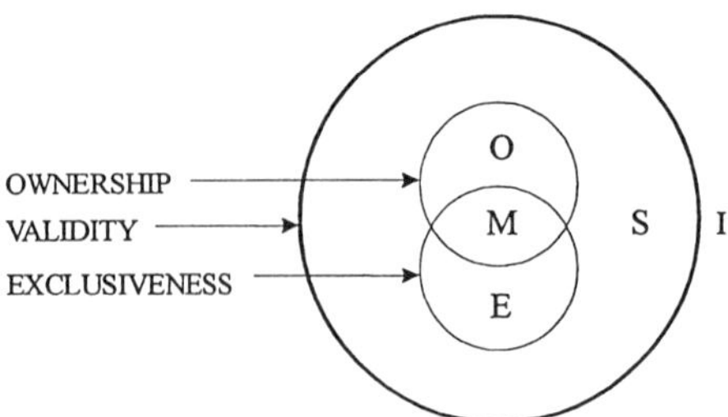


**Figure 7.3.** Explanation of the MOESI protocol. [*Source:* Tomešević and Milutinović (1993).]
*Comment:* Different protocols cover different parts of the symbolic field of shapes. Ownership, validity, and exclusiveness are represented as three different circles.

One needs six signal lines in order to implement the MOESI protocol on the backplane of Futurebus (or its descendants). Three lines are used by the master for the transactions, to indicate the intentions of the master; the other three are used for other units on the bus, to assert either status or control.

In reality, one more signal line is needed, called BS or busy, to abort a transaction. It is needed to implement *versions* of other existing protocols, introduced before MOESI. Therefore, the total number of needed lines is seven.

Earlier protocols include Write-Once, Illinois, Berkeley (in the write-invalidate domain), and Firefly or Dragon (in the write-update domain). For details, see Tomešević and Milutinović (1993). All these protocols are based on three or four states, and, conditionally speaking, each one can be treated as a subset of the MOESI protocol. However, note that, in the general case, with the seven control lines defined above, one can implement only *versions* of many existing protocols—not their exact algorithms.

For example, the states of the Berkeley protocol are MOSI, while the states of the Dragon protocol are MOSE. In principle, Berkeley is treated as the most mature protocol in the write-update group, and Dragon is treated as the most mature protocol in the write-invalidate group. Of course, in each of the two groups there exist newer protocols with better performance, but they have been derived either from Berkeley, or Dragon, or from a combination of the two.

As long as different modules of different architectures and different cache consistency maintenance protocols dynamically select only the actions permitted by the MOESI protocol, the coexistence is possible, and the system will be consistent.

### 7.1.1.4. MESI Protocol

A subset of the MOESI protocol called the MESI protocol is implemented in a number of 32-bit and 64-bit microprocessors. The MESI protocol includes only four states (M, E, S, and I) and represents a type of architectural support for incorporation of off-the-shelf microprocessors into the modern SMP systems.

The MESI protocol can be treated as a first step toward a future goal of having an entire SMP system on a single VLSI chip. This goal is believed to be implementable as soon as the on-chip transistor count crosses the 10 million threshold. Of course, with simpler nodes, the goal can be achieved sooner.

A partial list of machines implementing the MESI protocol or its supersets (e.g., MOESI) includes, but is not limited to, (1) AMD K5 and K6; (2) Cyrix 6x86; (3) the DEC Alpha series; (4) the HP Precision Architecture series; (5) the IBM PowerPC series; (6) Intel Pentium, Pentium Pro, Pentium II, and Merced; (7) SGI series; and (8) Sun UltraSPARC series.

### 7.1.1.5. SI Protocol

For instruction cache memory, there is no need to maintain M and E states, because these caches support only cache read. Consequently, the protocol built into the instruction cache controllers is SI. The controller for the SI protocol is relatively simple to design. As such, it is a good example for student homework assignments.

### 7.1.2. Directory Protocols

In the case of directory protocols, the responsibility for each cache coherence maintenance is typically built into a centralized controller. This controller is typically located in or next to the main memory (on the chip or on the board). One directory entry is associated with each cache block and contains all consistency maintenance related information.

Figure 7.4 includes an explanation of a typical directory protocol. In principle, reading again is not a problem; however, if a cache needs to write to a value, it first consults a centralized cache consistency maintenance related directory and performs the desired action, which is of either the write-invalidate type (most often) or the write-update type (less often).

Actions involved in cache consistency maintenance are of the following possible types: (1) unicast (one source to one destination), (2) multicast (one source to several but not all destinations), and (3) broadcast (one source to all destinations).

Since not only broadcast is involved, directory protocols are better suited for interconnection networks of a more general type (other than shared bus). However, in theory, directory protocols can be used on any type of interconnection network: (1) simplest, like bus, ring, or LAN (BRL in the rest of the text); (2) one of the reduced interconnection network types (hypercube, $N$-cube, PM2I, etc.); or (3) the most complex, like grid, mesh, or crossbar (GMC in the rest of the text).

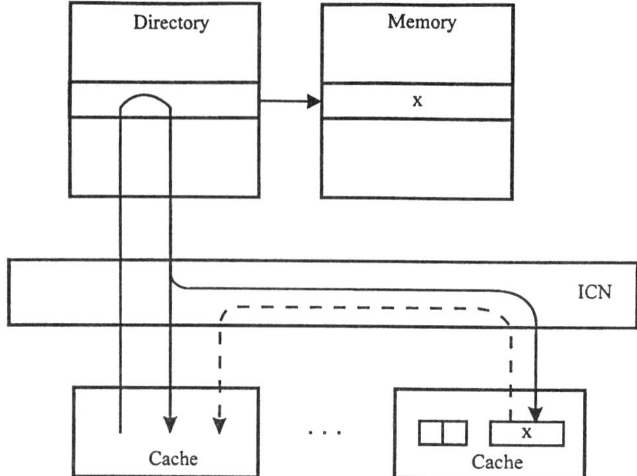



**Figure 7.4.** Explanation of a directory protocol (ICN—interconnection network). [*Source:* Tomešević and Milutinović (1993).] *Comment:* The directory is consulted before each and every consistency maintenance activity. After the appropriate knowledge is acquired from the directory, only updates or invalidations that are deemed necessary are performed. Consequently, broadcast is used very rarely; multicast and unicast to selected destinations are more appropriate solutions, on condition that the interconnection network permits multicast and unicast (shared bus permits only broadcast). Note that the directory approach can be used for both cache consistency maintenance (less often) and memory consistency maintenance (more often).

In this context, *bus* implies parallel transfer on an open-loop topology, *ring* implies parallel transfer on a closed-loop topology, and LAN implies serial transfer on any topology (open, close, star, etc.). Furthermore, *grid*, *mesh*, and *crossbar* imply the same topology of lines (a number of horizontal and a number of vertical lines); however, processing nodes are positioned differently. In the case of a grid, processing nodes are located next to line crossings (messages go by the processing nodes). In the case of a mesh, processing nodes are located on the line crossings (messages go through the processing nodes). In the case of a crossbar, processing nodes are located on horizontal and/or vertical line terminations.

Information stored in the directory can be organized in a number of different ways. This organization determines the type and characteristics of directory protocols. The three major types of directory protocols are (1) full-map directory, (2) limited directory, and (3) chained directory. All three of them are briefly discussed in the text to follow.

### 7.1.2.1. Full-Map Directory Protocols

The essence of full-map directory protocols is explained using Figure 7.5. Each directory entry (related to a cache block) includes $N$ presence bits

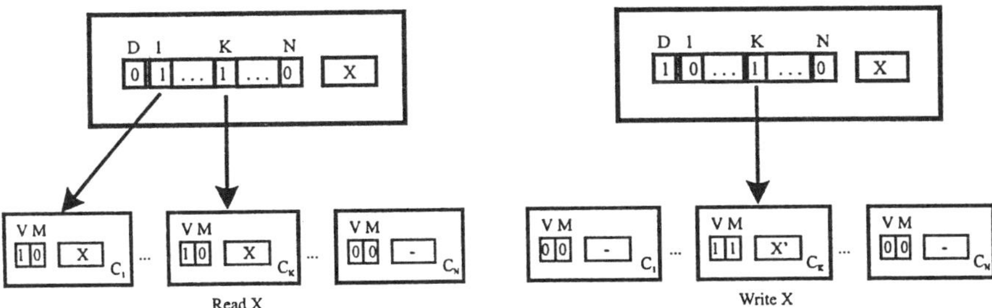



**Figure 7.5.** Explanation of a full-map directory scheme (V—validity bit; M—modified bit; D—dirty bit; K—processor that modifies the block). [*Source:* Tomešević and Milutinović (1993).]
*Comment:* The full-map directory approach is not scalable, and it is suitable only for systems of a relatively small size (e.g., up to 16 nodes, as in the case of the Stanford DASH multiprocessor).

corresponding to $N$ processors in the system, plus a single bit ("dirty" or D) that denotes if the memory is updated, or not (D = 0 means that the memory is updated, i.e., not dirty). This means that each directory entry includes $N + 1$ bits.

This type of protocol is denoted as Dir($N$)NB, which means that each entry includes $N$ presence related fields, and no broadcast operation is ever used.

Some of the consistency maintenance bits are kept in the local private cache memories; in the case of full-map protocols, 2 more bits are added per each cache block. Each cache block includes one validity bit (V) indicating whether the cached copy is valid (V = 1), and one modified bit (M) indicating whether the value in the cache compared to the value in memory is different/modified (M = 1).

In Figure 7.5 (left side) processor $K$ reads a value, which is shared by processors 1 and $K$. The sharing status is reflected through the fact that two out of the $N$ bits in the centralized directory entry are set to 1. Also, since the value in the memory and the values in the caches are the same, memory bit D is set to zero (D = 0), and cache bits M are set to zero (M = 0). Of course, the V bit for all shares is set to one (V = 1).

In Figure 7.5 (right side) processor $K$ writes a new value into its own cache. Consequently, all except one of the bits in the presence vector are set to zero. Since a writeback approach is assumed in the example of Figure 7.5, the dirty bit in the corresponding memory entry is set to one (D = 1), indicating that the memory is not up to date any more. Since a write-invalidate approach is assumed in the example of Figure 7.5, the V bits of other caches in the system are set to zero; under such a condition, the M bits in other caches become irrelevant (typically, designs are such that V = 0 causes M = 0, as well).

The full-map protocol is characterized with the best performance, compared to other directory protocols. This is because the coherence traffic is the smallest, compared to other directory protocols. However, this protocol is characterized by several drawbacks:

1. The size of the directory storage is $O(M - N)$, where $M$ refers to the number of blocks in the memory, and $N$ refers to the number of processors in the system. This is essentially an $O(N^2)$ complexity, which means that the protocol is not scalable. It is not able to support very large systems, due to a large cost that grows as $O(N^2)$.

2. Adding a new node requires system changes, like the widening of the centralized controller. Consequently, the protocol is not flexible for expansion; that is, the cost per added node is considerably larger than the cost of the node alone.

3. The centralized controller is a potential fault-tolerance bottleneck, as well as a performance degradation factor, especially if more nodes are present in the system.

The protocols to follow eliminate some or all of the drawbacks of full-map protocols, at the expense of performance reduction, due to an increased consistency-maintenance-related traffic.

### 7.1.2.2. Limited Directory Protocols

The essence of the limited directory protocols is explained using Figure 7.6, which includes a straightforward modification of a full-map protocol, called Dir($i$)NB. Each directory entry (related to a cache block) includes the same single dirty bit (D) plus only $i$ ($i < N$) presence fields; each presence field is of the length log $N$ bits. In total, this is $1 + i \log_2 N$ bits (remember that the

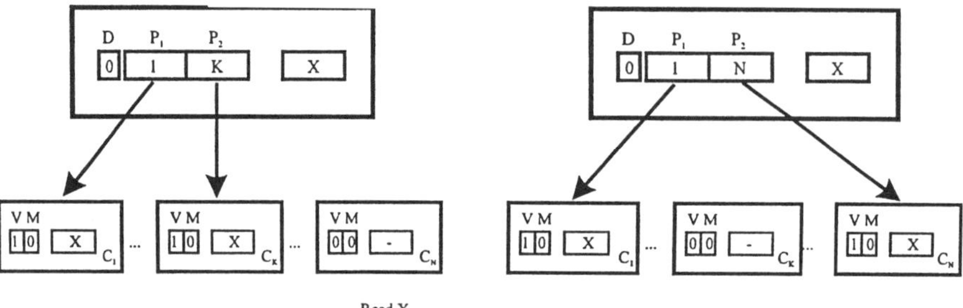


**Figure 7.6.** Explanation of the no-broadcast scheme. [*Source:* Tomešević and Milutinović (1993).]
*Comment:* Note that the coherence overhead exists even for read operations. The amount of overhead decreases if appropriate operating system support is incorporated.

full-map protocol entries include $1 + N$ bits). In reality, such an approach means fewer bits per entry. If $N$ is large enough, and $i$ is small enough, the size of one directory entry is less, compared to the full-map protocol. For example, if $N = 16$ and $i = 2$, a Dir($i$)NB limited directory protocol includes 9 bits, and a full-map directory protocol includes 17 bits. In other words, if N = 16, the Dir($i$)NB limited directory protocols are less costly if $i < 4$.

The approach described above is possible, because several studies have shown that typically a block is shared only by a relatively small number of processors. In many cases, it is only a producer and a consumer. The question is what happens if a need arises for a cache block to be shared by more than $i$ processors, that is, what type of mechanism is used to handle the presence vector overflow (which happens when the number of sharers becomes larger than $i$).

Actually, there are two types of limited directory protocols. The one described so far, as indicated, is denoted as Dir($i$)NB. Another one is denoted as Dir($i$)B, which means that each entry includes $i$ presence related fields, and that the broadcast operation is used in the protocol.

In both schemes, some of the consistency maintenance bits are kept in the local private cache memories. As in the case of full-map protocols, two more bits are added per each cache block. As before, each cache block includes one validity bit (V) indicating whether the cached copy is valid (V = 1), or not, and one modified bit (M) indicating whether the value in the cache compared to the value in the memory is different or modified (M = 1).

In both cases, the protocols are scalable, as far as the directory storage overhead, which is $O(M \log N)$, or essentially $O(N \log N)$. In both cases, protocols are less inflexible, because a new node does not always mean changes in the central directory (adding a new node may not change the prevalent pattern of sharing). However, in both cases, performance is affected by increased sharing. Also, the centralized directory continues to cause bottlenecks of many kinds.


*7.1.2.2.1. The Dir(i)NB Protocol.* In Figure 7.6, which explains Dir($i$)NB, there is a restriction for reading (which is a consequence of the fact that $i < N$), meaning that the number of copies for simultaneous read is also limited to $i$. For example (left side of the figure), processor $N$ reads a value, which is shared by processors 1 and $K$. After the read miss, the value will be brought into cache $N$ (right side of the figure), and the corresponding pointer field will be updated ($K$ is substituted by $N$), but the copy in another cache ($K$ in this example) has to be invalidated. In other words, the sharing status had to be changed, because two out of the $i = 2$ fields (in the centralized directory entry) have already been in use. Also, since the value in the memory and the values in the active caches continue to be the same, memory bit D continues to be set to zero (D = 0), and cache bits M continue to be set to zero (M = 0). Of course, the V bit for all active sharers is set to one (V = 1).

Because of the restriction on the number of simultaneously cached copies, there is an overhead even for read sharing. This results in performance degradation, due to an increased miss ratio, which may become especially dramatic in the case of intensive sharing of read-only and read-mostly data. The term intensive sharing denotes a situation in which more than $i$ nodes keep accessing the same value (for reading purposes) in some random order (the worst case is when they rotate their access order).

The scheme to follow eliminates the restrictions for reading, at some minimal cost increase, and a potential performance degradation in some applications.

*7.1.2.2.2. The Dir(i)B Protocol.* In Figure 7.7, which explains Dir($i$)B, there is no restriction on the number of read sharers. For example, processor $N$ reads a value (left-hand side of the figure), which is shared by processors 1 and $K$. Processor $N$ is again allowed to cache a copy. However, the sharing status will not be reflected through a change in the centralized directory. The two pointers remain unchanged and continue to be set to 1 and $K$, respectively. However, this protocol includes one more control bit in each directory entry, referred to as the *broadcast bit* (B), and that one is then set to one (B = 1), to reflect the fact that the number of sharers is now larger than $i$ (pointer overflow). The meaning of the two control bits in cache controllers is the same as before.

In this context, there are no restrictions on the number of simultaneous readers. However, if a write happens under conditions of pointer overflow (B = 1), an invalidation broadcast signal will be generated. Of course, the need for invalidation broadcast increases the write latency; moreover, some of the invalidation broadcasts are not necessary, which wastes a fraction of the communications bandwidth.

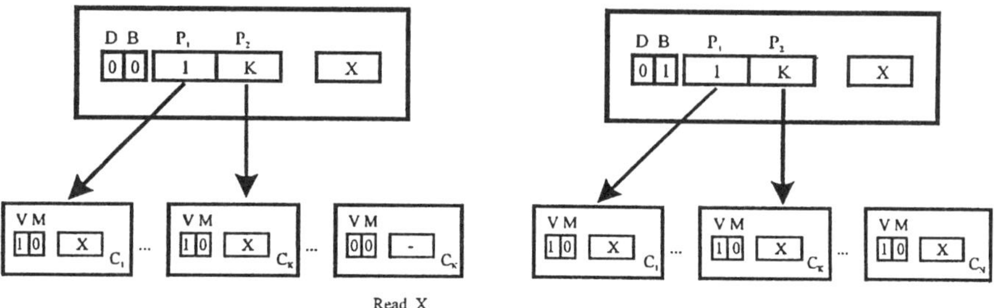


**Figure 7.7.** Explanation of the broadcast scheme (D—dirty bit; B—broadcast bit). [*Source:* Tomešević and Milutinović (1993).]
*Comment:* Note that the coherence overhead does not exist now for read operations; however, the system state-related information is not precise any more (it becomes fuzzy, unless appropriate operating system support is incorporated).

All protocols described so far share some common drawbacks: (1) too much is centralized, which creates a bottleneck; (2) adding a new node requires centralized changes, which is inflexible; and (3) the size of the central directory is still relatively large. The chained directory protocols, to be described next, remove these problems, at a price that is negligible in a number of common applications.

### 7.1.2.3. Chained Directory Protocols

In the case of chained directory protocols, directories are distributed across all caches, in the form of a chain of pointers (shared copies of a block are chained into a linked list). As indicated in Figure 7.8, only the head of the list (H) is in the memory and occupies $\log N$ bits, so that it can point to one of the $N$ nodes.

This means that the three above-mentioned problems have been solved. First, almost nothing is centralized (only the vector of link lists heads), and there is no central bottleneck. Next, the system is absolutely flexible for expansion, and no changes in the centralized directory are needed, if a new node has to be added. Finally, the size of the centralized directory is down to the minimum; that is, the storage cost is defined by $O(\log N)$.

Figure 7.8 gives an example based on the singly linked lists. After a new reader shows up (processor 1), it is added to the linked list, as the first node after the head of the list. All searches start from the head of the list; consequently, according to the locality principle, the most recently added read sharer is the most likely one to be reading in the near future. The end of the linked list is denoted with a terminator symbol (CT).

Performance can be improved at a minimal cost increase, if a doubly linked list is used. Research at MIT has shown that such an approach leads to performance that is close to the one provided by full-map directory protocols, at a cost that continues to be $O(\log N)$.

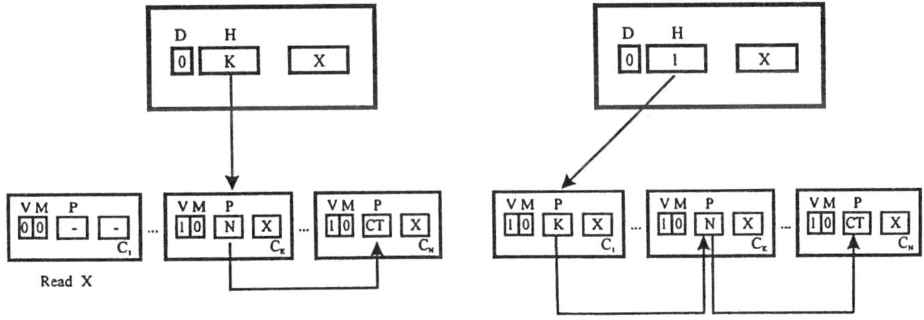


**Figure 7.8.** Explanation of chained directory schemes (H—header). [*Source:* Tomešević and Milutinović (1993).]
*Comment:* The chained directory protocols are very frequently used in new designs, especially those based on the **IEEE Scalable Coherent Interface** standard, which supports the chained directory approach.

## 7.2. ADVANCED ISSUES

This part contains the author's selection of research activities that, in his opinion, have made an important contribution to the field in recent time and are compatible with the overall profile of this book. Here we concentrate only on extended pointer schemes and protocols for efficient elimination of negative effects of false and passive sharing.

Other problems of interest include reduction of contention on shared resources. A good survey of techniques for reduction of contention in shared memory multiprocessors as given in Stenstrom (1988).

The directory protocol of Agarwal (1990, 1991) is described in Figure 7.9. It is referred to as the LimitLESS Directory Protocol and represents an effort to rationalize the storage problems of typical directory schemes, at a minimal performance deterioration. It is based on standard limited directory protocols, like Dir($i$)NB or Dir($i$)B, with a software mechanism to handle pointer overflow. On a pointer overflow, an interrupt is generated, and the hardware pointers (of the cache block that is about to become shared more than supportable by hardware) are then stored in memory, in order to free up directory space for lower-intensity sharers. Figure 7.9 (left side) shows that the directory structure is basically the same as in standard limited directory protocols, except that a trap bit (T) is included to indicate the overflow processing status. Figure 7.9 (right side) shows that some space has to be allocated in memory, for the software maintained table of overflow pointers.

Simoni's approach [Simoni 1990, 1992; Simoni and Horowitz 1991] is described in Figure 7.10. It describes an effort to improve the characteristics of chained directory protocols, by using dynamic pointer allocation. Each directory entry includes a short part and a long part. The short directory includes only a small directory header—one for each shared cache block. This header includes two fields: (1) a dirty state bit and (2) a head link field. The long directory includes two types of structures: (1) one linked list of free

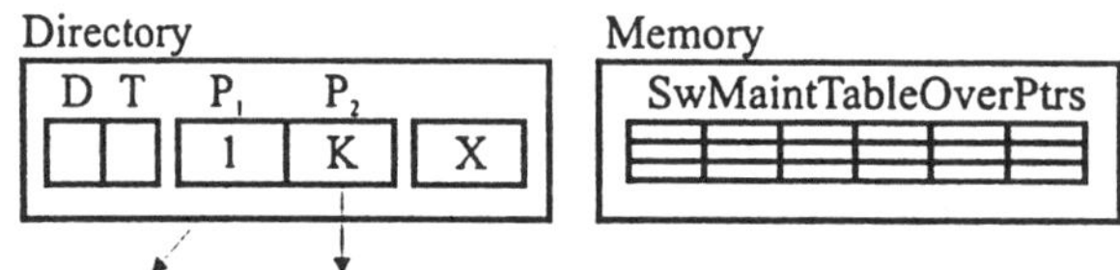


**Figure 7.9.** LimitLESS directory protocol (D—dirty bit; T—trap bit; SwMaint-TableOverPtrs—software maintained table of overflow pointers). [*Source:* Agarwal (1990, 1991).]

*Comment:* In essence, the LimitLESS directory protocol represents a synergistic interaction between a limited directory protocol and the operating system. Consequently, the information about the system state continues to be precise, even if the number of sharers exceeds the number of directory entries.

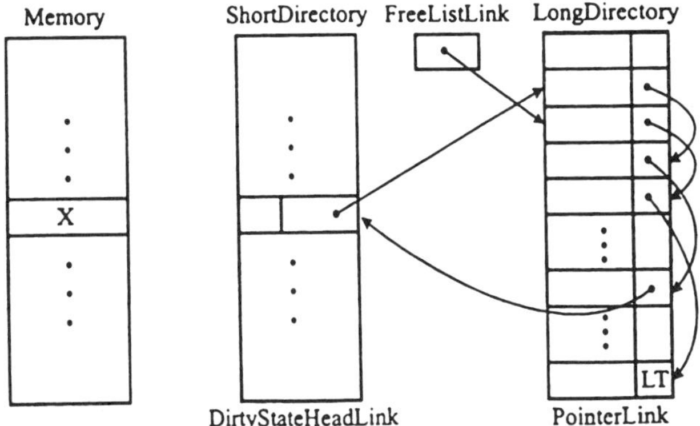


**Figure 7.10.** Dynamic pointer allocation protocol [XC—CacheBlock; LT—list terminator; DirtyState (1 bit); HeadLink (22 bits); Pointer (8 bits); Link (22 bits); P/L—Pointer/Link Store].
*Comment:* In essence, dynamic pointer allocation protocol relies on operating system support to maintain the absolute minimum of memory storage for specification of sharers.

entries of the so-called pointer/link store (referred to as the *free list link*) and (2) numerous linked lists for shared cache blocks, with two fields per link list element (processor pointer and forward link). On read, the new sharer is added at the head position (the space for the new sharer is obtained by reallocating an entry from the free list link). On write, the linked list is traversed for invalidation purposes (and the space is freed up by returning the entries into the free list link). Essentially, this protocol represents a hardware implementation of the LimitLESS protocol.

James's approach of [James et al. 1990] is described in Figure 7.11. This approach represents another effort to improve the characteristics of chained directory protocols, by using doubly linked lists. This effort resulted in the creation of a new IEEE SCI (Scalable Coherent Interface) standard, also referred to as IEEE 1592-1992 [IEEE 1993]. This protocol is based on the previously described chained directory protocols and specifies a number of details of interest for its incorporation into industrial products. Reading a shared value implies (1) finding the current head of the list in memory and (2) adding the specifiers of the reading node, if it is the first reading of that node. Writing to a shared value implies (1) sending the invalidation signal sequentially down the list and (2) creating the new list header for the modified block. Advantages of doubling the links are (1) on block replacement, the corresponding linked list element has to be extracted from the list, and that operation is much easier to execute if the linked lists are bidirectional; and (2) the possibility of a bidirectional traversal brings up a number of potential opportunities. Figure 7.11 indicates that the major fields in memory entries are (1) validity bit (V) and (2) memory header pointer

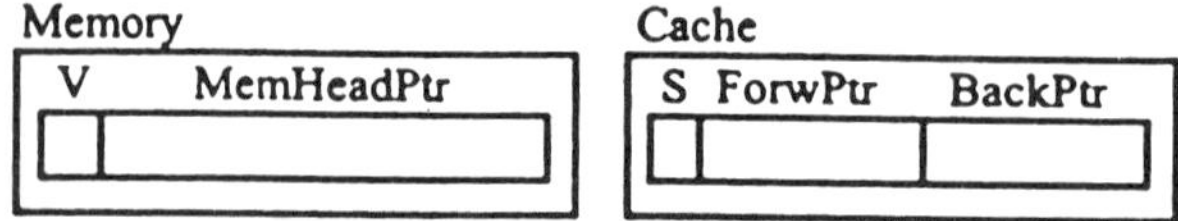

**Figure 7.11.** An explanation of the SCI protocol [V—validity (1 bit); MemHeadPtr (8 bits); State (5 bits); ForwPtr (8 bits); BackPtr (8 bits)]. [*Source:* James et al. (1990).] [*Sources:* Simoni (1990, 1992) and Simoni and Horowitz (1991).]
*Comment:* A detailed description of the IEEE Scalable Coherent Interface 1592-1992 standard can be found in IEEE (1993).

(MemHeadPtr). Major fields in cache entries are (1) status bit (S), (2) forward pointer (ForwPtr), and (3) backward pointer (BackPtr).

Prete (1991) described an effort to minimize the number of unnecessary writebacks related to the replacement of modified shared copies. This approach, referred to as reduced state transition (RST), represents a modification of a write-update protocol.

Prete later described [Prete et al. 1995] an effort to minimize the negative effects of false sharing, which occurs when cache blocks are relatively large, and several objects or processes share the same cache block, as a consequence of process migration or an unsophisticated compiler. This approach [Prete et al. 1995] is referred to as useless shared copies removal (USCR).

Prete also described [Prete et al. 1997] an effort to minimize the negative effects of passive sharing, which happens when a process moves from one processor to another and leaves behind private values that, after migration, formally become shared. This last approach [Prete et al. 1997] is referred to as useless private copies removal (UPCR).

The approaches of the University of Pisa were introduced in the early and mid-1990s; however, related experiences have been used lately in a number of state-of-the-art microprocessor designs intended for multimicroprocessor systems. This is why they have been included in the state-of-the-art group of protocols.

An important new trend in SMP research is related to hardware fault containment and efforts to understand better the effects of communication latency, overhead, and bandwidth [Teodosiu et al. 1997, Martin et al. 1997].

Keeton et al. (1998) treat performance evaluation of a Quad Pentium Pro SMP using the OLTP (online transaction processing) workloads. Commercial applications are an important market for SMP machines. Their paper shows that multiprocessors scale well on commerical applications, which opens the door for their more efficient penetration into commercial markets. This seems to be due mostly to the effectiveness of cache memory. Only 0.33% of database accesses and 0.47% of OS (operating system) accesses reach memory.

Mukherjee and Hill (1998) have investigated the usage of prediction to accelerate coherence protocols. The authors have developed a predictor for coherence messages (Cosmos), which predicts the source and type of the next coherence message, for a cache block. This prediction is based on logic similar to the one in the PAp branch predictor. If the prediction of the coherence message types and their parameters is good enough, negative effects of latencies related to misses on remotely cached blocks will be decreased. It has been shown that Cosmos can achieve prediction accuracies of about 62% to about 93%. The bottom line is that a coherence protocol can execute faster if future actions can be predicted and executed speculatively.

The main assumption of this book is that one of the major goals of future on-chip processor designs is to have an entire shared memory multiprocessor

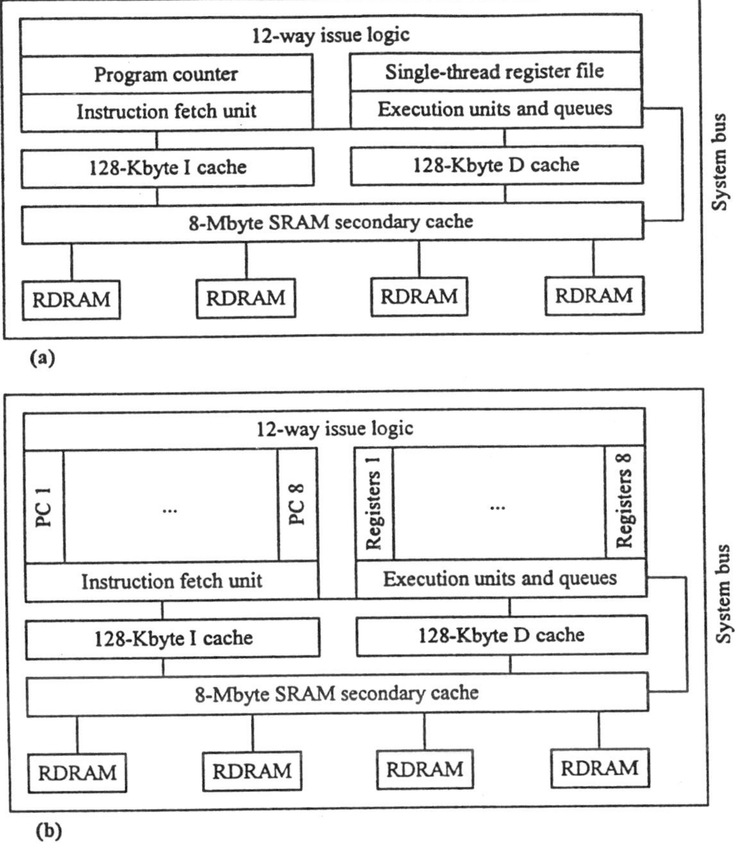


**Figure 7.12.** Comparing (*a*) superscalar, (*b*) simultaneous multithreading, and (*c*) chip multiprocessor architectures [Hammond et al. 1997] (I—Instruction; D—data; PC—program counter). These three architectures, each one of the same transistor count (one billion), are compared in Hammond et al. (1997).

*Comment:* All three architectures shown in Figure 7.12 are implementable with approximately the same number of transistors (one billion). Hammond et al. (1997) conclude that the third architecture (multiprocessor) is the best one to implement.

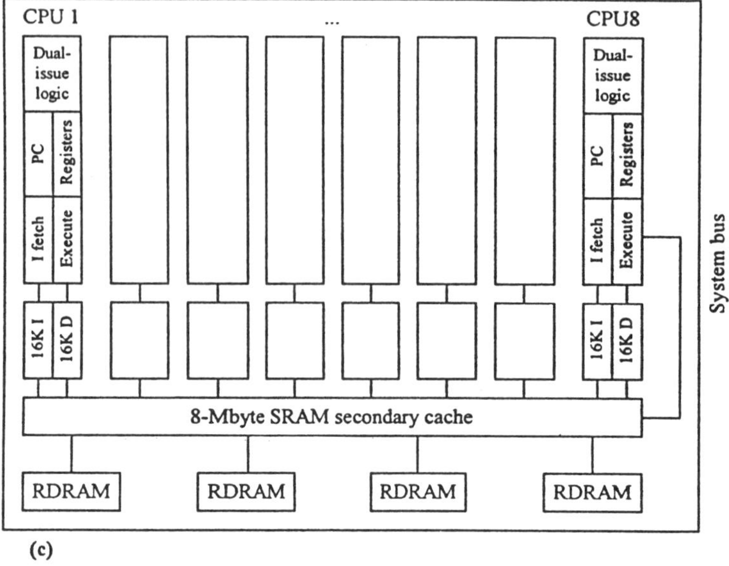


**Figure 7.12.** (*Continued*).

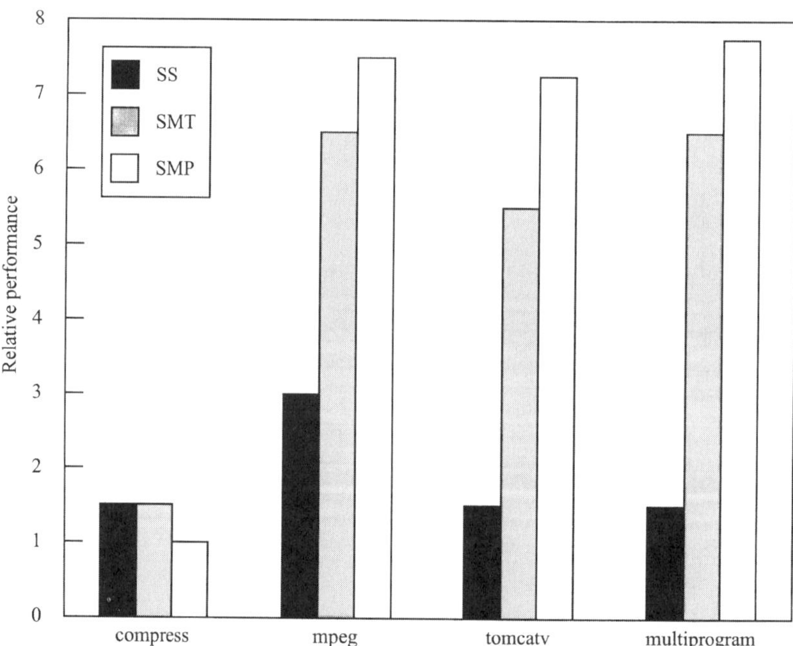


**Figure 7.13.** Relative performance of the simulated on-chip processor architectures: SS—superscalar; SMT—simultaneous multithreading; SMP—shared memory multiprocessor.
*Comment:* On average, SMP performs the best and is the easiest one to implement.

on the same silicon die. This assumption has been confirmed by several studies. Here, results from a simulation study [Hammond et al. 1997] will be presented in detail.

The three architectures compared by researchers from Stanford University are (1) superscalar, (2) simultaneous multithreading, and (3) shared memory multiprocessor. These three architectures are shown in Figure 7.12. They are characterized by approximately the same number of transistors (one billion). Characteristics of the three compared architectures are given in Table 7.1. The simulated performance is shown in Figure 7.13. On average, the shared

**TABLE 7.1. Compared Characteristics of the Superscalar, Simultaneous Multithreading, and Chip Multiprocessor Architectures**

| Characteristic | Superscalar | Simultaneous Multithreading | Chip Multiprocessor |
|---|---|---|---|
| Number of CPUs | 1 | 1 | 8 |
| CPU issue width | 12 | 12 | 2 per CPU |
| Number of threads | 1 | 8 | 1 per CPU |
| Architecture registers (for integer and floating point) | 32 | 32 per thread | 32 per CPU |
| Physical registers (for integer and floating point) | 32 + 256 | 256 + 256 | 32 + 32 per CPU |
| Instruction window size | 256 | 256 | 32 per CPU |
| Branch predictor table size (entries) | 32,768 | 32,768 | $8 \times 4096$ |
| Return stack size | 64 entries | 64 entries | $8 \times 8$ entries |
| Instruction (I) and data (D) cache organization | $1 \times 8$ banks | $1 \times 8$ banks | 1 bank |
| I and D cache size | 128 kB | 128 kB | 16 kB per CPU |
| I and D cache associativities | 4-way | 4-way | 4-way |
| I and D line sizes (bytes) | 32 | 32 | 32 |
| I and D cache access times (cycles) | 2 | 2 | 1 |
| Secondary cache organization (MB) | $1 \times 8$ banks | $1 \times 8$ banks | $1 \times 8$ banks |
| Secondary cache size (bytes) | 8 | 8 | 8 |
| Secondary cache associativity | 4-way | 4-way | 4-way |
| Secondary cache line size (bytes) | 32 | 32 | 32 |
| Secondary cache access time (cycle) | 5 | 5 | 7 |
| Secondary cache occupancy per access (cycles) | 1 | 1 | 1 |
| Memory organization (number of banks) | 4 | 4 | 4 |
| Memory access time (cycles) | 50 | 50 | 50 |
| Memory occupancy per access (cycles) | 13 | 13 | 13 |

memory multiprocessor architecture performs best and is the easiest to implement.

Results of the described simulation study can easily be explained. When the minimum feature size decreases, the transistor gate delay decreases linearly, while the wire delay stays nearly constant or increases. Consequently, as the feature size improves, on-chip wire delays improve 2–4 times more slowly than the gate delay. This means that resources on a processor chip that must communicate with each other must be physically close to each other. For this reason, the simulation study has shown (as expected) that the shared memory multiprocessor architecture exhibits (on average) the best performance, for the same overall transistor count. In addition, due to modular structure, the complexity of the shared memory multiprocessor architecture is much easier to manage at design time and test time. Because of the scalability of the multiprocessor architectures with logically shared memory address spaces, the same conclusion applies to a wider range of architectures with the logically shared memory address space.

The author and his associates were active in the field of shared memory multiprocessing and especially in the field of cache consistency for shared memory multiprocessors. For more details, see Ekmecic et al. (1995), Tartalja (1997), Tomešević and Milutinović (1992, 1993), and Tomešević (1992).

## PROBLEMS

**7.1.** Design a control unit for a write-invalidate snoopy protocol of your choice. What is the transistor count, and how does it depend on the cache size?

**7.2.** Design a control unit for a write-update snoopy protocol of your choice. What is the transistor count, and how does it depend on the cache size?

**7.3.** If one word in a block is dirty, the whole cache block is considered dirty. However, an improvement of the MOESI protocol can be devices (called, e.g., MOESI-Advanced, or MOESIA), in which a cache block is declared dirty only after the second word becomes dirty. Develop the details of such a cache consistency maintenance protocol. Explain for what applications such a protocol is potentially useful.

**7.4.** Design a control unit for a version of the full-map directory protocol. What is the transistor count?

**7.5.** Design a control unit for a version of the Dir($i$)NB limited directory protocol. What is the transistor count?

**7.6.** Design a control unit for a version of the Dir($i$)B limited directory protocol. What is the transistor count?

**7.7.** Compare single-linked-list and double-linked-list chained directory protocols, from the performance and complexity points of view. What are the pros and cons of each approach?

**7.8.** Consult the open literature (e.g., the Internet) and find out about the commercial SMP products using SCI (Scalable Coherent Interface). Construct a table that compares the major features of these products.

**7.9.** What cache consistency maintenance protocols are most efficient for elimination of false sharing? Why?

**7.10.** Compare different passive sharing oriented cache consistency maintenance protocols, and explain pros and cons of each one. In which applications does passive sharing become a problem?

# CHAPTER 8

# Distributed Shared Memory

The distributed shared memory (DSM) approach represents an effort to combine the best characteristics of two different approaches: (1) SMP—shared memory multiprocessing and (2) DCS—distributed computing systems. As already indicated, SMP has one major good characteristic (simple-to-use programming model, based on one shared address space), and one major bad characteristic (it is not scalable). As is well known, DCS has one major good characteristic (it is scalable), and one major bad characteristic (difficult-to-use programming model, based on a number of distinct address spaces, and message passing between these spaces).

The major characteristics of DSM are that (1) it is a simple-to-use programming model and (2) it is scalability. The simple programming model is a consequence of the fact that DSM is also based on a single logical address space. The scalability is a consequence of the fact that different portions of the address space (which is logically unique) are implemented in different processing nodes (which means physically distinct).

This approach has emerged relatively recently. However, some systems in the 1980s, and even in the 1970s, do include properties of today's DSM systems. Some examples dating back to the 1970s, and elaborated on in the early 1980s, include the Encore RMS (reflective memory system) system [Gould 1981], the Modcomp MMM (mirror memory multiprocessor) system [Modcomp 1983], the University of Belgrade Lola system [Protić 1985], and the University of Pisa's MuTeam system [Corsini and Prete 1987].

For two different points of view on DSM, see Bell and Van Ingen (1999) and Hagersten and Papadopulas (1999), both published in the March 1999 *Proceedings of IEEE* Special Issue on DSM.

## 8.1. BASIC ISSUES

The basic structure and organization of a DSM system are given in Figure 8.1. The backbone of the system is a system-level interconnection network (ICN) of any type. This means that some systems include a minimal ICN of the BRL (bus, ring, or LAN) type. Other systems include a maximal ICN

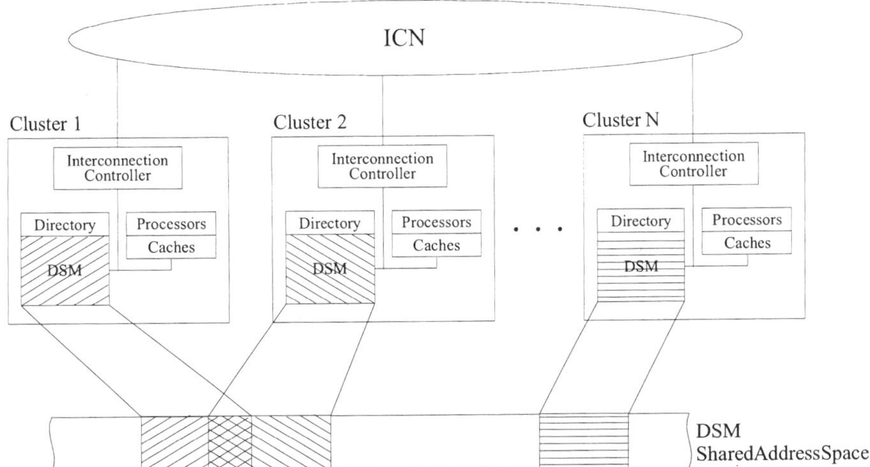


**Figure 8.1.** Structure and organization of a DSM system. [*Source:* Protić et al. (1996).] *Comment:* Note that each cluster (node) can be a uniprocessor system, or a relatively sophisticated multiprocessor system. In a number of designs, both clusters and the interconnection network are off-the-shelf systems. Only the DSM mechanisms are custom made, in hardware, software, or a combination of the two.

of the GMC (grid, mesh, or crossbar) type or some reduced ICN, which is in between, as far as cost and performance.

Nodes of a DSM system are referred to as "clusters" in Figure 8.1. In reality, a cluster can include a single microprocessor or an entire SMP system. Systems can be uniform (all clusters implemented as a single microprocessor or all clusters implemented as an SMP) or heterogeneous (different clusters based on different architectures).

Basic elements of a cluster are: (1) one or more processors with their level 1 (L1) and level 2 (L2) caches, (2) a part of memory with the memory consistency maintenance directory, and (3) an interconnection network interface, which is more or less sophisticated.

Finally, the logical DSM memory address space is obtained by combining the memory address spaces of physical modules placed in different processing nodes. Address spaces of different physical modules can logically overlap (which is the case in systems with replication), or they can be logically nonoverlapping (which is the case in systems with migration), or the overlap can be partial (as indicated in the bottom left corner of Figure 8.1).

### 8.1.1. The Mechanisms of a DSM System and Their Implementation

The term *mechanisms of a DSM system* covers all engineering aspects necessary for proper functioning of the concept. Typical DSM mechanisms include (1) internal organization of shared data and related system issues, (2) granularity of consistency maintenance and related system issues, (3) access algorithms and related issues, (4) property management and related issues, (5) cache consistency maintenance on the cluster level, and (6) memory consistency maintenance on the system level.

The DSM mechanisms can be implemented in three basic ways: (1) software, (2) hardware, and (3) a combination of the two.

First DSM implementations were in software, on platforms such as message passing multicomputers or clusters of workstations. Software implementation can be placed on various layers of the software system: (1) library routines, (2) optimizing compiler, (3) operating system outside the kernel, and (4) operating system inside the kernel. These implementations were easy to build and cheap to implement, but relatively slow and inefficient. Consequently, researchers turned to hardware implementations.

Hardware implementations were fast and efficient, but relatively complex and expensive. Therefore, the researchers had to turn to hybrid implementations, where some mechanisms are implemented in hardware and others in software.

According to many, hybrid DSM is the best way to go. Each particular mechanism is to be implemented in the layer (software or hardware) for which it is best suited. This means the best price/performance ratio. In some cases, hybrid implementations start from a full hardware approach and then migrate selected mechanisms into software (typically, those that are complex and expensive to implement in hardware). In other cases, hybrid implementations start from a full software approach and then migrate selected mechanisms into hardware (typically, those that are slow if implemented in software and frequently used in applications of importance).

The sections to follow give brief explanations of the major DSM mechanisms. For detailed explanations, interested readers are referred to Protić et al. (1997) and Protić (1998). These references include both the details and the references to original research publications.


## 8.1.2. The Internal Organization of Shared Data

In principle, shared data can belong to any data type. They can be nonstructured or structured. If structured, they can be structured as traditional complex data types or as objects in line with modern programming languages.

Shared data are being accessed during the program segments called *critical code*. These critical code segments have to be properly protected, which is the subject of a later section of this book.

Organization of shared data is an issue of interest for optimizing compiler design and system programming. It is not to be confused with the type and granularity of consistency maintenance units, which is the subject of the next section.

### 8.1.3. The Granularity of Consistency Maintenance

As far as cache memory consistency maintenance, the granularity is typically determined by the internal design of the microprocessors used in different clusters of the system. These microprocessors are typically off-the-shelf machines, and cache block sizes are in the range from 4 to 8 or 16 words.

As far as memory consistency maintenance is concerned, granularity is typically determined by the system designers, and it can take on a number of different sizes: (1) word, (2) block, (3) small page, (4) large page, (5) object, and (6) segment.

Hardware approaches typically utilize relatively small units of consistency maintenance (word, block, or small page). This is so because hardware can easily support smaller granularity, which is good for issues such as space utilization, data thrashing, and false sharing.

Software approaches typically utilize relatively large units of consistency maintenance (large page, objects, or segments). This is so because software cannot easily support smaller granularity; however, software approaches can easily support mechanisms that check and improve on issues such as space utilization, data thrashing, and false sharing. Also, software algorithms can support semantic data structures such as objects and/or segments, which are good for some applications.

Hybrid approaches typically utilize medium size units of consistency maintenance (small page or large page). This is so because hybrid approaches can easily combine the best of the two extreme solutions, for the best price/performance ratio given the technology/application prerequisites.

### 8.1.4. The Access Algorithms of a DSM System

There are three basic access algorithms used in DSM systems: (1) single reader, single writer (SRSW), (2) multiple reader, single writer (MRSW), and (3) multiple reader, multiple writer (MRMW). Each one is briefly discussed.

The SRSW algorithm implies that only one node is allowed to read during a given period of time, and only one node is allowed to write during the same period of time. This algorithm is well suited for typical producer–consumer relationships in two-node systems. If applied to multiple-node systems, the management mechanism (to be discussed in the next section) must allow the migration of memory consistency units (pages, in most cases). In that case, if another node needs to read or write, it first performs an action to migrate the memory consistency unit of interest into its own physical memory, which makes it become the owner of that memory consistency unit, and then it reads or writes. In DSM systems, the SRSW algorithm is used only in conjunction with the other two algorithms, but never as the only algorithm in the system. In such schemes, typically, SRSW is used only for some data types (which behave as SRSW), and the other two algorithms are used for other data types (which demonstrate a behavior unsuitable for SRSW).

The MRSW algorithm implies that several nodes are allowed to read during a given period of time, but only one node is allowed to write during a given period of time. In the case of reading, no migration of memory consistency units is needed. Migration is needed only in the case of writing. This approach is compatible with typical application scenarios, where multiple consumers use data generated by one producer. Consequently, this type of algorithm is used in most of the traditional DSM machines.

The MRMW algorithm implies that all or most nodes are allowed to both read and write during a given period of time. In systems with replication, in writethrough update-all systems [e.g., reflective memory systems (RMSs) or a mirror memory multiprocessor (MMM)], whenever a node writes to a shared variable, this writing is also sent to all other nodes that potentially need that value, and the remote nodes are updated after a network delay. If the network serializes all writes in the system, data consistency will be preserved, and each node can read shared variables from its own local physical memory. In systems without replication, in writeback systems, and in systems in which the interconnection network does not serialize the writes, additional action (on the software or hardware part) is needed, in order to preserve data consistency. In systems with migration, appropriate mechanisms have to be applied (in software or in hardware), to help preserve the data consistency.

### 8.1.5. The Property Management of a DSM System

Except in all-cache systems (COMA) (e.g., KSR or DDM) and many low-complexity systems ([uniform memory access (UMA)], like RMS or memory channel), with caches only and no main memory, each unit of consistency is typically associated with a home node [an important characteristic of cache consistent nonuniform memory access (CC-NUMA) systems]. The home node for a given unit of consistency (read "page" in most of the systems) is responsible for the management of ownership of that unit of consistency (different processing nodes can own it during different time intervals).

Nodes can be assigned to units of consistency in a number of different ways. One way is, at compile time, to assign unit 0 to node 0, unit 1 to node 1, and so on. In principle, it helps the application execution time, if the home node is the same node as the most frequent owner. Consequently, a good optimizing compiler always tries to do static assignment (compile time) of units of consistency to nodes, in such a way that the home for each unit of consistency is its most frequent owner. Another way is, at run time, to assign nodes according to a predefined algorithm. An approach that is sometimes used is to assign unit 0 to the first node that needs it, unit 1 to the first node that needs it, and so forth. This approach starts from the assumption that the first user is most likely to be the most frequent user.

In SRSW algorithms, when a node has to read or write, it first consults the home of the memory consistency unit to which the variable to be read and/or written belongs and requests that the ownership of the entire memory consistency unit be passed to it (i.e., to the node that is to read or write). After the ownership is granted, and the unit of consistency is migrated, the reading or writing can occur. In MRSW algorithms, the same procedure happens only on writing. In MRMW algorithms this procedure is not needed on either reading or writing, but other appropriate actions are needed, in order to preserve the data consistency.

The management mechanism can be either centralized or distributed. If distributed, the responsibility for different units of consistency can be fixed (at compile time or run time) or variable (which means changeable at run time). Knowledge about current nodes for different units of consistency can be kept in the form of traditional tables or linked lists. Knowledge about the concrete node that is the home or owner for a specific unit of consistency can be public, or only the node involved can have that knowledge. Different management strategies and more have been elaborated on in the Yale University Ph.D. thesis of Kai Li.

### 8.1.6. The Cache Consistency Protocols of a DSM System

As far as cache consistency maintenance is concerned, both research prototypes and industrial products rely heavily on the mechanisms supported by built-in off-the-shelf microprocessors. This means that either WI (write-invalidate), or WU (write-update), or TS (type-specific) protocols can be used. *Type-specific* means that WI is used for some data types, and WU for other data types.

As indicated earlier, snoopy protocols are used at the cache consistency maintenance level, and directory protocols are used at the memory consistency maintenance level. Consistency maintenance on the memory level is always based on one of the chosen memory consistency models; consequently, the next section sheds more light on different memory consistency models.

### 8.1.7. The Memory Consistency Protocols of a DSM System

Memory consistency models (MCMs) can be restricted (strict or sequential) or relaxed (processor, weak, release, lazy, lazy release, entry, AURC, scope, or generalized). The more relaxed a memory consistency model, the more difficult it is to program a DSM machine. However, the performance improves. Memory consistency models determine the timing when different nodes in the system have to become aware about changes on variables that they share and potentially need in the future.

In principle, memory consistency protocols can be based on either update or invalidation. In other words, in order to keep the memory consistent, protocols can use either update or invalidation of memory consistency units.

The strict MCM implies that all nodes become immediately aware of all shared data changes. This type of MCM has only a theoretical value, unless relative interconnection network (ICN) delays are negligible.

The sequential MCM implies that all nodes become aware of all shared data changes, after a delay introduced by the ICN. Sequential MCM is typically supported on systems that use BRL (bus, ring, or LAN) in the ICN. All processors see each particular data access stream, as well as the global data access stream, in the same order. For example, if processor P1 writes first to A and then to B, and after that processor P2 writes first to C and than to D, each and every processor in the system sees the same global order of writes: A, B, C, D.

The processor MCM implies that all nodes see each particular data access stream in the same order; however, each node sees the global data access stream in a different order. Processor MCM is typically supported by systems that use BRL and include ICN access buffers, which forward data to the ICN using different speeds and priorities. For example, in the same scenario as in the previous paragraph, possible orders of writes that different processors in the system can see are ABCD, ACBD, ACDB, CDAB, CADB, and CABD. Note that in each case A is before B, and C is before D. This means that each processor in the system sees the same order of writes of processor P1, which is A before B, and the same order of writes of processor P2, which is C before D.

The weak MCM (as well as all other MCMs to follow) implies that special synchronization primitives are incorporated into the program code. In between the synchronization points, the memory consistency is not maintained, and programmers have to be aware of that fact, to avoid semantic problems in the code. Synchronization points are referred to as either (1) *acquire*, at the entry point of a critical section; or (2) *release*, at the exit point of the critical section. In the weak MCM, consistency is maintained at both synchronization points, and the code is not allowed to proceed after the synchronization points, unless the memory consistency has been established. Synchronization points by themselves have to follow the sequential consistency rule (all nodes seeing the same global order of accesses).

The release MCM implies that memory consistency has to be established only at release points, and the release points by themselves have to follow the processor consistency rule. At each release point, all other processors and processes in the system get updated with the changes to variables made prior to the release point of the critical code section just completed. This generates some potentially unnecessary ICN traffic, which may have a negative impact on the speed of the application code.

The lazy release MCM implies that memory consistency has to be established only at the next acquire point, which means that the ICN traffic will include only the updated variables that might be used by the critical code section that is just about to start. This means less ICN traffic, but more buffering space at each processor, to keep the updated variables until the start of all critical code sections to follow. The better execution speed of the application code is paid for by more data buffering at all nodes. Note that some of the updated variables will not be used in the critical code section to follow. Nevertheless, such variables will contribute to the ICN traffic, because they might be used, and consequently have to be passed over to the next critical code section, which is about to start running on another processor.

The entry MCM implies that each shared variable (or each group of shared variables) is protected by a synchronization variable (a critical section is bounded by a pair of synchronization accesses to the synchronization variable). Consequently, updated variables will be passed over the network if and only when absolutely needed by the follow-up critical code sections. Such an approach leads to the best possible performance. However, concrete performance depends a lot on the details of the practical implementation. The first implementation of entry consistency (Midway project at CMU) requires the programmer to be the one to protect the specific shared variables (or the groups of shared variables). If such implementation of the entry MCM is compared with a lazy release MCM, differences are small; for some of the SPLASH-2 applications [Woo et al. 1995] the entry MCM works better, and for other SPLASH-2 applications the lazy release MCM works better [Adve et al. 1996]. This author believes that, with an appropriate implementation, the entry MCM is able to provide better performance for most SPLASH-2 applications.

In principle, implementations of various MCMs can be done predominantly in hardware or predominantly in software. Table 8.1 summarizes the discussion on MCMs. For each of the six different MCMs discussed so far, two projects (or two groups of projects) have been specified: (1) one that refers to the first predominantly hardware implementation of the given MCM and (2) one that refers to the first predominantly software implementation of the given MCM.

Table 8.1 also includes the names of the major researchers, as well as the names of universities/industries involved. Information in this table has to be considered conditionally, because different researchers classify different approaches in different ways.

Chronologically, software implementations of relaxed MCMs have come first. Here is an explanation. The basic idea behind software-implemented DSM, often referred to as *shared virtual memory* (SVM), is to emulate a cache coherence protocol at the granularity of pages rather than cache blocks. Under such conditions (coarser granularity of memory coherence units), major issues that affect performance are false sharing and fragmentation. False sharing happens when two semantically unrelated processors/processes share the same page (accessing unrelated data, which fall into the same page), and one of them writes data. Data of the other will be unnecessarily invalidated and it (the other one) will have to bring over the ICN the data that it already has. If data update is used, the ICN traffic will increase again, due to the updates, which are not needed because they refer to data, which are not being used by the processor or process being updated. Page fragmentation happens when a processor or process needs only a part

**TABLE 8.1. Memory Consistency Models**[a]

| | |
|---|---|
| Sequential | MEMNET+KSR1 vs. IVY+MIRAGE<br>Delp/Farber(Delaware)+Frank(KSR) vs. Li(Yale)<br>+Fleish/Popek(UCLA) |
| Processor | RM vs. PLUS<br>Gould/Encore/DEC/Compaq(USA) vs. Bisiani/<br>Ravishankar/Nowatzyk(CMU) |
| Weak | TSO vs. DSB<br>SUN/HAL/SGI(USA) vs. Dubois/Scheurich/Briggs(USC) |
| Release | Stanford-DASH vs. Eager-MUNIN<br>Gharachorloo/Gupta/Hennessy(SU) vs. Carter/<br>Bennett/Zwaenopoel(Rice) |
| LazyRelease | AURC vs. TREADMARKS<br>Iftode/Dubnicki/Li(Princeton) vs. Zwaenopoel/<br>Keleher/Cox(Rice) |
| Entry | SCOPE vs. MIDWAY<br>Iftode/Singh/Li(Princeton) vs. Bershad/Zekauskas/<br>Sawdon(CMU) |

[a]Implementations of MCMs (hw-mostly vs. sw-mostly): TSO—total store order; AURC—automatic update release consistency. It is important to know the names behind the major achievements in the field of interest. Conditionally speaking, each memory consistency model can be implemented mostly in hardware and/or mostly in software. For all major memory consistency models, for both hardware-mostly and software-mostly implementations, one can find here the name of the project, the names of major contributors, and the name of the university. In the case of industry, individual names have been omitted.

of the page, because there is no finer granularity available. Both problems are alleviated through the use of relaxed MCMs, which permit the system to delay the coherence activity from the time when the modification happens to the time of the next synchronization point of interest (release or acquire or both). Postponing the coherence activity means that some of the ICN traffic will not happen. On the other hand, hardware-implemented DSM typically uses cache blocks as memory coherence units, which means that impacts of false sharing and fragmentation are not so severe, and that there is no urgent need to do relaxed MCMs in hardware, except for uniformity reasons when everything is done in hardware (e.g., relaxed MCM in DASH), or when the speed-up of a software solution is needed (e.g., AURC and SCOPE in SHRIMP).

The major MCMs will now be revisited from another point of view, using pictures and specific examples (definitions of all mnemonics used in the pictures have been given in figure captions).

•Dash:

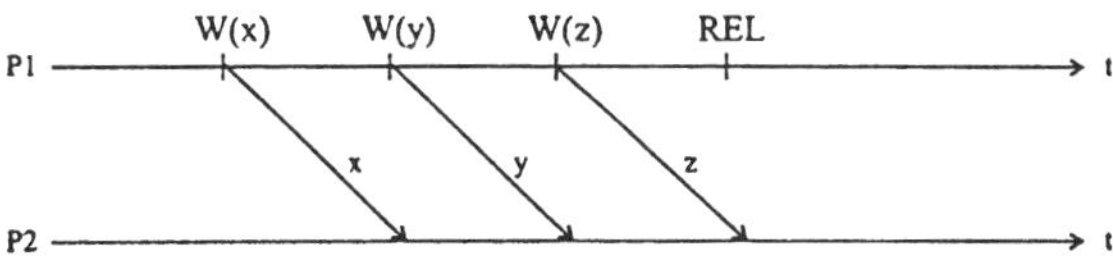


•Munin:

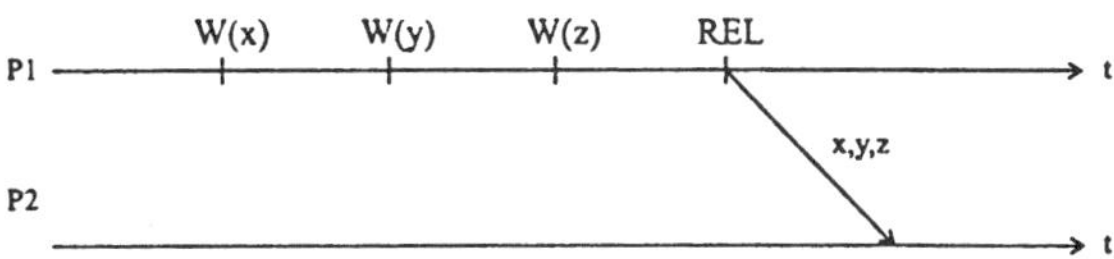


•Repeated updates of cached copies:

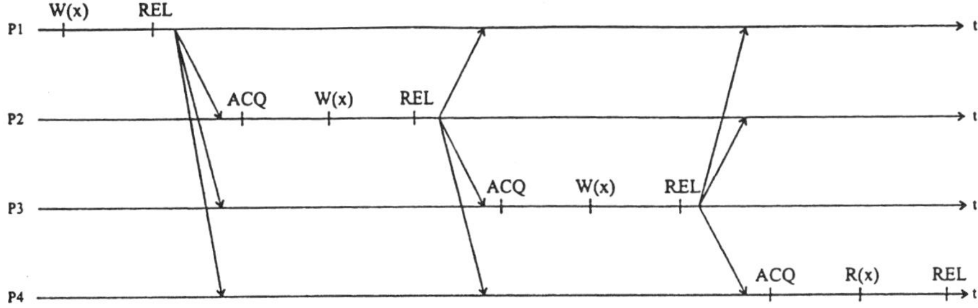


**Figure 8.2.** Release consistency—explanation (W—write; R—read; REL—release; ACQ—acquire). [*Sources:* Lenoski et al. (1992) and Carter et al. (1991).]

*Comment:* Note that DASH (hardware implementation) and Munin (software implementation) create different interconnection network traffic patterns. DASH accesses the network after each write, which means a more uniform traffic. Munin accumulates the information from all writes within the critical section and accesses the network at the release point, which means bursty traffic. This difference is due to inherent characteristics of hardware (low overhead for single data broadcast, multicast, or unicast) and software (high overhead for single data broadcast, multicast, or unicast). Note that release consistency, in general, no matter if implemented in hardware or software, updates and invalidates more processors than needed, which is a drawback to be overcome by the lazy release consistency, at the cost of extra buffering in the system. In theory, release consistency can be based on either update or invalidate approaches.

### 8.1.7.1. Release Consistency

The Stanford DASH project includes a predominantly hardware implementation of the release consistency, which is depicted in the upper part of Figure 8.2. As soon as a value is generated (which is potentially needed by another process on another processor), the update is performed. Consequently, values $W(x)$, $W(y)$, and $W(z)$ are being updated at different times.

The Rice Munin project includes a predominantly software implementation of the release consistency, which is depicted in the middle part of Figure 8.1. All updated values are being collected in a special-purpose buffer, and the update is done only at the release point, which is more convenient in software implementations. Obviously, the software implementations create a bursty ICN traffic.

The example in the lower part of Figure 8.2 underlines the fact that release consistency broadcasts the updated values to all potentially interested processors, in spite of the fact that many of them will not need the data. This picture is provided to serve as a contrast to the picture that explains lazy release consistency and follows next.

### 8.1.7.2. Lazy Release Consistency

The first predominantly software implementation of lazy release consistency was done as a part of the TreadMarks project at Rice [Keleher et al. 1994]. Lazy release consistency is best explained using the example in Figure 8.3. This figure covers the same example as Figure 8.2.

It is seen immediately that lazy release consistency generates considerably less ICN traffic. As indicated earlier, this potential performance improvement is paid for with extra buffering in each processor—a price that is fully justified by the improved performance.

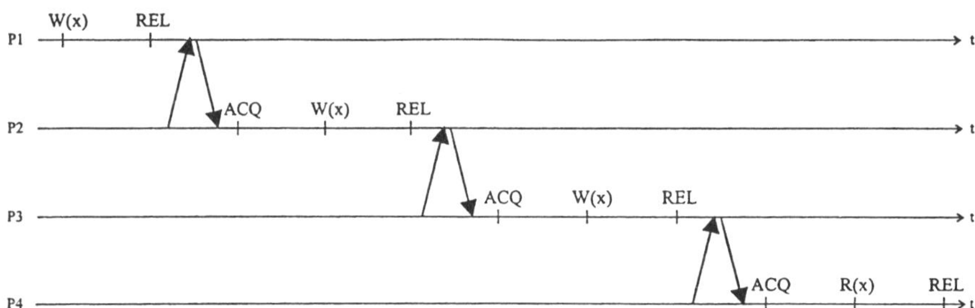


**Figure 8.3.** Lazy release consistency—explanation. [*Source:* Keleher et al. (1994]).]
*Comment:* Updates and invalidates happen at the next acquire point. Only the shared variables related to the acquiring process are being updated/invalidated, which means less interconnection network traffic, at the cost of extra buffering in each processor (remaining updates/invalidates related to other shared variables changed by the previous process have to be held in a buffer, until the time comes when they are needed by a future acquiring process).

The LRC (lazy release consistency) implementation of TreadMarks is a multiple writer protocol using invalidation (other implementations are possible as well). Each writer records all changes that it makes on all shared pages, since the last acquire point. This is done by comparing a copy of the original page (*twin*) and the modified original page; this comparison generates a data structure called *diff*. At the next acquire point, pages modified by other processes get invalidated. At the first access to an invalidated page, diffs are collected and applied in the proper causal order, to rebuild the coherent page.

Figure 8.3 is related to a software implementation of lazy release (TreadMarks); software implementation is noticeable when a single update/invalidate burst happens at the next acquire point. In theory, lazy release consistency can be based on either update or invalidate approaches.

In the meantime, there were two more developments:

1. Researchers at the University of Maryland (Keleher) showed that LRC can be implemented in the single writer mode. No diffs are necessary, but performance is poor if the application exposes larger amount of write–write false sharing.
2. Researchers at Rice University proposed an adaptive LRC, which can switch dynamically between single-writer and multiple-writer modes; this seems to be a good price/performance compromise.

### 8.1.7.3. Entry Consistency

The first predominantly software implementation of entry consistency was done as a part of the Midway project at CMU [Bershad et al. 1993]. Entry consistency is best explained using the example in Figure 8.4. This figure covers the same example as Figures 8.3 and 8.2.

It is immediately seen that (in this idealized example) entry consistency generates even less ICN traffic than lazy release consistency. As indicated earlier, this potential performance improvement is paid for (in the Midway implementation) with extra effort on the programmer side.

The EC (entry consistency) implementation of Midway is a single writer protocol using updating (other implementations are possible as well). Entry consistency guarantees that shared data become consistent only when the processor acquires a synchronization object. Designers of the Midway system were aware of the fact that advantages of entry consistency may be difficult to obtain if limited programming effort is possible; consequently, in addition to entry consistency, Midway also supports release and processor consistency.

### 8.1.7.4. Automatic Update Release Consistency

The automatic update release consistency (AURC) was developed at Princeton, as a part of the SHRIMP project [Blumrich et al. 1994, 1998; Iftode

• Message traffic:

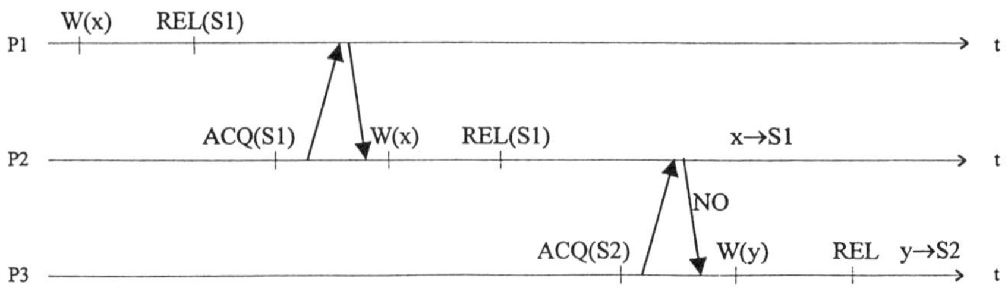


**Figure 8.4.** Entry consistency—explanation (→ —protected by: x → S1—data variable x protected by synchronous variable S1; NO—not needed). [*Source:* Bershad et al. (1993).]

*Comment:* This figure is related to a software implementation of entry consistency (Midway). With the help of a programmer, at the next acquire point, updates and invalidates are done only for the shared variables related to a given lock (S1 in Fig. 8.4), which means even less interconnection network traffic, on average. In theory, entry consistency can be based on either update or invalidate approaches, although some people believe that entry consistency is inherently oriented to update, because the Midway implementation uses the update approach.

1996a]. Some researchers consider it an implementation of lazy release consistency, which includes elements of hardware support. Others treat it as a new MCM. In fact, AURC is a home-based LRC, which uses automatic update hardware support (fine-grain remote writes). According to this author, it is definitely a new protocol, but not a new MCM.

Figure 8.5 shows an example depicting the difference between the classical lazy release consistency (TreadMarks) and the automatic update release consistency (SHRIMP-2). The upper part of the figure includes a classical lazy release example. The middle part of the figure includes an AURC example based on the Copyset-2 mechanism. The lower part of the figure includes an AURC example based on the Copyset-*N* mechanism. Differences between Copyset-2 and Copyset-*N* are explained in Figure 8.5.

Automatic update release consistency implemented in the SHRIMP-2 project is best treated as a hardware-supported implementation of lazy release. Hardware supports the point-to-point communications between a writing node and the owner node for the page being written into (Copyset-2); however, this point-to-point mechanism, enhanced by software, can be used as a building block for a more general communications pattern (Copyset-*N*).

As indicated in Table 8.2, the Copyset-2 mechanism implies automatic hardware-based update of the tables in the home processor, as soon as a value is updated. The Copyset-*N* mechanism is a logical extension of the Copyset-2 mechanism (this means that the Copyset-2 mechanism is directly

•Classical LRC:

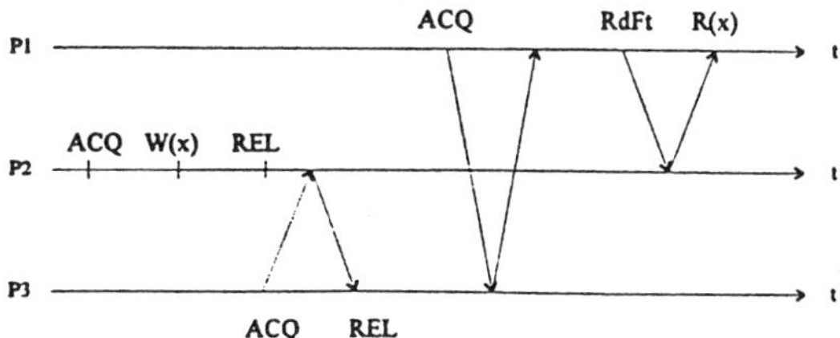

•Copyset-2 AURC:

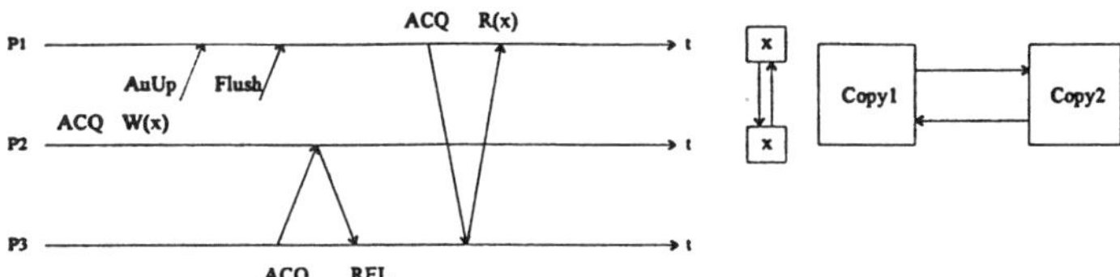

•Copyset-2 AURC:

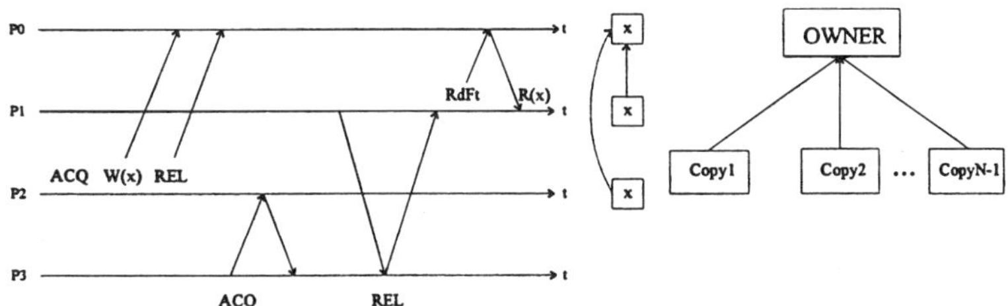

**Figure 8.5.** AURC—explanation (RdFt—read fault; AuUp—automatic update). [*Source:* Iftode et al. (1996a).]

*Comment:* Automatic update supports only point-to-point communication; therefore, pages shared by only two nodes can be kept coherent using automatic update only (Copyset-2). For pages shared by more than 2 nodes, the coherence is maintained by a combination of hardware and software (Copyset-$N$). A node is designated as home and automatic update is used to propagate the updates from nonhome copies to the home copy. The nonhome copies of the shared page are updated on demand by transferring the whole page from home.

supported in hardware, and Copyset-$N$ is synthesized out of the appropriate number of actions of the Copyset-2 mechanism). Details of AURC are given in Table 8.2, using a version of AURC (the AURC idea has undergone several modifications). Details not covered in Table 8.2 are left to the interested reader, to be developed as homework.

**TABLE 8.2. AURC Revisited[a]**

| | |
|---|---|
| 1. | The hardware update mechanism works on the point-to-point basis. On a write to a location, only one remote node is updated—the home—no matter how many different nodes share that same data item. |
| 2. | The home for a specific page is determined at run time, as the node, which writes first into a page. It is assumed that the first writer will become the most frequent writer. |
| 3. | On the acquire point, the acquiring node sends a message to the last owner of the lock (indirect through the home of the lock, which is statically assigned), as in LRC. |
| 4. | From the last owner, along with the lock, the acquiring node gets the "write notices" indicating all pages of that acquire that were updated in the past, according to the happen-before partial ordering (as in LRC). |
| 5. | Pages indicated in the write notices are invalidated unless the versions of the local copies are posterior to the versions indicated in the write notices. Versions are represented as a timestamp vector and can be constructed from the write notices received. |
| 6. | At the faulting time the process sends the request for the page to the home indicating the version. |
| 7. | Home keeps the copy set (list of sharers) and the updated "version vector" telling which locks were writing to a particular version of each page, in the set of pages for which it is the home. Each node keeps just one element of the "version vector." This is the vector element related to the pages replicated in that specific node. The home replies as to whether it has at least the required version; otherwise it waits until it reaches that version. Along with the page, the home also sends the current version of the page, which can be greater than the one requested. This acts as a prefetching that could save future invalidations. |
| 8. | Before page fault is resumed, the faulting node write-protects the page in order to detect the update. |
| 9. | At the first write on the page following an acquire, a write fault occurs and the faulting node adds the page to the update list and unprotects the page for writing such that future writes will not generate faults. |
| 10. | Each write will be performed into the local copy and propagated as an automatic update to the home and also performed into the home's copy. Others will be updated after they acquire the lock and after they contact the home to obtain the last update of the page. This decreases the ICN traffic, in comparison with some other reflective (writethrough-update) schemes, which update *all* sharers using a hardware reflection mechanism (RM and/or MC). The major issue here is to minimize the negative effects of false sharing, in cases when page sizes are relatively large and several processes (locks) share the same page. |
| 11. | At the release point, the processor increments its local timestamp and sends it to all homes of the pages in its update list. In this way, the home timestamp vector (version of the page) is updated and the links between the releaser and the homes of the shared pages are flushed. |

**TABLE 8.2.** (*Continued*)

| | |
|---|---|
| 12. | A new interval starts. All pages from the update list are write-protected and the update list is emptied. |
| 13. | This scheme implements the lazy release consistency model through a joint effort of hardware and software (unlike TreadMarks, which does the entire work in software). The AURC scheme has been implemented in the SHRIMP-2 project at Princeton. |

[a]The AURC idea was going through a number of different versions; the version selected for presentation here is the one that seems to be the most suitable from the point of view of the final goal of this book, which is to present the facts of importance for implementation of a DSM on a chip, together with a number of dedicated accelerators for the most frequently used application functions.

The AURC is a special case of a more general class of LRC protocols that some call *home-based LRC* protocols (HLRC). Researchers at Princeton have implemented an HLRC on a 64-node Intel Paragon. Later, they implemented an HLRC on Typhoon-0.

The basic difference between HLRC and the canonical LRC is how the updates are collected. In LRC (which some call "homeless" or distributed or standard LRC) the diffs are kept everywhere and are fetched on demand and applied to the local copy in order to bring it up-to-date.

In HLRC, the updates are propagated to the home using diffs (instead of automatic update) at the release time and applied to the home's copy. Later, at page fault time, the whole page is fetched from the home instead of fetching diffs potentially from multiple nodes.

The advantage of HLRC is simplicity and scalability. The memory overhead (which can be huge in standard LRC as a result of diff storage and may require garbage collection) is negligible in HLRC since diffs are discarded after they are sent home. The disadvantages of HLRC are that (1) it may send more data sometimes (pages instead of diffs, but diffs can also accumulate, and bandwidth is not a problem now) and (2) its home choice can create hot spots sometimes (or twice as many messages as LRC for migratory data patterns).

However, running HLRC and LRC up to 64 nodes confirmed that HLRC scales better than does the standard LRC. Moreover, AURC can be viewed now as an HLRC protocol, which uses the AU mechanisms to propagate the updates.

### 8.1.7.5. SCOPE Consistency

The scope consistency (ScC) was developed at Princeton, also as a part of the SHRIMP project [Iftode et al. 1996b]. Some researchers consider it to be an implementation of entry consistency, which includes elements of hardware support and excludes programmer interaction (which represents a very important improvement). Others treat it as a new MCM. It is definitely a new protocol.

The scope consistency is based on the same basic hardware support as the AURC, except that more detailed bookkeeping is maintained, which brings it conceptually closer to entry consistency, but avoids the need for programmer support. Details of scope consistency are given in Table 8.3, using a version of scope consistency (this idea has undergone several modifications). Details not covered in Table 8.3 are left to the interested reader, to be developed as homework.

Actually, ScC does not necessarily require hardware support. It can be implemented totally in software (using a home-based protocol). There is a system called *Brazos* (Rice University) that implements ScC on top of Windows NT, also in software.

Actually, the bookkeeping is simpler in ScC. It assumes an implicit (and dynamic) binding between data and locks (scopes). A local scope corresponds, for instance, to all critical sections protected with the same lock. Barriers form a global scope. Binding between data (page) and lock occurs at the first write to that page and it is reset at each barrier. A page can be bound to multiple scopes either because of false sharing or because binding can be ambiguous (if there are nested scopes, the protocol cannot determine to which lock to bind the page and conservatively binds it to all scopes open at that time).

### 8.1.8. A Special Case: Barriers and Their Treatment

A closer look at some of the typical parallel applications (e.g., those found in the SPLASH-2 suite) reveals that many shared reads and writes frequently occur outside explicit critical sections. We will briefly discuss the treatment of such memory references by the more relaxed memory consistency models mentioned above.

The aforementioned applications (frequently of scientific and engineering nature) often use *barriers* as a means of synchronization, in addition to using ordinary locks. Of $N$ processes participating in a parallel computation, each time one of them encounters a barrier, it will block until all the other $N - 1$ processes reach the same barrier. At that point, all $N$ of them are released. Barriers are used to separate phases in parallel computation; each such phase is called an *epoch*. A barrier can be intuitively understood as a point where total consistency is established, including the data written outside explicit critical sections (with an exception that this is not necessarily true for the entry consistency model). It may be helpful to view a barrier as a release–acquire pair, that is, as if a process performed a *release* at barrier arrival and an *acquire* at barrier departure.

To illustrate the treatment of shared accesses both inside and outside (explicit) critical sections, let us consider the following sequence of actions performed by three processors, P1, P2, and P3, in parallel:

**TABLE 8.3. SCOPE Revisited**[a]

| | |
|---|---|
| 1. | The SCOPE is the same as the AURC, except for the following differences. |
| 2. | Write notices are kept per lock rather than per processor. |
| 3. | The version update vector at the home includes one more field, telling which variables (addresses) were written into, by a given lock; this is repeated for each version of the page. At the acquire time, the acquiring node receives from the last releaser only the write notices corresponding to that lock. |
| 4. | Pages are invalidated as in LRC/AURC. Pages are fetched from the home at page fault time as in AURC, then write-protected. |
| 5. | The home is not checked at the next acquire point, but after that point, when the variable is invoked, which corresponds to the ENTRY consistency model. |
| 6. | However, unlike with ENTRY, the activity invoking a variable is related to the entire page; thus, the entire page is brought over, if so necessary. |
| 7. | In ScC a node keeps a list of update lists, one for each lock (scope) that is open. This means page-based maintenance and more traffic, rather than object-based maintenance and less traffic (as in Midway). In principle, Midway can also do page-based maintenance, but it does not. |
| 8. | At the release point, the processor increments its local timestamp and the epoch number of the lock. The epoch number is used at the acquire time to determine which write notices to send to the next acquirer. |
| 9. | At the barrier time, all write notices since the last barrier (global scope) are performed. This ensures that, when the barrier is reached, the entire address space is made coherent and the local scopes can be redefined. |
| 10. | In comparison with Midway, the SCOPE is faster, although it is page-based rather than object-based, since it does a part of the protocol in hardware, using the same automatic update hardware as the AURC. |
| 11. | Unlike Midway/EC, ScC takes advantage of the data to lock binding using a page-based protocol. It does not require compiler assistance and it seems simpler to program than EC (although some precautions must be taken). Many (but not all) RC applications are ScC without any modifications. Some applications require changes (or even explicit scopes to be added) in order to comply with ScC. |
| 12. | In conclusion, SCOPE can be treated as a hw/sw (hardware/software) implementation of ENTRY. Its advantage over ENTRY is in complete avoidance of compiler assistance. |
| 13. | An implementation of ScC on SHRIMP is in progress at Princeton. ScC was implemented in Brazos at Rice on a cluster of Pentium dual-processor PCs running Windows NT and the results confirm the simulations reported in the ScC paper. The SCOPE was not implemented in any of the Princeton DSM machines. |
| 14. | The SCOPE research brings up a number of new ideas: merge, diff-avoid. |

[a]The SCOPE idea was going through a number of different versions, and the one selected for presentation here may not correspond to some of the implementations of ScC.

```
              P1                    P2                   P3
              BARRIER               BARRIER              BARRIER
              A := 1
t₁            acquire S1            ...                  ...
              B := 2
t₂            release S1
t₃                                  acquire S1           acquire S2
              ...                   b := B               a := A
t₄                                  release S1           release S2
C := 3                              ...                  ...
t₅            BARRIER               BARRIER              BARRIER
```

Assume that variables $A$, $B$, and $C$ are shared and the others are private;
assume further that all reads and writes not shown are local. Symbol $t_i$
denotes a point of time at which a synchronization operation (acquire,
release, barrier) occurs. Different models treat this code in different ways,
and they are ordered here by the amount of extra programming effort
required:

- *Release Consistency.* $A = 1$ and $B = 2$ are guaranteed to be visible to P2
  and P3 after $t_2$, and no later than immediately after $t_3$. $C = 3$ will be
  visible to P2 and P3 immediately after $t_5$.
- *Lazy Release Consistency.* $A = 1$ and $B = 2$ must be seen by P2 immediately after $t_3$. However, this is not the case with P3, as it uses a different
  lock: while the interval $t_1 - t_2$ on P1 clearly must precede the interval
  $t_3 - t_4$ on P2, nothing can be claimed about the relative order of the
  interval $t_1 - t_2$ on P1 and the interval $t_3 - t_4$ on P3. Therefore, $A = 1$
  and $B = 2$ will be visible to P3 only after $t_5$. Had P3 tried to acquire lock
  S1 sometime between $t_4$ and $t_5$, the values of $A$ and $B$ would have been
  propagated. $C = 3$, as for the release consistency, will be visible after $t_5$.
- *ENTRY Consistency.* As in the previous case, $B = 2$ is guaranteed to be
  visible to P2 immediately after $t_3$ (provided that the lock S1 guards $B$),
  but no further assumptions are made, not even as to what will be visible
  *after* the epoch. In order to ensure visibility of shared variables updated
  outside critical sections, the programmer must either (1) use read-only
  locks to read the shared but not propagated values when they are
  needed, (2) explicitly bind them to the barrier, or (3) bind the entire
  address space to the barrier (the latter two when visibility in the next
  epoch is the goal).
- *SCOPE Consistency.* $B = 2$ is guaranteed to be visible to P2 immediately
  after $t_3$, but no further commitments exist within the epoch. The model
  does *not* ensure visibility of $A = 1$ before $t_5$. $A = 1$ and $C = 3$ will be
  visible to P2 and $A = 1$, $B = 2$, $C = 3$ to P3 only after $t_5$.

## 8.1.9. Producer–Consumer Communication in DSM Multiprocessors

Conventional DSM implies consumer-driven communication. When data are needed, they get retrieved from the global shared memory. In cache-coherent DSM systems, data reside either in a remote memory module or in the producer's cache memory.

Some of the recent research focuses on producer-initiated mechanisms. Producer-initiated mechanisms [Byrd and Flynn 1999] reduce communication latency by sending data to potential consumers, as soon as data are produced. Producer-initiated mechanisms are either implicit (if precise identity of the consumer is not known) or explicit (if precise identity of the consumer is known).

Consumer-initiated mechanisms are (1) invalidation and (2) prefetch. Producer-initiated mechanisms are (3) data forwarding, (4) message passing, (5) update, (6) lock, and (7) selective update. An excellent simulation-based comparison of all these schemes can be found in Byrd and Flynn (1999).

Consumer-initiated communication in a cache-coherent system is illustrated in Figure 8.6. Producer-initiated communication, in the same scenario, is illustrated in Figure 8.7 [Byrd and Flynn 1999].

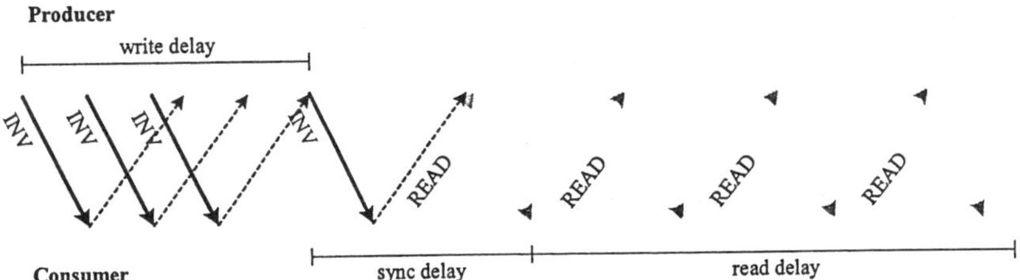


**Figure 8.6.** Illustration of consumer-initiated communication (INV—invalidation). [*Source:* Byrd and Flynn (1999).]
*Comment:* The read operation is triggered by a cache miss.

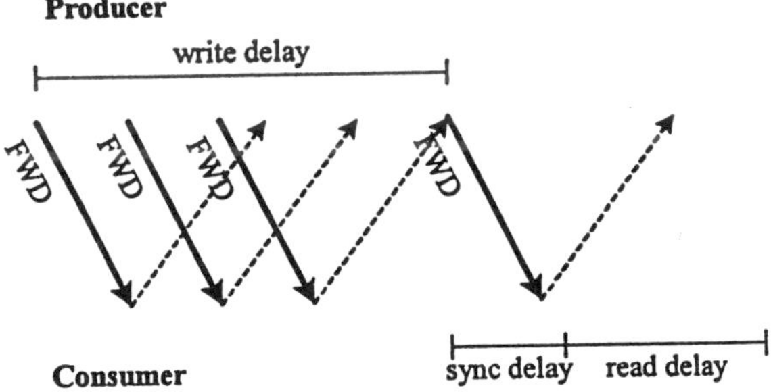


**Figure 8.7.** Illustration of producer-initiated communication (the same scenario as in Fig. 8.6) (FWD—forward). [*Source:* Byrd and Flynn (1999).]
*Comment:* Write delay stays the same. Synchronization delay and read delay become shorter.

Consumer-initiated communications (for an example with invalidations) work as follows. Produced data are written into the cache memory of the processor executing the producer process. If these data are shared, remote copies have to be invalidated. The write is completed, and the write delay is over, after the invalidations are acknowledged. After the write is completed, the producer writes to a synchronization variable, to notify the consumer that data are ready. The consumer reads the synchronization variable repeatedly (through its cache) until it realizes that an updated value is ready to be retrieved from the producer's cache into the consumer's cache. At that moment, the synchronization delay is over. Finally, during the read delay interval, the consumer's cache is updated.

Producer-initiated communications (for the same scenario) work as follows. As the data are produced in the producer's cache, data are forwarded directly to the consumer's cache. As before, the write is completed, and the write delay is over, after the last write is acknowledged. The write to the synchronization variable also is forwarded from producer to consumer (consumer reads, rather than fetches), and the synchronization delay is shorter. After the synchronization delay is over, data are already in the consumers' caches, and the consumer reads them from its cache. Because the consumer reads data from the cache (no cache miss involved), the read delay is much shorter.

These two simple examples demonstrate the potentials. However, the issue is how much these potentials would materialize given a real application, for different implementations of producer-initiated mechanisms. One classification of DSM communication mechanisms is given in Table 8.4.

**TABLE 8.4. Classification of DSM Communications Mechanisms**[a,b]

| | Consumer-Initiated | | Producer-Initiated | | | | |
|---|---|---|---|---|---|---|---|
| | Invalidate | Prefetch | Data Fwd | Message Passing | Update | Lock | Sel Update |
| Transfer targets | — | — | Explicit | Explicit | Implicit | Implicit | Implicit |
| Transfer initiation | Implicit | Explicit | Explicit | Explicit | Implicit | Explicit | Explicit |
| Transfer granularity | Line | Line word | Line/ | Message word | Line/ | Line word | Line/ |
| Sync granularity | Block/ word | Block/ word | Block/ word | Block | Block/ word | Line | Block/ word |
| Write delay | — | = | > or < | > | > or = | < | > or < |
| Sync. delay | — | = | < or = | < | < | < | < or = |
| Read delay | — | < | ≪ | ≪ | ≪ | ≪ | ≪ |

*Source:* Byrd and Flynn (1999).

[a]*Key:* > increase, < decrease, ≪ greatly decrease, = no change.

[b]The last three lines show the expected effects relative to the invalidate mechanism.

**TABLE 8.5. Base System Parameters Used in the Simulation Study**[a]

| Processors | 64 |
|---|---|
| L1 cache size | 8 kB |
| L1 write buffer | 8 lines |
| L2 cache size | Unlimited |
| L2 outstanding accesses | 16 |
| L2 line size | 8 words (64 bytes) |
| L2 access | 3 cycles |
| Network latency | 4 cycles/hop |
| Network bandwidth | 4 bytes/cycle |
| Network topology | hypercube |
| Memory access | 24 cycles |
| Memory bandwidth | 1 word/cycle |

*Source:* Byrd and Flynn (1999).

[a]Cache and network parameters have been chosen to be consistent with the current-generation technology; the release consistency memory model was chosen (as before, L1 and L2 denote levels 1 and 2 cache memory).

Simulation was done for concrete values of major design parameters as specified in Table 8.5. The Byrd–Flynn simulation study has shown that producer-initiated mechanisms typically improve performance but consumer-initiated mechanisms are less expensive to implement.

### 8.1.10. Some of the Existing Systems

The best way to give a brief overview of existing systems is through a table with itemized characteristics. Software implementations are summarized using Table 8.6. Hardware implementations are summarized using Table 8.7. Hybrid software/hardware implementations are summarized using Table 8.8.

In software implementations, the majority of newer systems support OS (operating system) level and OC (optimizing compiler) level approaches. Most of the systems use the MRSW algorithms; however, new systems typically explore the MRMW algorithms. Sequential MCM is still represented a lot, in spite of its lower performance; newer approaches are typically oriented to more sophisticated MCMs. Granularity of the memory consistency unit tends to be larger (pages of 1 kB are rarely used). Invalidate consistency maintenance protocols seem to be more frequently used, compared to update protocols. Of course, the conclusions from tables of this sort will change, as time passes, and new research ideas find their way into research prototypes and/or industrial products.

In hardware implementations, the majority of systems support clusters with multiple processors (typically SMP). As for the type of ICN, the variety is large. The MRSW algorithm is more frequently used than the MRMW algorithm. Again, sequential consistency prevails. The granularity unit is at

**TABLE 8.6. A Summary of Software-Implemented DSM[a]**

| Reference | Type of Implementation | Type of Algorithm | Consistency Model | Granularity Unit | Coherence Policy |
|---|---|---|---|---|---|
| IVY [Li 1988] | User-level library + OS[b] modification | MRSW | Sequential | 1 kB | Invalidate |
| Mermaid [Zhou et al. 1990] | User-level library + OS modifications | MRSW | Sequential | 1 kB, 8 kB | Invalidate |
| Munin [Carter et al. 1991] | Run-time system + linker + library + preprocessor + OS modifications | Type-specific (SRSW, MRSW, MRMW) | Release | Variable-size objects | Type-specific (delayed update, invalidate) |
| Midway [Bershad et al. 1993] | Run-time system + compiler | MRMW | Entry, release, processor | 4 kB | Update |
| TreadMarks [Keleher et al. 1994] | User-level | MRMW | Lazy release | 4 kB | Update, invalidate |
| Blizzard [Schoinas et al. 1994] | User-level + OS kernel modification | MRSW | Sequential | 32–128 bytes | Invalidate |
| Mirage [Fleisch and Popek 1989] | OS kernel | MRSW | Sequential | 512 bytes | Invalidate |
| Clouds [Ramachandran and Khalid 1991] | OS, out of kernel | MRSW | Inconsistent, sequential | 8 kB | Discard segment when unlocked |
| Linda [Ahuja et al. 1986] | Language | MRSW | Sequential | Variable (tuple size) | Implementation-dependent |
| Orca [Bal and Tanenbaum 1988] | Language | MRSW | Synchronization-dependent | Shared data object size | Update |

*Source:* Protić (1996).

[a]Note that the bulk of software DSM research starts in the mid to late 1980s. This author especially likes the selective approach of Munin, which uses different approaches for different data types, in almost all major aspects of DSM, and introduces the release memory consistency model. TreadMarks introduces the lazy release memory consistency model. Midway introduces the entry memory consistency model.

[b]Operating system.

**TABLE 8.7. A Summary of Hardware-Implemented DSM**[a, b]

| Reference | Cluster Configuration | Type of Network | Type of Algorithm | Consistency Model | Granularity Unit (bytes) | Coherence Policy |
|---|---|---|---|---|---|---|
| Memnet [Delp et al. 1991] | Single processor, Memnet device | Token ring | MRSW | Sequential | 32 | Invalidate |
| Dash [Lenoski et al. 1992] | SGI 4D/340 (4 PEs, L2 caches), local memory | Mesh | MRSW | Release | 16 | Invalidate |
| SCI [James 1994] | Arbitrary | Arbitrary | MRSW | Sequential | 16 | Invalidate |
| KSR1 [Frank et al. 1993] | 64-bit custom PE, I + D caches, 32M local memory | Ring-based hierarchy | MRSW | Sequential | 128 | Invalidate |
| DDM [Hagersten et al. 1992] | 4 MC88110s, L2 caches, 8–32 local memory | Bus-based hierarchy | MRSW | Sequential | 16 | Invalidate |
| Merlin [Maples and Wittie 1990] | 40-MIPS computer | Mesh | MRMW | Processor | 8 | Update |
| RMS [Lucci et al. 1995] | 1–4 processors, caches, 256M local memory | RM bus | MRMW | Processor | 4 | Update |

*Source:* Protić (1996).

[a]*Key:* RM—reflective memory; RMS—reflective memory system; SCI—scalable coherent interface; DDM—data diffusion machine.
[b]Note that the bulk of hardware DSM research starts around the year 1990, except for some pioneering work like RMS (reflective memory system) at Gould and Encore; Stanford Dash introduces the first hardware implementation of the release memory consistency model.

**TABLE 8.8. A Summary of Hybrid Hardware / Software-Implemented DSM[a, b]**

| Name and Reference | Cluster Configuration + Network | Type of Algorithm | Consistency Model | Granularity Unit | Coherence Policy |
|---|---|---|---|---|---|
| PLUS [Bisani and Ravishankar 1990] | M88000, 32K cache, 8–32M local memory, mesh | MRMW | Processor | 4 kB | Update |
| Galactica Net [Wilson et al. 1994] | 4 M88110s, L2 caches 256M local memory, mesh | MRMW | Multiple | 8 kbytes | Update / invalidate |
| Alewife [Chaiken et al. 1994] | Sparcle PE, 64K cache, 4M local memory, CMMU, mesh | MRSW | Sequential | 16 bytes | Invalidate |
| FLASH [Kuskin et al. 1994] | MIPS T5, I + D caches, MAGIC controller, mesh | MRSW | Release | 128 bytes | Invalidate |
| Typhoon [Reinhardt et al. 1994] | SuperSPARC, L2 caches, NP controller | MRSW | Custom | 32 bytes | Invalidate custom |
| Hybrid DSM [Chandra et al. 1993] | FLASH-like | MRSW | Release | Variable | Invalidate |
| SHRIMP [Iftode et al. 1996a] | 16 Pentium PC nodes, Intel Paragon routing network | MRMW | AURC, SCOPE | 4 kB | Update / invalidate |

*Source:* Protić (1996).

[a]*Key:* CMMU—cache memory management unit; NP—network protocol; K—thousand; M—million.

[b]Note that the bulk of hybrid DSM research starts in the early to mid-1990s. The Princeton SHRIMP-2 project introduces the AURC and the SCOPE memory consistency models.

most 128 bytes. The consistency maintenance protocols are usually of the invalidate type. Again, the conclusions from tables of this sort will change in time.

In hybrid implementations, the majority of systems is based on SMP multiprocessor clusters, using off-the-shelf microprocessors. Since these machines are of a newer date, the MRMW and MRSW algorithms are used about equally often. For the same reason, a variety of MCMs are used. The granularity units vary from as low as 16 bytes to as high as 8 kB. Update and invalidate consistency protocols are being used about equally often.

These three tables assume the classification of DSM systems according to the type of implementation (hardware, software, or hybrid), as the major classification criterion. Other classifications are also possible. One widely used classification is based on the access pattern characteristics, as the major classification criterion. In that classification, the major classes are (1) uniform memory access (UMA), (2) nonuniform memory access (NUMA), and (3) cache-only memory access (COMA). Each class can be further subdivided into a number of subclasses: (1) flat COMA (F-COMA) versus hierarchical COMA (H-COMA), (2) cache-coherent NUMA (CC-NUMA) versus non-cache-coherent NUMA (NCC-NUMA), and so on.

### 8.1.11. Case Studies

Two different case studies are now presented. One is related to a UMA approach—the reflective memory system (RMS)—which originated at Gould Corporation and was later reincarnated in various forms in industry and academia. The other is related to a CC-NUMA approach—the DASH that originated at Stanford University (in the group of professor John Hennessy) and was reincarnated in various forms later on, mostly in industry. These two case studies cover two promising hardware DSM approaches (since this book is oriented toward future on-chip implementation of DSM).

#### *8.1.11.1. Case Study: RMS*

The RMS approach was patented by Gould Electronics in 1985. Later on, the rights on RMS passed to Encore, DEC, and Compaq. In essence, this is a writethrough update approach, and the first RMS multiprocessors were oriented to real-time markets. At Encore, the RMS approach was upgraded to support block updates, which resulted in the reflective memory/memory channel (RM/MC) approach. In the literature, the acronym RMS sometimes refers to the initial RMS, and sometimes to RM/MC. The description to follow is related to RM/MC and is about the Encore system introduced in 1990 [Jovanovic and Milutinović 1999, Jovanovic 2000].

The system consists of up to eight processing nodes (Fig. 8.8) connected by means of a multiplexed synchronous 64-bit wide RM/MC bus that supports both single word and block update transfers. In order to expand connection bandwidth, more parallel connected RM/MC buses can be used. The arbitration on the RM/MC bus is centralized and a round-robin synchronous

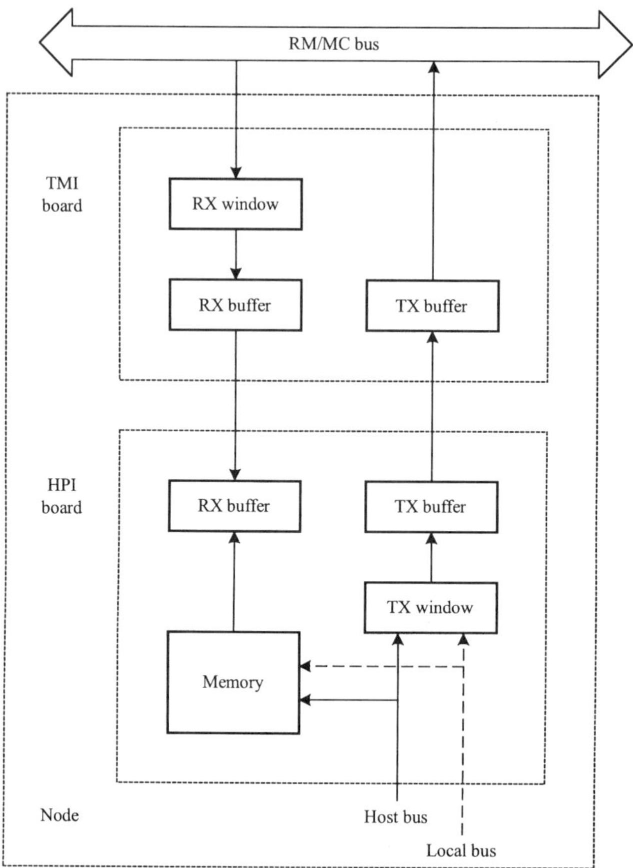


**Figure 8.8.** An RM/MC node (RX—receive; TX—transmit; TMI—transition module interface; HPI—host port interface).
*Comment:* Each shared memory write (determined by its 48-bit address that falls into an open transmit window) is written into the local memory, and after the address translation, it is propagated to the RM/MC bus. Other nodes snoop the bus. If the receive window of some particular node is open for a particular write, after the address translation, the local memory of that node is updated.

arbitration algorithm is used. Supported distances are 80 ft with copper cables, 240 ft with coaxial cables, and 15,000 ft with a fiberoptic cable.

The challenge was to use commodity components in open-system environments. To accomplish this goal, the node was divided into two parts: a transition module interface (TMI) board that implements the interface to the RM/MC bus, and a host port interface (HPI) board that provides the interface to the host system bus. There are realizations for the VME bus, the EISA bus, and the PCI bus.

Local memory pages can be configured as reflective (shared) or private (nonshared) by means of translation widows on transmit and receive sides.

These windows are used for determination of whether data from a particular page have to be transmitted to and/or received from the RM/MC bus, as well as for address mapping (copies of a shared page could be at different addresses on different nodes). Receive and transmit FIFO buffers are provided to make the transfers to/from the RM/MC bus going asynchronously from/to the host.

The RMS supports numerous real-time applications such as vehicle simulation, telemetry, instrumentation, and nuclear power plant simulation and control. The RM/MC system was initially designed to satisfy OLTP (online transaction processing) application needs. Besides that, projects employing this system have included mainframe-class data storage, a database distributed lock manager, and decision support applications.

The RM/MC system combines potentials for a high bandwidth with a low latency time. This system has some advantages because it is bus-based. A broadcast mechanism is easy for implementation. Update messages are small, consisting basically of virtual address and data (there are no special headers and trailers, etc.). Propagation time for update messages is small, because the interconnection medium spends only one bus cycle (in the case of RMS where the bus is nonmultiplexed) to propagate an update message to all nodes. The system tolerates node failures without service disruption.

The main disadvantage is the low system scalability. Also, the caching of RM is disabled, because the HPI board, as a slave on the system bus, cannot initiate update or invalidate messages on receiving an update from the RM/MC bus, in order to keep the cache memory consistent. Short (word) messages and long (block) messages share the same FIFO buffers, so the short messages always have to wait after the long ones.

Differences between various reflective memory systems (see list in Table 8.9) are summarized in Table 8.10 [Jovanović and Milutinović 1999], while their strengths and weaknesses are compared in Table 8.11.

### 8.1.11.2. Case Study: Stanford DASH

The major motivation to select Stanford DASH for presentation in this book is that it is the first fully operational hardware DSM multiprocessor of the CC-NUMA type [Hennessy et al. 1999], and among the closest to what may appear one day as DSM on a single chip.

The Stanford DASH architecture is given in Figure 8.9. Each node includes four processors organized in the form of the bus-based SMP using snoopy cache coherence. The prototype with 16 nodes included 64 processors in total. Directory structure was implemented using a bit vector representation described earlier in the book.

Access delays in DASH were as follows: (1) *local fill*—29 processor clocks, (2) *fill from home node*—101 processor clocks, and (3) *fill from remote node*—132 processor clocks. The length of the processor clock was 30 ns, which reflects the technology capabilities of the time of DASH implementation.

**TABLE 8.9. A List of Relevant Reflective Memory Systems[a]**

| System | Academy (A) Industrial (I) | University/ Company | Year Introduced | Interconnection Network | URL |
|---|---|---|---|---|---|
| RMS/Multimax | I | Encore | 1991 | Bus | `http://www.encore.com` |
| RMS for PC | A + 1 | Belgrade/Encore | 1993 | Bus | `http://galeb etf.bg.ac.yu/ ~ dsm` |
| RM/MC | I | Encore | 1993 | Bus | `http://www.encore.com` |
| RM/MC ++ | A + 1 | Belgrade/Encore | 1994 | Bus | `http://galeb.etf.bg.ac.yu/ ~ dsm` |
| LAM | I + A | Florida/Encore | 1996 | Bus | `http://www.encore.com` |
| MMM | I | Modcomp | 1993 | Bus | `http://www.modcomp.com` |
| RSM | A | Tokyo | 1994 | Bus | `http://www.sail.t.u-tokyo.ac.yp/ ~ oguchi` |
| MC | I | DEC | 1996 | Crossbar | `http://www.digital.com` |
| NSM | I | ATC | 1994 | Ring | `http://www.atcorp.com` |
| SCRAMNet | I | Systran | 1991 | Ring | `http://www.systran.com` |
| VMIC | I | VMIC | 1995 | Ring | `http://www.vmic.com` |
| Sesame | A | Stony Brook | 1991 | Ring | `ftp://ftp.cd.sunyb.edu/pub/techreport/wittie` |
| SHRIMP | A | Princeton | 1994 | Ring | `http://www.cs.princeton.edu/shrimp` |

[a]For detailed explanations, see WWW.

**TABLE 8.10. Various Reflective Memory Systems**[a, b]

| System | Processor Count | Sharing Granularity | RM Mapping | Update Granularity | MCM Protocol | Caching Included | Main Applications |
|---|---|---|---|---|---|---|---|
| Encore RMS | 8 | Page | Dynamic | Word | PC | No | Real-time |
| RMS for PC | 16, 32, more | Page | Dynamic | Word | PC | Yes | Real-time |
| Encore RM/MC | 8 | Page | Dynamic | Word, block | PC | No | OLTP |
| RM/MC ++ | 8 | Page | Dynamic | Word, block | PC | No | OLTP |
| Encore LAM | 8 | Segment | Dynamic | Word, block | EC | No | OLTP |
| Modcomp MMM | 8 | Page | Dynamic | Word | SC | No | Real-time |
| RSM | 8 | Segment | Static | Word | RC | No | Real-time |
| DEC MC | 16 | Page | Dynamic | Word, block | PC | Yes | Client–server |
| NSM | 60 | Segment | Static | Word | SC | No | S&E |
| SCRAMNet + | 256 | Segment | Static | Word | PC | No | Real-time |
| VMIC Network | 256 | Segment | Static | Word | PC | No | Real-time |
| Sesame | > 1000 | Page | Dynamic | Word | PC | No | S&E |
| SHRIMP | > 1000 | Page | Dynamic | Word, block | AURC | Yes | Client–server |

*Source:* Jovanovic (1999).

[a]*Key:* SC—sequential consistency; PC—processor consistency; RC—release consistency; EC—entry consistency; AURC—automatic update release consistency; S&E—scientific and engineering.

[b]A traditional drawback of reflective memory systems is scalability. However, with appropriate system modifications, the scalability problem can be overcome, and processor count can go above 1000, as pointed out in this table.

**TABLE 8.11. Strengths and Weaknesses of the Reflective Memory Systems Presented**[a,b]

| System | Advantages | Disadvantages | Possible Improvement Avenues |
|---|---|---|---|
| Encore RMS | Simple compared to others | Low scalability, high cable complexity, limited number of nodes | Caching, data filtering, better control of FIFO overflows |
| RAMS for PC | Better scalability, data filtering | High cable complexity | More relaxed MCMs lazy propagation of updates |
| Encore RM/MC | Widely applicable | Low scalability, high cable complexity, limited number of nodes | Caching, hierarchy of RM buses, prioritization |
| RM/MC ++ | Widely applicable, prioritization, better control of word and block streams | Low scalability, compiler has to be modified | Caching, hierarchy of RM buses |
| Encore LAM | Widely applicable, entry consistency, shared data are organized as segments | One memory pool | Caching, multiple memory pools, prioritization, hierarchy of RM buses |
| Modcomp MMM | Hard and soft real-time applications | Poor scalability, dual function of the VME bus | Bus hierarchy, system bus |
| RSM | Release MCM, lazy updates | RM concept in software, only word updates | Dynamic RM mapping, block update granularity |
| DEC MC | Extremely high bandwidth, caching, update acknowledgments, remote read primitive, sender could bypass local memory | Homogeneous nodes, low number of nodes | Heterogeneous computing, relaxed MCM, inclusion of some kind of hierarchy, segment sharing granularity |
| NSM | Hardware support for synchronization | No overlapping computation with communication, each node keeps a copy of the entire RM space | Relaxed MCM model, dynamic RM mapping |
| SCRAMNet + | Data filtering, modular media interface, heterogeneous computing, merging word updates | RM is limited to 8 MB, each node keeps a copy of the entire RM space | Dynamic RM mapping, block update granularity |
| VMIC Network | Onboard RM is SRAM, heterogeneous computing | Each node keeps a copy of the entire RM space; use of interrupts for data synchronization | Dynamic RM mapping, block update granularity, data filtering |
| Sesame | High scalability, heterogeneous computing, merging of word updates, distributed synchronization, dynamic sharing | No real block updates | Block update granularity |

**TABLE 8.11.** (*Continued*)

| System | Advantages | Disadvantages | Possible Improvement Avenues |
|---|---|---|---|
| SHRIMP | High scalability, caching, sophisticated relaxed MCMs, deliberate updates, word and block updates, merging of word updates | No broadcast mechanism | Hardware broadcast mechanism |

*Source:* Jovanovic (1999).

[a]*Key:* MCM—memory consistency model.

[b]See Jovanovic and Milutinović (1999) for a discussion of possible improvements of the RM approach.

The DASH prototype was used to experiment on three different latency hiding techniques: (1) memory consistency modules, (2) prefetching, and (3) remote access cache (RAC).

DASH supports the relaxed memory consistency models. See details in the section on memory consistency models earlier in this book. DASH supports software prefetch. It allows the processor to specify the address of a data item to be fetched before it is actually needed.

DASH introduced the remote access cache approach, to allow remote accesses to be combined and buffered within the individual nodes. It stores remote data that were accessed recently. If a remote data item is requested, and is included in RAC, it comes from RAC. RAC is useful when two different processors of the same four-processor node use the same data, or when data are not "competitive" enough to be captured in the regular cache, but are "competitive" enough to be captured in RAC.

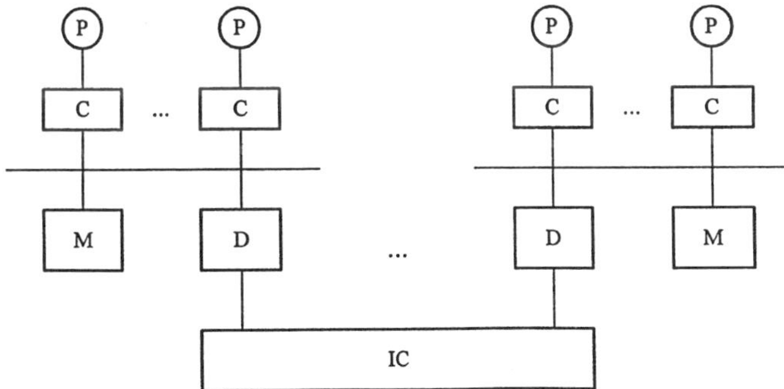

**Figure 8.9.** The Stanford DASH architecture (P—processor; C—cache; D—directory; M—memory; IC—interconnection network). [*Source:* Hennessy et al. (1999).] *Comment:* Remember that the chosen directory structure provides limited scalability.

A reincarnation of the concept appeared in SGI Origin, except that each node includes two processors, and many improvements were made, to reflect the lessons learned from the DASH research and prototype implementation. For lessons learned from the DASH project and for new challenges inspired by the DASH project, see Hennessy et al. (1999).

Multiprocessors such as HP Exemplar, Sequent NUMA-Q, DG NUMA-Line, and HAL S-1 can be treated conditionally as reincarnations of DASH, with less or more departures from the baseline. Finally, the FLASH project of Stanford University can be treated as an effort to move important features from the hardware layer (as in DASH) to the software layer (to achieve a more efficient hardware/software codesign, for a better price/performance ratio).

**TABLE 8.12.  A List of Modern Post-DASH Systems**[a]

| System | Academic/ Industrial | University or Company | Year Introduced | Interconnection Network | URL |
|---|---|---|---|---|---|
| SGI Origin | I[b] | Silicon Graphics | 1996 | Cray Link Spider | http://www.sgi.com |
| HP Exemplar | I | Hewlett-Packard | 1996 | SCI-based dual rings | http www.hp.com |
| Sequent NUMA-Q | I | Sequent | 1997 | SCI-based single ring | http://www.seauent.com |
| DG NUMA-Line | I | Data General | 1997 | SCI-based dual ring | http://www.dg.com / numaline |

[a]For detailed explanations, see WWW.

[b]Industrial.

**TABLE 8.13.  Differences Between Modern DSM Systems**[a]

| System | Processor Count | Sharing Granularity | Mapping | Update Granularity | MCM Protocol | Caching Included | Main Application |
|---|---|---|---|---|---|---|---|
| SGI Origin | 1024 $(512 \times 2)$ | Page | Dynamic/ static | Block | SC/RC[b] | Yes | Scientific/ commercial |
| HP Exemplar | 512 $(8 \times 4$ torus) | Page | Dynamic/ static | Block | SR/RC | Yes | Scientific/ commercial |
| Sequent NUMA-Q | 252 $(63 \times 4)$ | Page | Dynamic/ static | Block | SC/RC | Yes | Commercial |
| DG NUMA-Line | Logically $\leq 1024$ | Page | Dynamic/ static | Block | SC/RC | Yes | Commercial |

[a]For detailed explanations, see WWW.

[b]SC—sequential consistency; RC—release consistency.

Table 8.12 lists a number of modern post-DASH DASH-like DSM systems. Table 8.13 compares the systems presented in Table 8.12. Table 8.14 discusses the strengths and weaknesses of the presented modern DSM post-DASH DASH-like systems.

### 8.1.12. Other Research

The field is already well established and new developments are not expected to be very exciting. Researchers work mostly on fine improvements in performance, and on implementations that are better suited for new technologies. The major crown of all these efforts will be the birth of a DSM system on a single VLSI chip, which will happen after the on-chip transistor count reaches one billion, or even before (if simpler processors are used within the SMP clusters).

Section 8.2 summarizes some of the recent developments and gives references to the original literature, so that interested readers can get a closer insight. The selection of examples to follow has been created to be in line with the major message of this book (DSM on a single VLSI chip, as an ultimate goal of microprocessor industry, according to this author).

The first commercial microprocessor with large on-chip DRAM [Geppert 1997] was announced in 1996 by Mitsubishi Electronics America in Sunnyvale, California (M32R/D). It is a 32-bit machine with a 16-MB DRAM on the same chip. This DRAM can be used to implement a part of distributed shared memory. Such a development is an important milestone on the way to complete DSM on a single chip, together with appropriate accelerators (like MMX, or similar).

**TABLE 8.14. Strengths and Weaknesses of Modern DSM Systems**[a]

| System | Advantages | Disadvantages | Possible Improvement Avenues |
|---|---|---|---|
| SGI Origin | High-speed nonlocal memory access, page migration | High-cost specialized processors | More sophisticated MCM[b] |
| HP Exemplar | Based on high-performance processors | High-cost specialization processors | More sophisticated MCM |
| Sequent NUMA-Q | Based on commodity processors | Software support | More sophisticated MCM |
| DG NUMA-Line | Based on commodity processors | Software support | More sophisticated MCM |

[a]For detailed explanations see WWW.

[b]Memory consistency model.

## 8.2. ADVANCED ISSUES

This section contains the author's selection of research activities that, in his opinion, have made an important contribution to the field in recent time and are compatible with the overall mission of this book.

Two papers [Nowatzyk et al. 1993, Saulsbury et al. 1996] describe efforts at Sun Microsystems to come up with a DSM architecture that represents a good candidate for future porting to a DSM-on-a-single-chip environment. A major conclusion of their study is that a siliconless motherboard, as a first next step toward the final goal, is achievable once the feature size drops down to 0.25 $\mu$m. Major highlights are summarized in Table 8.15.

Lovett and Clapp (1996) describe an effort at Sequent Corporation to come up with a commercially successful CC-NUMA machine based on the SCI standard chips. The major conclusion of their study is that once Intel

**TABLE 8.15. S3.mp and Beyond**

**Origin and Environment**
Nowatzyk et al. (1993)
Sun Microsystems

**Major Highlights**
Going toward the DSM-based workstation (S3.mp)
Going toward the siliconless motherboard (LEGO)[a]
Using many less powerful CPUs, rather than a few brainiacs,
   since the performance is limited by the "memory wall"
Simulation studies oriented to 0.25-$\mu$m 256-Mbit DRAM

[a]The LEGO project from Sun Microsystems can be treated as the earliest effort toward a DSM on a chip, with a number of on-chip accelerators.

**TABLE 8.16. The Sequent STiNG[a]**

**Origin and Environment**
Lovett and Clapp (1966)
Sequent Computer Systems, Beaverton, Oregon, USA
A CC-NUMA for the commercial market (1996)

**Major Highlights**
Combines 4 quads using SCI
Quad is based on Intel P6
Quad includes up to 4 GB of system memory, 2 PCI buses for I/O,
   and a Lynx board for SCI interface and systemwide cache consistency
Architecture similar to Encore Mongoose (1995)
Processor consistency MCM
Application: OLTP

[a]The STiNG product from Sequent can be treated as one of the first and the most successful implementations based on the chips that support the SCI standard.

becomes able to place four P6 machines on a single die (Quad Pentium Pro), it will be possible to have a much more efficient implementation of their STiNG architecture (small letter "i" comes from "iNTEL inside"). Major highlights are summarized in Table 8.16.

Savic et al. (1995) describe an effort at Encore Computer Systems to come up with a board that one can plug into a PC (personal computer), in order to

---

**TABLE 8.17. The IFACT RM / MC for Networks of PCs[a]**

**Origin and Environment**

    Milutinović, Tomašević, Savic, Jovanovic, Grujic, Protić,
    Aral, Gertner, Natale, Gupta, Tran, Grant
    Supported by Encore on a contract for DEC

**Major Highlights**

    Basic research and design in 1993
    Board implementation and testing in 1994
    Five different concept improvements for higher node counts
        Efficient integration of RM and MC
        Write filtering
        Transmit FIFO priorities
        Caching of reflective memory regions
        Duplication of critical resources

---

[a]The IFACT RM/MC for PC environments project at Encore can be treated as the first effort to implement a board that turns a PC into a DSM node (based on the reflective memory approach).

---

**TABLE 8.18. The DEC MC for NOWs[a, b]**

**Origin and Environment**
    Gillett (1996)
    A follow-up on the "Digital/Encore MC team" (1994/95)

**Major Highlights**
    A PCI version of the IFACT RM/MC board
    Digital UNIX cluster team: better advantage of MC
    Digital HPC team: optimized application interfaces (including PVM)
    Reason for adoption
        Performance potentials over 1000 times the conventional NOW
        No compromise in cost per added node
        Computer architecture for availability
        Error handling at no cost to the applications

---

[a]*Key:* NOW—network of workstations; HPC—high-performance computing; PVM—parallel virtual machine.

[b]The DEC memory channel product is treated as one of the most successful market-oriented products based on the reflective memory approach, done as a follow-up effort, after a contract with Encore.

enable it to become a node in DSM systems of the RMS type. The major conclusion of their study is that the board (implemented using FPGA VLSI chips and fully operational) can be ported into a single standard-cell VLSI chip, which means that the RMS approach might be the first one to fit within the single chip boundaries. Major highlights are summarized in Table 8.17.

Gillett (1996) describes an effort at DEC to come up with a support product for their client–server systems, using the principles of RMS [this can be treated as a follow-up effort after Savic et al. (1995), done in addition to a contract with Encore]. A major conclusion of their study is that RMS still represents a successful way to go, even though the concept has been around for such a long time (as long as appropriate innovations are incorporated, like those selected by DEC). Major highlights are summarized in Table 8.18.

Milutinović (1996) describes another effort at Encore Computer Systems to come up with further improvements of the RMS concept, for better exploitation in the I/O environment (in order for their Infinity SP "I/O pump" to continue to be, in their words, the "Fastest I/O pump on [the] Planet"). A major conclusion of the study is that the RMS can be further improved if it is combined with more sophisticated MCMs, and if an appropriate layer is added, which can be viewed as distributed shared I/O on top of distributed shared memory. Major highlights are summarized in Table 8.19.

Reinhardt et al. (1996) describe an effort at the University of Wisconsin to come up with an approach based on NOWs and additional commodity components. A major conclusion of their study is that the level of success of a new product depends on the level of utilization of commodity components in both hardware and software domains; however, appropriate architectural changes have to be done first (decoupling of functional hardware). Major highlights are summarized in Table 8.20.

TABLE 8.19.  **The IFACT RM / MC for Infinity SP**[a]

**Origin and Environment**

Milutinović, Protić, Milenković, Raskovic, Jovanovic, Denton, Aral
Supported by Encore, on contract for IBM

**Major Highlights**

Basic research in 1996
Goal: continuing to be the highest performance I/O processor on planet
Five different ideas introduced for higher performance
Separation of temporal and spatial data in DSM
Direct cache injection mechanisms in DSM
Distribution shared I/O on top of DSM
Moving to more sophisticated memory consistency models

[a]The major goal of the IFACT RM/MC for Infinity SP project at Encore is to make the reflective memory approach more competitive in the performance race with other approaches acquired by industry.

An important new trend in DSM research implies the building of commercially successful machines, as well as the usage of the IRAM approach to achieve energy-efficient architectures or replication/migration trade-offs to achieve performance-efficient architectures [Fromm et al. 1997, Laudon and Lenoski 1997, Soundararajan et al. 1998].

One of the companies most active lately in the DSM arena is Convex (the Examplar family). The initial SPP1000 was introduced in 1994; the SPP2000, in 1997. The latter uses a superscalar processor with out-of-order execution and nonblocking caches; in addition, it includes more nodes, richer interconnection topology, and a better optimized protocol for improved memory latency and lower bandwidth requirements.

Both SPP1000 and SPP2000 connect their nodes using multiple rings. An SPP1000 node has four pairs of processors connected by a crossbar. An SPP2000 node has eight processor pairs also connected by a crossbar. Each processor pair has an agent that connects it to a crossbar port. The SPP1000 uses the HP PA 7100 (a two-way superscalar), while the SPP2000 uses the HP PA 8000 (a four-way superscalar). Figure 8.10 shows the nodes, and Figure 8.11 shows the internode communication topologies of the two machines [Abandah 1998].

The most recent advances in DSM and related problems can be found in some of the papers of *IEEE Transactions on Computers*, Special Issue on DSM, February 1999. For example, Heinrich et al. (1999) give an excellent quantitative analysis of scalability of DSM cache coherence protocols, and Zhang et al. (1999) discuss the novel Excel-NUMA approach. The first paper [Heinrich et al. 1999] stresses the importance of the impact of cache coherence protocols on the overall performance of DSM machines and illustrates the point for the case of the Stanford FLASH project. The second paper [Zhang et al. 1999] stresses the importance of adequate data layout for

---

**TABLE 8.20. Typhoon$^a$ and Beyond**

**Origin and Environment**
Reinhardt, Pfile, Wood
University of Wisconsin
Hardware-supported software DSM in NOW

**Major Highlights**
Decoupling the functional hardware, for higher off-the-shelf utilization
Typhoon-0: an off-the-shelf protocol processor + network interface
Typhoon-1: higher level of hardware integration
Basic DSM functions to decouple:
    Messaging and networking (doing internode communications)
    Access control (detecting memory references for nonlocal action)
    Protocol processing (maintaining the global consistency)
    Commodity components; now also for parallel processing (FPGAs + $\mu$Ps)

---

$^a$Typhoon is a university project oriented to DSM based on off-the-shelf components in both hardware and software domains.

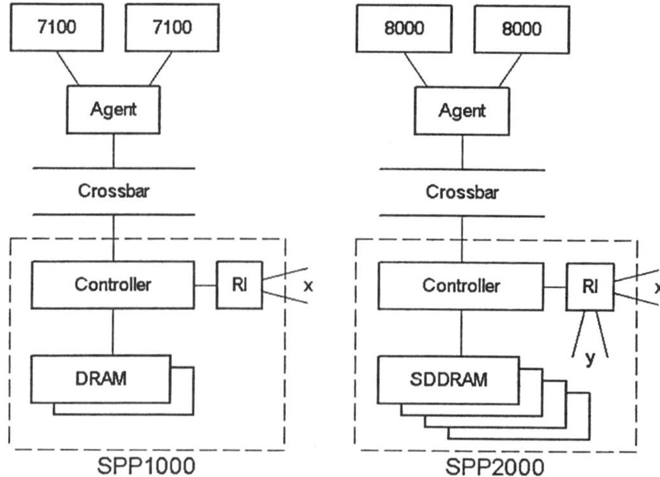



**Figure 8.10.** Processing nodes of the SPP machines (RI—ring interface).

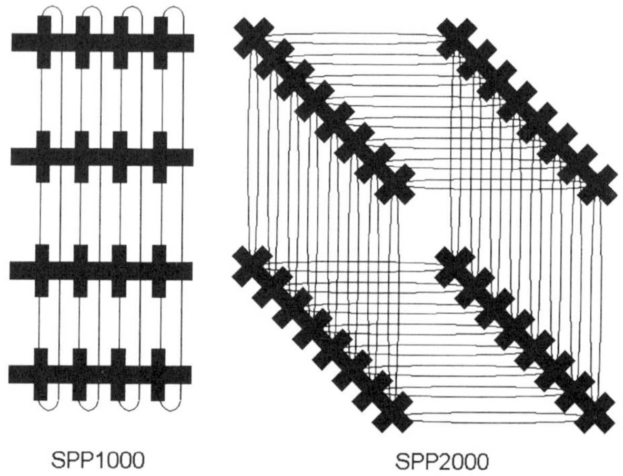



**Figure 8.11.** Interconnection topologies of the SPP machines. In SPP1000, four 8-processor nodes are connected in one dimension, using 4 rings. In SPP2000, four 16-processor nodes are connected in two dimensions, using 32 rings.

dynamic locality and proposes that after a memory line is written and cached, the storage that kept that line in memory remain unutilized, so that it can be used to hold remote data displaced from local caches (the approach referred to as the *Excel-NUMA*).

Other papers of special importance are Luk and Mowry (1999), Lai and Falsafi (1999), Bilir et al. (1999), and Jiang and Singh (1999). The first paper [Luk and Mowry 1999] elaborates the concept of data forwarding or remote write, where the goal is to forward data closer to the user, before the data are actually needed, which minimizes the latencies related to data fetching. The second paper [Lai and Falsafi 1999] extends the prediction concept into the message prediction domain; if one can predict concrete messages used to maintain coherence in DSM, the performance of DSM systems can be maximized, because much of the remote access latency can be efficiently hiden. The third paper [Bilir et al. 1999] introduces a hybrid approach that combines broadcast snooping and directory protocols; it includes the so-called multicast mask that minimizes the amount of unnecessary bandwidth utilization, and consequently increases the performance. The fourth paper [Jiang and Singh 1999] examines the scalability of hardware DSM and concludes that application restructuring can help considerably in achieving better scalability.

An interesting project for the reader is to estimate the needed transistor count for all DSM systems mentioned so far, having in mind both the transistor count of the actual components used, as well as the transistor count if only the needed resources are used. Such an exercise can bring a better understanding of the future referred to as "DSM on a VLSI chip."

Another type of exercise for students, used in the author's classes, for the above-mentioned systems and for new ones to come, is to prepare one-academic-hour lectures, to explain the details to their colleagues, using the presentation strategy explained in Milutinović and Petkovic (1995) and Milutinović (1996b): (1) problem being attacked by the chosen research project; (2) essence of the existing solutions, and what is to be criticized with them, from the point of view defined in the problem statement; (3) essence of the proposed approach, and why it is expected to be better under the conditions of interest, which are defined as a part of the problem statement; (4) details that deserve attention; (5) performance evaluation results; (6) complexity evaluation results; and (7) final conclusion.

The author and his associates were active in the DSM architecture research, and especially in the field of reflective memory. For more details, see Grujic et al. (1996), Jovanovic and Milutinović (1999), Jovanovic (2000), Milenković and Milutinović (1996a, 1996b), Milutinović (1992, 1995, 1996), Protić (1996, 1998), Protić and Milutinović (1996), Protić et al. (1997), and Savic et al. (1995).

## PROBLEMS

**8.1.** Compare the DSM systems of the 1970s, 1980s, and 1990s. Which issues stayed essentially the same? What changes have been driven by advances in technology and applications?

**8.2.** Compare the modern DSM systems as far as the type of interconnection network used on various levels of the system. Which type prevails?

**8.3.** Discuss pros and cons of various granularity levels for maintenance of memory consistency in DSM systems. Do that from both software and hardware implementation points of view.

**8.4.** The MRMW algorithm is potentially the most efficient one; however, write conflicts are possible, if multiple writers have to write to the same location. Check how this problem is solved in the DSM systems listed in Tables 8.6–8.8, and propose improved solutions.

**8.5.** Using a tool such as Limes (see Appendix B at the end of this book), compare two alternative-COMA approaches of the same complexity.

**8.6.** Using a tool such as Limes (see Appendix B), compare two alternative-RMS approaches of the same complexity.

**8.7.** Using a tool such as Limes (see Appendix B), compare two alternative CC-NUMA approaches of the same complexity.

**8.8.** Compare hardware and software implementations of the release and the lazy release memory consistency models, from the implementation point of view. Show a block scheme of one hardware implementation and a flowchart of one software implementation.

**8.9.** Develop details of AURC and SCOPE, and show differences on the example of one short program of your choice. Discuss the complexity of specific mechanisms of SCOPE, which make it superior to AURC.

**8.10.** Develop relevant details for one consumer-initiated scheme and one producer-initiated scheme, and compare the time delays: write delay, sync delay, and read delay. What is the ratio of write, sync, and read delays?

# REFERENCES

The list to follow includes the references used in this book plus some references of the author and his associates that have influenced the author's thought, directly or indirectly. For information on ongoing research in the field, the reader is referred to a WWW presentation that includes a number of interesting and useful pieces of information related to the general field of computer architecture; see `http://www.cs.wisc.edu/~arch/www/#Participation`.

[Abandah98] Abandah, G., Davidson, E., "Effects of Architectural and Technological Advances on the HP/Convex Exemplar's Memory and Communications Performance," *Proc. ISCA-98*, Barcelona, Catalonia, Spain, June 27–July 1, 1998, pp. 318–329.

[Adve96] Adve, S. V., Cox, A. L., Dwarkadas, S., Rajamony, R., Zwaenopoel, W., "A Comparison of Entry Consistency and Lazy Release Consistency Implementations," *Proc. IEEE HPCA-96*, San Jose, CA, Feb. 1996, pp. 26–37. (See `http://www-ece.rice.edu/~sarita/publications.html` for related work.)

[Agarwal90] Agarwal, A., Lim, B.-H., Kranz, D., Kubiatowicz, J., "April: A Processor Architecture for Multiprocessing, *Proc. of the ISCA-90*, Seattle, WA, May 1990, pp. 104–114.

[Agarwal91] Agarwal, A., Lim, B.-H., Kranz, D., Kubiatowicz, J., "LimitLESS Directories: A Scalable Cache Coherence Scheme," *Proc. ASPLOS-91*, Apr. 1991, pp. 224–234.

[Ahuja86] Ahuja, S., Carriero, N., Gelernter, D., "Linda and Friends," *IEEE Comput.* **19**(8), 26–34 (May 1986).

[Alvarez97] Alvarez, G. A., Burkhard, W. A., Cristian, F., "Tolerating Multiple Failures in RAID Architectures with Optimal Storage and Uniform Declustering," *Proc. ISCA-24*, Denver, CO, June 1997, pp. 62–72.

[Alvarez98] Alvarez, G. A., Burkhard, W. A., Stockmeyer, L. J., Cristian, F., "Declustered Disk Array Architectures with Optimal and Near-Optimal Parallelism," *Proc. ISCA-98*, Barcelona, Catalonia, Spain, June 27–July 1, 1998, pp. 109–120.

[AMD97] *AMD K6 MMX Processor Data Sheet* (`http://www.amd.com/K6/k6docs/k6ds/index.html`), Advanced Micro Devices, Menlo Park, CA, 1997.

[August98]   August, D. I., Connors, D. A., Mahlke, S. A., Sias, J. W., Crozier, K. M., Cheng, B.-C., Eaton, P. R., Olaniran, Q. B., Hwu, W.-M., "Integrated Predicated and Speculative Execution in the IMPACT EPIC Architecture," *Proc. ISCA-98*, Barcelona, Catalonia, Spain, June 27–July 1, 1998, pp. 227–237.

[Bal88]   Bal, H. E., Tanenbaum, A. S., "Distributed Programming with Shared Data," *Proc Int. Conf. Computer Languages 88*, Oct. 1988, pp. 82–91.

[Becker98]   Becker, J., Hartenstein, R., Herz, M., Nageldinger, U., "Parallelization and Co-Compilation for Configurable Accelereators: A Host/Accelerator Partitioning Compilation Method," *Proc. ASP-DAC 98* (Asian and South Pacific Design Automation Conf. 98), Yokohama, Japan, Feb. 10–13, 1998, pp. 23–33.

[Bekerman99]   Bekerman, M., et al., "Correlated Load-Address Predictors," *Proc. ISCA-99*, Atlanta, May 1999, pp. 54–63.

[Bell99]   Bell, G., van Ingen, K., "DSM Perspective: Another Point of View," *Proc. IEEE* **87**(3), 412–417 (Mar. 1999).

[Bershad93]   Bershad, B. N., Zekauskas M. J., Sawdon, W. A., "The Midway Distributed Shared Memory System," *Proc. COMPCON 93*, Feb. 1993, pp. 528–537.

[Bierman77]   Bierman, G. J., *Factorization Methods for Discrete Sequential Estimation*, Academic Press, New York, 1977.

[Bilir99]   Bilir, E., et al., "Multicast Snooping: A New Coherence Method Using a Multicast Address Network," *Proc. ISCA-99*, Atlanta, May 1999, pp. 294–304.

[Bisani90]   Bisani, R., Ravishankar M., "PLUS: A Distributed Shared-Memory System," *Proc. 17th Annual Int. Symp. Computer Architecture*, **18**(2), 115–124 (May 1990).

[Black99]   Black, B., Rychlik, B., Shen, J., "The Block-Based Trace Cache," *Proc. ISCA-99*, May 1999, pp. 196–207.

[Blumrich94]   Blumrich, M. A., Li, K., Alpert, R., Dubnicki, C., Felten, E. W., Sandberg, J., "The Virtual Memory Mapped Network Interface for the Shrimp Multicomputer," *Proc. 21st Annual Int. Symp. on Computer Architecture*, Chicago, Apr. 1994, pp. 142–153

[Blumrich98]   Blumrich, M., Alpert, R., Chen, Y., Clark, D., Damianakis, S., Dubnicki, C., Felten, E., Iftode, L., Li, K., Martonosi, M., Shillner, R., "Design Choices in the SHRIMP System: An Empirical Study," *Proc. ISCA-98*, Barcelona, Catalonia, Spain, June 27–July 1, 1998, pp. 330–341.

[Brauch98]   Brauch, J., Fleischman, J., "Design of Cache Test Hardware on the HP8500," *IEEE Design Test Comput.* **15**(3), 58–63 (July–Sept. 1998).

[Burger97]   Burger, D., Goodman, J. R., "Billion Transistor Architectures," *IEEE Comput.* **30**(9), 46–48 (Sept. 1997).

[Byrd99]   Byrd, G. T., Flynn, M. J., "Producer-Consumer Communication in Distributed Shared Memory Multiprocessors," *Proc. IEEE* **87**(3), 456–466 (Mar. 1999)

[Calder99]       Calder, B., Reinman, G., Tullsen, D., "Selective Value Prediction," *Proc. ISCA-99*, Atlanta, May 1999, pp. 64–74.

[Carter91]       Carter, J. B., Bennet, J. K., Zwaenepoel, W., "Implementation and Performance of Munin," *Proc. 13th ACM Symp. Operating Systems Principles*, Oct. 1991, pp. 152–164.

[Chaiken94]      Chaiken, D., Kubiatowicz, J., Agarwal, A., "Software-Extended Coherent Shared Memory: Performance and Cost," *Proc. 21st Annual Int. Symp. Computer Architecture*, Apr. 1994, pp. 314–324.

[Chan96]         Chan, K. K., Hay, C. C., Keller, J. R., Kurpanek, G. P., Schumacher, F. X., Zheng, J., "Design of the HP PA7200 CPU," *Hewlett-Packard J.* 1–12 (Feb. 1996).

[Chandra93]      Chandra, R., et al., *Performance Evaluation of Hybrid Hardware and Software Distributed Shared Memory Protocols*, Technical Report CSL-TR-93-597, Stanford Univ., Palo Alto, CA, Dec. 1993.

[Chang95]        Chang, P., Banerjee, U, "Profile-guided Multiheuristic Branch Prediction," *Proc. Int. Confer. Parallel Processing*, July 1995.

[Chang97]        Chang, P.-Y., Hao, E., Patt, Y., "Target Prediction for Indirect Jumps," *Proc. ISCA-24*, Denver, June 1997, pp. 274–283.

[Chappell99]     Chappell, R., et al., "Simultaneous Subordinate Microthreading (SSMT)," *Proc. ISCA-99*, Atlanta, May 1999, pp. 186–195.

[Cho99]          Cho, S., Yew, P., Lee, G., "Decoupling Local Variable Accesses in a Wide-Issue Superscalar Processor," *Proc. ISCA-99*, Atlanta, May 1999, pp. 100–110.

[Corsini87]      Corsini, P., Prete, C. A., "Architecture of the MuTeam System," *IEE Proc.* **134**, (Part E)(5), 217–227 (Sept. 1987).

[Davidovic97]    Davidovic, G., Ciric, J., Ristic-Djurovic, J., Milutinović, V., Flynn, M., "A Comparative Study of Adders: Wave Pipelining vs. Classical Design," *IEEE TCCA Newsl.* 64–71 (June 1997).

[DeJuan87]       DeJuan, E., Casals, O., Labarta, J., "Cache Memory with Hybrid Mapping," *Proc. IASTED* **87**, Grindelwald, Switzerland, 1987, pp. 27–30.

[DeJuan88]       DeJuan, E., Casals, O., Labarta, J., "Management Algorithms for an Hybrid Mapping Cache Memory," *Proc. MIMI-88*, Sant Feliu, Spain, June 1988, pp. 368–372.

[Delp91]         Delp, G., Farber, D., Minnich, R., "Memory as a Network Abstraction," *IEEE Network*, July 1991, pp. 34–41.

[Digital96a]     Digital Equipment Corp., *Alpha 21064A Microprocessors Data Sheet* (http://ftp.europe.digital.com/pub/Digital/info/semiconductor/literature/21064Ads.pdf), Digital Equipment Corporation, Maynard, MA, 1997.

[Digital96b]     Digital Equipment Corp., *Alpha 21064A Microprocessor Product Brief* (http://www.digital.com/info/semiconductor/a264up1/index.html), Digital Equipment Corporation, Maynard, MA, 1996.

[Digital97a]     Digital Equipment Corp., *Digital Semiconductor Alpha 21164 Microprocessor Data Sheet* (http://com/pub/Digital/

info/semiconductor/literature/21264pb.pdf), Digital Equipment Corporation, Maynard, MA, 1997.

[Digital97b] Digital Equipment Corp., *Digital Semiconductor Alpha 21264 Microprocessor Product Brief* (http://ftp.europe.digital.com/pub/Digital/info/semiconductor/literature/21264pb.pdf), Digital Equipment Corporation, Maynard, MA, 1997.

[Driesen98] Driesen, K., Holzle, U., "Accurate Indirect Branch Prediction," *Proc. ISCA-98*, Barcelona, Catalonia, Spain, June 27–July 1, 1998, pp. 167–178.

[Eickmeyer96] Eickmeyer, R. J., Johnson, R. E., Kunkel, S. R., Liu, S., Sqillante, M. S., "Evaluation of Multithreaded Uniprocessor for Commercial Application Environments," *Proc. ISCA-96*, Philadelphia, May 1996, pp. 203–212.

[Ekmecic95] Ekmecic, I., Tartalja, I., Milutinović, V., "A Taxonomy of Heterogeneous Computing," *IEEE Comput.* **28**(12), 68–70 (Dec. 1995).

[Ekmecic96] Ekmecic, I., Tartalja, I., Milutinović, V., "A Survey of Heterogeneous Computing: Concepts and Systems," *Proc. IEEE*, Aug. 1996, pp. 1127–1144.

[Espasa97] Espasa, R., Valero, M., "Multithreaded Vector Architectures," *Proc. HPCA-3*, San Antonio, TX, Feb. 1997, pp. 237–249.

[Evers96] Evers, M., Chang, P.-Y., Patt, Y., "Using Hybrid Branch Predictors to Improve Branch Prediction Accuracy in the Presence of Context Switches," *Proc. ISCA-96*, Philadelphia, May 1996, pp. 3–11.

[Evers98] Evers, M., Patel, S., Chappell, R., Patt, Y., "An Analysis of Correlation and Predictability: What Makes Two-Level Branch Predictors Work," *Proc. ISCA-98*, Barcelona, Catalonia, Spain, June 27–July 1, 1998, pp. 52–61.

[Fetherston98] Fetherston, R. S., Shaik, I. P., Ma, S. C., "Testability Features of the AMD-K6 Microprocessor," *IEEE Design Test Comput.* **15**(3), 64–69 (July–Sept. 1998).

[Fleisch89] Fleisch, B., Popek, G., "Mirage: A Coherent Distributed Shared Memory Design," *Proc. 14th ACM Symp. on Operating System Principles*, ACM, New York, 1989, pp. 211–223.

[Flynn95] Flynn, M. J., *Computer Architecture: Pipelined and Parallel Processor Design*, Jones and Bartlett, Boston, 1995.

[Forman94] Forman, G. H., Zahorjan, J. "The Challenges of Mobile Computing," *IEEE Comput.* **27**(4), 38–47 (Apr. 1994).

[Fortes86] Fortes, J., Milutinović, V., Dock, R., Helbig, W., Moyers, W., "A High-Level Systolic Architecture for GaAs," *Proc HICSS-86*, Honolulu, Jan. 1986, pp. 253–258.

[Frank93] Frank, S., Burkhardt III, Rothnie, J., "The KSR1: Bridging the Gap Between Shared Memory and MPPs," *COMPCON 93*, Feb. 1993, pp. 285–294.

[Fromm97]       Fromm, R., Perissakis, S., Cardwell, N., Kozyrakis, C., McGaughy, B., Patterson, D., Anderson, T., Yelick, K., "The Energy Efficiency of IRAM Architectures," *Proc. ISCA-24*, Denver, June 1997, pp. 327–337.

[Gabbay98]      Gabbay, F., Mendelson, A., "The Effect of Instruction Fetch Bandwidth on Value Prediction," *Proc. ISCA-25*, Barcelona, Catalonia, Spain, June 27–July 1, 1998 pp. 272–281.

[Geppert97]     Geppert, L., "Technology 1997 Analysis and Forecast: Solid State," *IEEE Spectrum*, **34**(1), 55–59 (Jan. 1997).

[Gillett96]     Gillett, R. B., "Memory Network for PCI," *IEEE MiCRO*, 12–18 (Feb. 1996).

[Gloy96]        Gloy, N., Young, C., Chen, J. B., Smith, M., "An Analysis of Branch Prediction Schemes on System Workloads," *Proc. ISCA-96*, Philadelphia, May 1996, pp. 12–21.

[Gonzalez95]    Gonzalez, A., Aliagas, C., Valero, M., "A Data Cache with Multiple Caching Strategies Tuned to Different Types of Locality," *Proc. Int. Conf. Supercomputing* (*ICS 95*), Barcelona, Spain, July 1995, pp. 338–347.

[Gonzalez96]    Gonzalez, J., Gonzalez, A., "Identifying Contributing Factors to ILP," *Proc. 22nd Euromicro Conf.*, Prague, Czech Republic, Sept. 1996, pp. 45–50.

[Gonzalez97]    Gonzalez, J., Gonzalez, A., "Speculative Execution via Address Prediction and Data Prefetching," *Proc. 11th ACM 1997 Int. Conf. Supercomputing*, Vienna, Austria, July 1997, pp. 196–203.

[Gould81]       Gould, Inc., *Reflective Memory System*, Gould, Inc., Fort Lauderdale, FL, Dec. 1981.

[Grujic96]      Grujic, A., Tomašević, M., Milutinović V., "A Simulational Study of Hardware-Oriented DSM Approaches," *IEEE Parallel Distrib. Technol.* **4**(1), 74–83.

[Gwennap97]     Gwennap, L., Digital 21264 Sets New Standard (http://www.chipanalyst.com/report/articles/21264/21264.html), A Micro Design Resources, Sebastopol, USA, 1997a.

[Gwennap97]     Gwennap, L., "Intel and HP Make EPIC Disclosure," *Microproc Rep.* **11**(14), 1–9 (Oct. 1997b).

[Hagersten92]   Hagersten, E., Landin, A., Haridi, S., "DDM—A Cache-Only Memory Architecture," *IEEE Comput.* **25**(9), 44–54 (Sept. 1992).

[Hagersten99]   Hagersten, E., Papadopulos, G., "Parallel Computing in the Commercial Marketplace: Research and Innovations at Work," *Proc. IEEE* **87**(3), 405–411 (Mar. 1999).

[Hammond97]     Hammond, L., Nayfeh, B., Olokotun, K., "Single-Chip Multiprocessor," *IEEE Comput.* **30**(9), 79–85 (Sept. 1997).

[Hank95]        Hank, R. E., Hwu, W.-M. W., Rau, B. R., "Region-Based Compilation: An Introduction and Motivation," *Proc. MICRO-28*, Ann Arbor, MI, Nov.–Dec. 1995, pp. 158–168.

[Hartenstien97] Hartenstein, R., "The Microprocessor Is No More General Purpose: Why Future Reconfigurable Platforms Will Win," *Proc.*

*ISIS'97* (Int. Conf. Innovative Systems in Silicon 97), Austin, TX, Oct. 8–10, 1997, pp. 2–12.

[Heinrich99]    Heinrich, M., Sounderarajan, V., Hennessy, J., Gupta, A., "A Quantitative Analysis of Performance and Scalability of Distributed Shared Memory Cache Coherence Protocols," *IEEE Trans. Comput.* **48**(2), 205–217 (Feb. 1999).

[Helbig89]    Helbig, W., Milutinović, V., "The RCA's DCFL E/D MESFET GaAs 32-bit Experimental RISC Machine," *IEEE Trans. on Comput.* **36**(2), 263–274 (Feb. 1989).

[Helstrom68]    Helstrom, G., *Statistical Theory of Signal Detection*, Pergamon Press, Oxford, 1968.

[Hennessy96]    Hennessy, J. L., Patterson, D. A., *Computer Architecture: A Quantitative Approach*, Morgan Kaufmann, San Francisco, 1996.

[Hennessy99]    Hennessy, J. L., Heinrich, M., Gupta, A., "CC-DSM: Perspectives on Its Development and Future Challenges," *Proc. IEEE* **87**(3), 418–429 (Mar. 1999).

[Hill88]    Hill, M., "A Case for Direct-Mapped Caches," *IEEE Comput.* **21**(12), 25–40 (Dec. 1988).

[Holland92]    Holland, M., Gibson, G., "Parity Declustering for Continuous Operations on Redundant Disk Arrays," *Proc. ASPLOS-5*, Boston, Oct. 1992, pp. 23–35.

[Hu96]    Hu, Y., Yang, Q., "DCD—Disk Caching Disk: A New Approach for Boosting I/O Performance," *Proc. ISCA-96*, Philadelphia, May 1996, pp. 169–178.

[Hunt97]    Hunt, D., *Advanced Performance Features of the 64-bit PA-8000* (`http:// hpcc920.external.hp.com / computing / framed / technology / micropro / pa- 8000 / docs /advperf.html`), Hewlett-Packard Company, Fort Collins, CO, 1997.

[Hwu95]    Hwu, W. W., et al., "Compiler Technology for Future Microprocessors," *Proc. IEEE* **83**(12), 1625–1640 (Dec. 1995).

[Iannucci94]    Iannucci, R. A., Gao, G. R., Halstead, R. H. Jr., Smith, B., *Multithreaded Computer Architecture: A Summary of the State of the Art*, Kluwer, Boston, 1994.

[IBM93]    IBM, *PowerPC 601 RISC Microprocessor Technical Summary* (`http:// www.chips.ibm.com / products / ppc / DataSheets/PPC/tech_sum/601TSr1.pdf`), IBM Microelectronics Division, New York, 1993.

[IBM96a]    IBM, *PowerPC 604e RISC Microprocessor Family: PID9V-604e Hardware Specifications* (`http:// www.chips.ibm.com / products/ppc/DataSheets/PPC/hw_spec/604eHSr0.pdf`), IBM Microelectronics Division, New York, 1996.

[IBM96b]    IBM, *PowerPC 620 RISC Microprocessor Technical Summary* (`http:// www.chips.ibm.com / products / ppc / DataSheets/PPC/tech_sum/620TS.pdf`), IBM Microelectronics Division, New York, 1996.

[IEEE93]        *IEEE Standard for Scalable Coherent Interface*, IEEE, New York, 1993.

[Iftode96a]     Iftode, L., Dubnicki, C., Felten, E., Li, K., "Improving Release-Consistent Shared Virtual Memory Using Automatic Update," *Proc. HPCA-2*, San Jose, CA, Feb. 1996, pp. 14–25.

[Iftode96b]     Iftode, L., Singh, J. P., Li, K., "Scope Consistency: A Bridge Between RC and EC," *Proc. 8th Annual ACM Symp. Parallel Algorithms and Architectures*, Padua, Italy, June 1996, pp. 277–287.

[Iftode96c]     Iftode, L., Singh, J. P., Li, K., "Understanding Application Performance in Shared Virtual Memory Systems," *Proc. ISCA-96*, Philadelphia, May 1996, pp. 122–133.

[Intel93]       *Intel Pentium Processor Users Manual*, Intel, Santa Clara, CA, 1993.

[Intel96]       Intel Corporation. *Product Proceedings.* (`http://www.intel.com/procs/p6/p6white/index.html`), Intel, Santa Clara, CA, 1997.

[Intel97a]      *Intel Pentium II Processor Home* (`http://www.intel.com/english/PentiumII/zdn.htm`), Intel, Santa Clara, CA, 1997a.

[Intel97b]      *Intel Architecture Software Developer's Manual*, Vol. 2: *Instruction Set Reference* (`http://developer.intel.com/design/MMX/manuals/24319101.PDF`), Intel, Santa Clara, CA, 1997.

[Iseli95]       Iseli, C., Sanchez, E., "Spyder: A SURE (SUperscalar and REconfigurable) Processor," *J. Supercomput.* **9**, 231–252 (1995).

[Jacobson97]    Jacobson, Q., Bennett, S., Sharma, N., Smith, J. E., "Control Flow Speculation in Multiscalar Processors," *Proc. HPCA-3*, San Antonio, TX, Feb. 1997, pp. 218–229.

[James90]       James, D., Laundrie, A., Gjessing, S., Sohi, G., "Distributed-Directory Scheme: Scalable Coherent Interface," *IEEE Comput.* **23**(6), 74–77 (June 1990).

[James94]       James, D. V, "The Scalable Coherent Interface: Scaling to High-Performance Systems," *COMPCON 94*: *Digest of Papers*, Mar. 1994, pp. 64–71.

[Jiang99]       Jiang, D., Singh, J., "Scaling Application Performance on a Cache-Coherent Multiprocessor," *Proc. ISCA-99*, Atlanta, May 1999, pp. 305–316.

[Johnson91]     Johnson, M., *Superscalar Microprocessor Design*, Prentice-Hall, Englewood Cliffs, NJ, 1991.

[Johnson97a]    Johnson, T. L., Hwu, W. W., "Run-Time Adaptive Cache Hierarchy Management via Reference Analysis," *Proc. of 24th Int. Symp. Computer Architecture*, Denver, June 1997.

[Johnson97b]    Johnson, T. L., Merten, M. C., and Hwu, W. W., "Run-Time Spatial Locality Detection and Optimization," *Proc. Micro-30*, Research Triangle Park, NC, Dec. 1997.

[Jouppi90]      Jouppi, N., "Improving Direct-Mapped Cache Performance by the Addition of a Small Fully-Associative Cache and Prefetch Buffers," *Proc. ISCA-90*, May 1990, pp. 364–373.

[Jourdan96]    Jourdan, S., Sainrat, P., Litaize, D., "Exploring Configurations of Functional Units in an Out-of-Order Superscalar Processor," *Proc. ISCA-95*, Santa Margherita Ligure, Italy, May 1995, pp. 117–125.

[Jovanovic99]    Jovanovic, M., Milutinović, V., "An Overview of Reflective Memory Systems," *IEEE Concurrency* **7**(2), 50–58 (1999).

[Jovanovic00]    Jovanovic, M., *Advanced RMS for PC Environments*, Ph.D. thesis, School of Electrical Engineering, Univ. of Belgrade, Belgrade, Serbia, Yugoslavia, 2000.

[Juan96]    Juan, T., Lang, T., Navarro, J. J., "The Difference-Bit Code," *Proc. 23rd Annual Int. Symp. Computer Architecture*, May 1996, pp. 114–120.

[Juan98]    Juan, T., Sanjeevan, S., Navarro, J., "Dynamic History-Length Fitting: A Third Level of Adaptivity for Branch Prediction," *Proc. ISCA-98*, Barcelona, Catalonia, Spain, June 27–July 1, 1998, pp. 155–166.

[Kalamatianos99]    Kalamatianos, J., Khalafi, A., Kaeli, D. R., Meleis, W., "Analysis of Temporal-Based Program Behavior for Improved Instruction Cache Performance," *IEEE Trans. Comput.* **48**(2), 168–175 (Feb. 1999).

[Kandemir99]    Kandemir, M., Ramanujam, J., Choudhary, A., "Improving Cache Locality by a Combination of Loop and Data Transformation," *IEEE Trans. Comput.* **48**(2), 159–167 (Feb. 1999).

[Keeton98]    Keeton, K., Patterson, D. A., He, Y.-Q., Raphael, R. C., Baker, W. E., "Performance Characterization of a Quad Pentium Pro SMP Using OLTP Workloads," *Proc. ISCA-98*, Barcelona, Catalonia, Spain, June 27–July 1, 1998, pp. 15–26.

[Keleher94]    Keleher, P., Cox, A. L., Dwarkadas, S., Zwaenepoel, W., "TreadMarks: Distributed Shared Memory on Standard Workstations and Operating Systems," *Proc. USENIX Winter 1994 Conf.*, Jan. 1994, pp. 115–132.

[Kennedy97]    Kennedy, A. R., et al., "*A G3 PowerPC Superscalar Low-Power Microprocessor*," Digest of Papers Comlon-97, Feb. 1997, pp. 315–324.

[Kim96]    Kim, J. H., Chien, A. A., "Rotating Combined Queueing," *Proc. ISCA-96*, Philadelphia, May 1996, pp. 226–236.

[Klauser98]    Klauser, A., Painthankar, A., Grunwald, D., "Selective Eager Execution on the PolyPath Architecture," *Proc. ISCA-98*, Barcelona, Catalonia, Spain, June 27–July 1, 1998, pp. 250–259.

[Kozyakis97]    Kozyrakis, C. E., Patterson, D. A., Anderson, T. Asanovic, K., Cardwell, N., From, R., Golbus, J., Gribstad, B., Keeton, K., Thomas, R., Treuhaft, N., Yelick, K., "*IRAM*," *IEEE Comput.* **30**(9), 75–78 (Sept. 1997).

[Kumar97]    Kumar, A., "The HP PA-8000 RISC CPU," *IEEE Micro* **17**(2), 27–32 (Mar.–Apr. 1997).

[Kumar98]    Kumar, S., Wilkerson, C., "Exploiting Spatial Locality in Data Caches Using Spatial Footprints," *Proc. ISCA-98*, Barcelona, Catalonia, Spain, June 27–July 1, 1998, pp. 357–368.

[Kung88]    Kung, S. Y., *VLSI Array Processors*, Prentice-Hall, Englewood Cliffs, NJ, 1988.

[Kuskin94]    Kuskin, J., Ofelt, D., Heinrich, M., Heinlein, J., Simoni, R., Gharachorloo, K., Chapin, J., Nakahira, D., Baxter, J., Horowitz, M., Gupta, A., Rosenblum, M., Hennessy, J., "The Stanford FLASH Multiprocessor," *Proc. 21st Annual Int. Symp. Computer Architecture*, Apr. 1994, pp. 302–313.

[Kwak99]    Kwak, H., Lee, B., Hurson, A. R., Yoon, S.-H., Hahn, W.-J., "Effects of Multithreading on Cache Performance," *IEEE Trans. Comput.* **48**(2), 176–184 (Feb. 1999).

[Lai99]    Lai, A., Falsafi, B., "Memory Sharing Predictor: The Key to a Speculative Coherent DSM," *Proc. ISCA-99*, Atlanta, May 1999, pp. 172–183.

[Laudon97]    Laudon, J., Lenoski, D., "The SGI Origin: A ccNUMA Highly Scalable Server," *Proc. ISCA-24*, Denver, CO, June 1997, pp. 241–251.

[Lenoski92]    Lenoski, D., Laudon, J., et al., "The Stanford DASH Multiprocessor," *IEEE Comput.* **25**(3), 63–79 (Mar. 1992).

[Lesartre97]    Lesartre, G., Hunt D., *PA-8500: The Continuing Evolution of the PA-8000 Family* (`http://hpcc920.external.hp.com/computing/framed/technology/micropro/pa-8500/docs/8500.html`), Hewlett-Packard Company, Fort Collins, CO, 1997.

[Li88]    Li, K., "IVY: A Shared Virtual Memory System for Parallel Computing," *Proc. 1988 Int. Conf. Parallel Processing*, Aug. 1988, pp. 94–101.

[Lipasti96]    Lipasti, M. H., Shen, J. P., "Exceeding the Dataflow Limit via Value Prediction," *Proc. 29th Int. Symp. Microarchitecture—MICRO-29*, Paris, Dec. 1996, pp. 226–237.

[Lipasti97]    Lipasti, M. H., Shen, J. P., "*Superspeculative Microarchitecture for Beyond AD 2000*," *IEEE Comput.* **30**(9), 59–66 (Sept. 1997).

[Lo98]    Lo, J. L., Barroso, L. A., Eggers, S. J., Gharachorloo, K., Levy H. M., Parekh, S. S., "An Analysis of Database Workload Performance on Simultaneous Multithreading Processors," *Proc. ISCA-98*, Barcelona, Catalonia, Spain, June 27–July 1, 1998, pp. 39–50.

[Lovett96]    Lovett, T., Clapp, R., "Sting," *Proc. IEEE/ACM ISCA-96*, Philadelphia, May 1996, pp. 308–317.

[Lucci95]    Lucci, S., Gertner, I., Gupta, A., Hedge, U., "Reflective-Memory Multiprocessor," *Proc. 28th IEEE/ACM Hawaii International Conf. on System Sciences*, Maui, Hawaii, Jan. 1995, pp. 85–94.

[Luk99]        Luk, C., Mowry, T., "Memory Forwarding: Enabling Aggressive Layout Optimizations by Guaranteeing the Safety of Data Relocation," *Proc. ISCA-99*, Atlanta, May 1999, pp. 88–99.

[Manzingo80]   Manzingo, R., Miller, T., *Introduction to Adaptive Arrays*, Wiley, New York, 1980.

[Maples90]     Maples, C., Wittie, L., "Merlin: A Superglue for Multicomputer Systems," *COMPCON90*, Mar. 1990, pp. 73–81.

[Maquelin96]   Maquelin, O., Gao, G. R., Hum, H. H. J., Theobald, K., Tian, X., "Polling Watchdog: Combining Polling & Interrupts for Efficient Message Handling," *Proc. ISCA-96*, Philadelphia, May 1996, pp. 12–21.

[Martin97]     Martin, R. P., Vahdat, A. M., Culler, D. E., Anderson, T. E., "Effects of Communication Latency, Overhead, and Bandwidth in a Cluster Architecture," *Proc. ISCA-24*, Denver, June 1997, pp. 85–97.

[McCormack99]  McCormack, J., et al., "Implementing Neon: A 256-Bit Graphics Accelarator," *IEEE Micro* **19**(2), 58–69 (Mar.–Apr. 1999).

[McFarling95]  McFarling, S., "Technical Report on gshare" (http://www. research.digital.com/wrl), 1996.

[Milenkovic96a] Milenković, A., Milutinović, V., *Cache Injection Control Architecture for RM/MC*, Technical Report, Encore, Fort Lauderdale, FL, 1996a.

[Milenkovic96b] Milenković, A., Milutinović, V., *Memory Injection Control Architecture for RM/MC*, Technical Report, Encore, Fort Lauderdale, FL, 1996b.

[Milutinovic78] Milutinović, V., *Microprocessor-Based Modem Design*, Product Documentation, Michael Pupin Institute, Belgrade, Serbia, Yugoslavia, Dec. 1978.

[Milutinovic79] Milutinović, V., "A MOS Microprocessor-Based Medium-Speed Data Modem," *Microproc. Microprog.* 100–103 (Mar. 1979).

[Milutinovic80a] Milutinović, V., "One Approach to Multimicroprocessor Implementation of Modem for Data Transmission over HF Radio," *Proc. EUROMICRO-80*, London, Sept. 1980, pp. 107–111.

[Milutinovic80b] Milutinović, V., "Suboptimum Detection Procedure Based on the Weighting of Partial Decisions," *IEE Electron. Lett.* **16**(6), 237–238, (Mar. 13, 1980).

[Milutinovic80c] Milutinović, V., "Comparison of Three Suboptimum Detection Procedures," *IEE Electron. Lett.* **16**(17), 683–685 (Aug. 14, 1980).

[Milutinovic84] Milutinović, V., "Performance Comparison of Three Suboptimum Detection Procedures in Real Environment," *IEE Proc. Part F* **131**(4), 341–344 (July 1984).

[Milutinovic85a] Milutinović, V., "A 4800 bit/s Microprocessor-Based CCITT Compatible Data Modem," *Microproc. Microprog.*, 57–74 (Feb. 1985).

[Milutinovic85b]   Milutinović, V., "Generalized W.P.D. Procedure for Micropro-cessor Based Signal Detection," *IEEE Proc. Part F* **132**(1), 27–35 (Feb. 1985).

[Milutinovic85c]   Milutinović, V., *Avenues to Explore in GaAs Multimicroprocessor Research and Development*, RCA Internal Report (Solicited Expert Opinion), RCA, Moorestown, NJ, Aug. 1985.

[Milutinovic86a]   Milutinović, V., Fortes, J., Jamieson, L., "A Multimicroprocessor Architecture for Real-Time Computation of a Class of DFT Algorithms," *IEEE Trans. ASSP*, 1301–1309 (Oct. 1986).

[Milutinovic86b]   Milutinović, V., Special Issue Guest Editor, "GaAs Microprocessor Technology," *IEEE Comput.* **19**(10) (Oct. 1986).

[Milutinovic86c]   Milutinović, V., Silbey, A., Fura, D., Keirn, K., Bettinger, M., Helbig, W., Heagerty, W., Zeiger, R., Schellack, B., Curtice, W., "Issues of Importance in Designing GaAs Microcomputer Systems," *IEEE Comput.* **19**(10), 45–59 (Oct. 1986).

[Milutinovic87a]   Milutinović, V., Lopez-Benitez, N., Hwang, K., "A GaAs-Based Microprocessor Architecture for Real-Time Applications," *IEEE Trans. Comput.*, 714–727 (June 1987).

[Milutinovic87b]   Milutinović, V., "A Simulation Study of the Vertical-Migration Microprocessor Architecture," *IEEE Trans. Software Eng.* 1265–1277 (Dec. 1987).

[Milutinovic88a]   Milutinović, V., "A Comparison of Suboptimal Detection Algorithms Applied to the Additive Mix of Orthogonal Sinusoidal Signals," *IEEE Trans. Commun.* **COM-36**(5), 538–543 (May 1988).

[Milutinovic88b]   Milutinović, V., Crnkovic, J., Houstis, C., "A Simulation Study of Two Distributed Task Allocation Procedures," *IEEE Trans. Software Eng.* **SE-14**(1), 54–61 (Jan. 1988).

[Milutinovic92]   Milutinović, V., *Avenues to Explore in PC-Oriented DSM Based on RM*, ENCORE Internal Report (Solicited Expert Opinion), ENCORE, Fort Lauderdale, FL, Dec. 1992.

[Milutinovic95a]   Milutinović, V., *A New Cache Architecture Concept: The Split Temporal/Spatial Cache Memory*, Technical Report, (UBG-ETF-TR-95-035), School of Electrical Engineering, Univ. Belgrade, Belgrade, Serbia, Yugoslavia, Jan. 1995.

[Milutinovic95b]   Milutinović, V., Petkovic, Z., "Ten Lessons Learned from a RISC Design," *Computer*, 120 (Mar. 1995).

[Milutinovic95c]   Milutinović, V., *New Ideas for SMP/DSM*, Technical Report, School of Electrical Engineering, Univ. Belgrade, Belgrade, Serbia, Yugoslavia, 1995.

[Milutinovic96a]   Milutinović, V., Markovic, B., Tomašević, M., Tremblay, M., "The Split Temporal/Spatial Cache Memory: Initial Performance Analysis," *Proc. IEEE SCIzzL-5*, Santa Clara, CA, Mar. 1996, pp. 63–69.

[Milutinovic96b]   Milutinović, V., Markovic, B., Tomašević, M., Tremblay, M., "The Split Temporal/Spatial Cache Memory: Initial Complexity Analysis," *Proc. IEEE SCIzzL-6*, Santa Clara, CA, Sept. 1996, pp. 89–96.

[Milutinovic96c]    Milutinović, V., "Some Solutions for Critical Problems in Distributed Shared Memory," *IEEE TCCA Newsl.* (Sept. 1996).

[Milutinovic96d]    Milutinović, V., "The Best Method for Presentation of Research Results," *IEEE TCCA Newsl.* (Sept. 1996).

[Milutinovic99]    Milutinović, V., Valero, M., "Cache Memory and Related Problems: Enhancing and Exploiting the Locality," *IEEE Trans. Comput.* **48**(2), 97–99 (Feb. 1999).

[MIPS96]    *MIPS R10000 Microprocessor User's Manual*, Version 2.0, (ftp: // sgigate.sgi.com / pub / doc / R10000 / User_Manual / t5.ver.2.0.book.pdf), MIPS Technologies, Mountain View, CA 1996.

[Modcomp83]    Modcomp, Inc, *Mirror Memory System*, Internal Report, Modcomp, Fort Lauderdale, FL, Dec. 1983.

[Montoye90]    Montoye, R. K., Hokenek, E., Runyon, S. L., "Design of the IBM RISC System/6000 Floating-Point Execution Unit," *IBM J. Res. Dev.* **34**(1), 59–70 (Jan. 1990).

[Moshovos97]    Moshovos, A., Breach, S. E., Vijaykumar, T. N., Sohi, G. S., "Dynamic Speculation and Synchronization of Data Dependencies," *Proc. ISCA-24*, Denver, June 1997, pp. 181–193.

[Mukherjee98]    Mukherjee, S., Hill, M., "Using Predictors to Accelerate Coherence Protocols," *Proc. ISCA-98*, Barcelona, Catalonia, Spain, June 27–July 1, 1998, pp. 179–190.

[Nair97]    Nair, R., Hopkins, M., "Exploiting Instruction Level Parallelism in Processors by Caching Scheduled Groups," *Proc. ISCA-24*, Denver, June 1997, pp. 13–25.

[Nowatzyk93]    Nowatzyk, M., Monger, M., Parkin, M., Kelly, E., Browne, M., Aybay, G., Lee, D., "S3.mp: A Multiprocessor in Matchbox," *Proc. PASA*, 1993.

[Oberman99]    Oberman, S., Favor, G., Weber, F., "AMD 3DNow! Technology: Architecture and Implementations," *IEEE Micro* **19**(2), 37–48 (Mar.–Apr. 1999).

[Palacharla97]    Palacharla, S., Jouppi, N., Smith, J. E., "Complexity-Effective Superscalar Processors," *Proc. ISCA-24*, Denver, June 1997, pp. 206–218.

[Papworth96]    Papworth, D. B., "Tuning the PentiumPro Microarchitecture," *IEEE Micro*, 8–16 (Apr. 1996).

[Patel98]    Patel, S. J., Evers, M., Patt, Y. N., "Improving Trace Cache Effectiveness with Branch Promotion and Trace Packing," *Proc. ISCA-98*, Barcelona, Catalonia, Spain, June 27–July 1, 1998, pp. 262–271.

[Patel99]    Patel, S. J., Friendly, D. H., Patt, Y. N., "Evaluation of Design Options for the Trace Cache Fetch Mechanism," *IEEE Trans. Comput.* **48**(2), 193–204 (Feb. 1999).

[Patt94]     Patt, Y. N., "The I/O Subsystem—a Candidate for Improvement," *IEEE Comput.* **27**(3), (Mar. 1994) (special issue).

[Patt97]     Patt, Y. N., Ptel, S. J., Evers, M., Friendly, D. H., Stark, J., "One Billion Transistors, One Uniprocessor, One Chip," *IEEE Comput.* **30**(9), 51–57 (Sept. 1997).

[Patterson94]     Patterson, D. A., Hennessy, J. L., *Computer Organization and Design: The Hardware/Software Interface*, Morgan Koufmann, San Mateo, CA, 1994.

[Peir99]     Peir, J.-K., Hsu, W. W., Smith, A. J., "Functional Implementation Techniques for CPU Cache Memories," *IEEE Trans. Comput.* **48**(2), 100–110 (Feb. 1999).

[Peleg94]     Peleg, A., Wiser, V., *Dynamic Flow Instruction Cache Memory Organized around Trace Segments Independent of Virtual Address Line*, U.S. Patent 5,381,533, Washington, DC, 1994.

[Petterson96]     Petterson, L. L., Davie, B. S., *Computer Networks*, Morgan Kaufmann, San Francisco, CA, 1996.

[Pinkston97]     Pinkston, T. M., Warnakulasuriya, S., "On Deadlocks in Interconnection Networks," *Proc. ISCA-24*, Denver, June 1997, pp. 38–49.

[Prete91]     Prete, C. A., "RST: Cache Memory Design for a Tightly Coupled Multiprocessor System," *IEEE Micro*, 16–19, 40–52 (Apr. 1991).

[Prete95]     Prete, C. A., Riccardi, L., Prina, G., "Reducing Coherence-Related Overhead in Multiprocessor Systems," *Proc. IEEE/ Euromicro Workshop on Parallel and Distributed Processing*, San Remo, Italy, Jan. 1995, pp. 444–451.

[Prete97]     Prete, C. A., Prina, G., Giorgi, R., Ricciardi, L., "Some Considerations about Passive Sharing in Shared-Memory Multiprocessors," *IEEE TCCA Newsl.*, 34–40 (Mar. 1997).

[Protic85]     Protić, J., *System LOLA-85*, Technical Report (in Serbian), Lola Industry, Belgrade, Serbia, Yugoslavia, Dec. 1985. (jeca@etf.bg.ac.yu).

[Protic96a]     Protić, J., Tomašević, M., Milutinović, V., "Distributed Shared Memory: Concepts and Systems," *IEEE Parallel Distrib. Technol.* **4**(2), 63–79 (1996).

[Protic96b]     Protić, J., Milutinović, V., *Combining LRC and EC: Spatial Versus Temporal Data*, Technical Report, Encore, Fort Lauderdale, FL, 1996 (jeca@etf.bg.ac.yu).

[Protic97]     Protić, J., Tomašević, M., Milutinović, V., *Tutorial on Distributed Shared Memory (Lecture Transparencies)*, IEEE CS Press, Los Alamitos, CA, 1997.

[Protic98]     Protić, J., *A New Hybrid Adaptive Memory Consistency Model*, Ph.D. thesis, School of Electrical Engineering, Univ. Belgrade, Belgrade, Serbia, Yugoslavia, 1998.

[Prvulovic97] Prvulovic, M., "Microarchitecture Features of Modern RISC Microprocessors—An Overview," *Proc. SinfoN'97*, Zlatibor, Serbia, Yugoslavia, Nov. 1997 (prvul@computer.org).

[Prvulovic99a] Prvulovic, M., Marinov, D., Dimitrijevic, Z., Milutinović, V., "Split Temporal/Spatial Cache: A Survey and Reevaluation of Performance," *IEEE TCCA Newsl.* (1999).

[Prvulovic99b] Prvulovic, M., Marinov, D., Dimitrijevic, Z., Milutinović, V., "Split Temporal/Spatial Cache: A Performance and Complexibility Evaluation," *IEEE TCCA Newsl.* (1999).

[Pyron98] Pyron, C., Prado, J., Golab, J., "Test Strategy for the PowerPC 750 Microprocessor," *IEEE Design Test Comput.*, 90–97 (July–Sept. 1998).

[Ramachandran91] Ramachandran, U., Khalidi, M. Y. A., "An Implementation of Distributed Shared Memory," *Software Pract. Experience* **21**(5), 443–464 (May 1991).

[Raskovic95] Raskovic, D., Jovanov, E., Janicijevic, A., Milutinović, V., "An Implementation of Hash Based ATM Router Chip," *Proc. IEEE/ACM HICSS-95*, Maui, Hawaii, Jan. 1995.

[Reinhardt94] Reinhardt, S., Larus, J., Wood, D., "Tempest and Typhoon: User-Level Shared Memory," *Proc. 21st Annual Int. Symp. Computer Architecture*, Apr. 1994, pp. 325–336.

[Reinhardt96] Reinhardt, S. K., Pfile, R. W., Wood, D. A., "Decoupled Hardware Support for DSM," *Proc. IEEE/ACM ISCA-96*, Philadelphia, May 1996, pp. 34–43.

[Rexford96] Rexford, J., Hall, J., Shin, K. G., "A Router Architecture for Real-Time Point-to-Point Networks," *Proc. ISCA-96*, Philadelphia, May 1996, pp. 237–246.

[Rivers96] Rivers, J. A., Davidson, E. S., "Reducing Conflicts in Direct-Mapped Caches with a Temporality Based Design," *Proc. Int. Conf. Parallel Processing*, 1996.

[Rotenberg97] Rotenberg, E., Bennett, S., Smith, J. E., "Trace Cache: A Low Latency Approach to High Bandwidth Instruction Fetching," *Proc. 29th Annual ACM/IEEE Int. Symp. Microarchitecture*, Paris, Dec. 1997, pp. 24–34.

[Rotenberg99] Rotenberg, E., Bennett, S., Smith, J. E., "A Trace Cache Microarchitecture and Evaluation," *IEEE Trans. Comput.* **48**(2), 111–120.

[Sahuquillo99] Sahuquillo, J., Pont, A., "The Split Data Cache in Multiprocessors Systems: An Initial Hit Ratio Analysis," *Proc. 7th Euromicro Workshop on Parallel and Distributed Processing*, Madeira, Portugal, Feb. 1999.

[Sanchez97] Sanchez, F. J., Gonzalez, A., Valero, M., "Software Management of Selective and Dual Data Caches," *IEEE TCCA Newsl.*, 3–10 (Mar. 1997).

[Saulsbury96] Saulsbury, A., Pong, F., Nowatzyk, A., "Missing the Memory Wall: The Case for Processor/Memory Integration," *Proc. ISCA-96*, 1996, pp. 90–101.

[Savell99]      Savell, T. C., "The EMU10K1 Digital Audio Processor," *IEEE Micro* **19**(2), 49–57 (Mar.–Apr. 1999).

[Savic95]       Savic, S., Tomašević, M., Milutinović, V., Gupta, A., Natale, M., Gertner, I., "Improved RMS for the PC Environment," *Microproc. Microsyst.* **19**(10), 609–619 (Dec. 1995).

[Schoinas94]    Schoinas, I., Falsafi, B., Lebeck, A. R., Reinhardt, S. K., Larus, J. R., Wood, D. A., "Fine-Grain Access Control for Distributed Shared Memory," *Proc. 6th Int. Conf. Architectural Support for Programming Languages and Operating Systems*, Nov. 1994, pp. 297–306.

[Sechrest96]    Sechrest, S., Lee, C. C., Mudge, T., "Correlation and Aliasing in Dynamic Branch Predictors," *Proc. ISCA-96*, Philadelphia, May 1996, pp. 21–32.

[Seznec96]      Seznec, A., "Don't Use the Page Number, but a Pointer to It," *Proc. ISCA-96*, Philadelphia, June 1996.

[Sheaffer96]    Sheaffer, G., *Trends in Microprocessing*, Keynote Address, YU-INFO-96, Brezovica, Serbia, Yugoslavia, Apr. 1996.

[Shriver98]     Shriver, B., Smith, B., *The Anatomy of a High-Performance Microprocessor: A System Perspective*, IEEE Computer Society Press, Los Alamitos, CA, 1998.

[Silc98]        Silc, J., Robic, B., Ungerer, T., "Asynchrony in Parallel Computing: From Dataflow to Multithreading," *Parallel Distrib. Comput. Pract.* **1**(1), 3–30 (1998) (`http://goethe.ira.uka.de/people/ungerer/JPDCPdataflow.pdf`).

[Simha96]       Simha, P., *R4400 Microprocessor Product Information* (`ftp://sgigate.sgi.com/pub/doc/R4400/Prod_Overview/R4400_Overview.ps.Z`), MIPS Technologies, Mountain View, CA, 1996.

[Simoni90]      Simoni, R., *Implementing a Directory-Based Cache Coherence Protocol*, Stanford Univ., Technical Report, (CSL-TR-90-423), Palo Alto, CA, Mar. 1990.

[Simoni91]      Simoni, R., Horowitz, M., "Dynamic Pointer Allocation for Scalable Cache Coherence Directories," *Proc. Int. Symp. Shared Memory Multiprocessing*, Stanford Univ., Palo Alto, CA, Apr. 1991, pp. 72–81.

[Simoni92]      Simoni, R., *Cache Coherence Directories for Scalable Multiprocessors*, Ph.D. thesis, Stanford Univ., Palo Alto, CA, 1992.

[Slegel99]      Slegel, T. J., et al., "IBM's S/390 G5 Microprocessor," *IEEE Micro* **19**(2), 12–23 (Mar.–Apr. 1999).

[Smith95]       Smith, J. E., Sohi, G., "The Microarchitecture of Superscalar Processors," *Proc. IEEE* **83**(12), 1609–1624 (Dec. 1995).

[Smith97]       Smith, J. E., Vajapeyam, S., "Trace Processors," *IEEE Comput.* **30**(9), 68–73 (Sept. 1997).

[Sodani97]      Sodani, A., Sohi, G. S., "Dynamic Instruction Reuse," *Proc. ISCA-97*, Denver, June 1997, pp. 194–205.

| | |
|---|---|
| [Sohi95] | Sohi, G. S., Breach, S. E., Vijaykumar, T. N., "Multiscalar Processors," *Proc. ISCA-95*, Santa Margerita Ligure, Italy, June 1995. |
| [Soundararajan98] | Soundararajan, V., Heinrich, M., Verghese, B., Gharachorloo, K., Gupta, A., Hennessy, J., "Flexible Use of Memory for Replication/Migration in Cache-Coherent DSM Multiprocessors," *Proc. ISCA-98*, Barcelona, Catalonia, Spain, June 27–July 1, 1998, pp. 342–355. |
| [Sprangle97] | Sprangle, E., Chappell, R. S., Alsup, M., Patt, Y., "The Agree Predictor: A Mechanism for Reducing Negative Branch History Interference," *Proc. ISCA-24*, Denver, June 1997, pp. 284–291. |
| [Stallings96] | Stallings, W., *Computer Organizations and Architecture*, 4th ed., Prentice-Hall, Upper Saddle River, NJ, 1996. |
| [Stenstrom88] | Stenstrom, P., "Reducing Contention in Shared-Memory Multiprocessors," *IEEE Comput.*, 26–37 (Nov. 1988). |
| [Stiliadis97] | Stiliadis, D., Varma, A., "Selective Victim Caching: A Method to Improve the Performance of Direct-Mapped Caches," *IEEE Trans. Comput.* **46**(5), 603–610 (May 1997). |
| [Stojanovic95] | Stojanovic, M., *Advanced RISC Microprocessors*, Technical Report, School of Electrical Engineering, Univ. Belgrade, Belgrade, Serbia, Yugoslavia, Dec. 1995. |
| [Sun95] | Sun Microelectronics, *SuperSPARC Data Sheet: Highly Integrated 32-Bit RISC Microprocessor* (http://www.sun.com/sparc/stp1020a/datasheets/stp1020a.pdf), Sun Microelectronics, Mountain View, CA, 1995. |
| [Sun96] | Sun Microelectronics, *UltraSPARC-I High Performance, 167 & 200 MHz, 64-bit RISC Microprocessor Data Sheet* (http://www.sun.com/sparc/stp1030a/datasheets/stp1030a.pdf), Sun Microelectronics, Mountain View, CA, 1996. |
| [Sun97] | Sun Microelectronics, *UltraSPARC-II High Performance, 250 MHz, 64-bit RISC Processor Data Sheet* (http://www.sun.com/sparc/stp1031/datasheets/tp1031lga.pdf), Sun Microelectronics, Mountain View, CA, 1997. |
| [Sweazey86] | Sweazey, P., Smith, A. J., "A Class of Compatible Cache Consistency Protocols and Their Support by the IEEE Futurbus," *Proc. ISCA-86*, Tokyo, June 1986, pp. 414–423. |
| [Tabak98] | Tabak, D., "Special Issue on Instruction Level Parallelism: Guest Editor Introduction," *Microproc. Microsyst.* **22**(6), 291–292 (Nov. 1998). |
| [Tanenbaum90] | Tanenbaum, A. S., *Structured Computer Organization*, Prentice-Hall, Englewood Cliffs, NJ, 1990. |
| [Tartalja97] | Tartalja, I., *The Balkan Schemes for Software Based Maintenance of Cache Consistency is Shared Memory Multiprocessors*, Ph.D. thesis, Univ. Belgrade, Belgrade, Serbia, Yugoslavia, 1997. |

[Temam99]      Temam, O., "An Algorithm for Optimally Exploiting Spatial and Temporal Locality In Upper Memory Levels," *IEEE Trans. Comput.* **48**(2), 150–158 (Feb. 1999).

[Teodosiu97]    Teodosiu, D., Baxter, J., Govil, K., Chapin, J., Rosenblum, M., Horowitz, M., "Hardware Fault Containment in Scalable Shared-Memory Multiprocessors," *Proc. ISCA-24*, Denver, June 1997, pp. 73–84.

[Thompson94]   Thompson, T., Ryan, B., "PowerPC 620 Soars," *Byte* (Nov. 1994).

[Thornton64]    Thornton, J. E., "Parallel Operation on the Control Data 6600," *Proc. Fall Joint Computer Conf.*, Oct. 1964, pp. 33–40.

[Tomasevic92a]  Tomašević, M., Milutinović, V., "A Simulation Study of Snoopy Cache Coherence Protocols," *Proc. HICSS-92*, Koloa, Hawaii, 1992, pp. 427–436.

[Tomasevic92b]  Tomašević, M., *A New Snoopy Cache Coherence Protocol*, Ph.D. thesis, School of Electrical Engineering, Univ. Belgrade, Belgrade, Serbia, Yugoslavia, 1992.

[Tomasevic93]   Tomašević, M., Milutinović, V., *Tutorial on the Cache Coherence Problem in Shared Memory Multiprocessors: Hardware Solutions* (Lecture Transparencies; the 1996 update), IEEE CS Press, Los Alamitos, CA, 1993.

[Tomasko97]    Tomasko, M., Hadjiyannis, S., Najjar, W. A., "Experimental Evaluation of Array Caches," *IEEE TCCA Newsl.*, 11–16 (Mar. 1997).

[Tomasulo67]    Tomasulo, R. M., "An Efficient Algorithm for Exploiting Multiple Arithmetic Units," *IBM J. Res. Dev.*, 25–33 (Jan. 1967).

[Tredennick86]  Tredennick, N., "Microprocessor Based Computers," *IEEE Comput.* **29**(10), 27–37 (Oct. 1996).

[Tremblay96]    Trembay, M. O'Connor, J. M., "UltraSPARC I: A Four Issue Processor Supporting Multimedia," *IEEE Micro* **16**(4), 42–50 (Apr. 1996).

[Tse98]         Tse, J., Smith, A. J., "CPU Cache Prefetching: Timing Evaluation of Hardware Implementations," *IEEE Trans. Comput.*, 509–526 (May 1998).

[Tullsen95]     Tullsen, D. M., Eggers, S. J., Levy, H. M., "Simultaneous Multithreading: Maximizing On-Chip Parallelism," *Proc. ISCA-95*, Santa Margherita Ligure, Italy, June 1995, pp. 392–403.

[Tullsen96]     Tullsen, D. M., Eggers, S. J., Emer, J. S., Levi, H. M., Lo, J. L., Stamm, R. L., "Exploiting Choice: Instruction Fetch and Issue on an Implementable Simultaneous Multithreading Processor," *Proc. ISCA-96*, Philadelphia, May 1996, pp. 191–202.

[Tullsen99]     Tullsen, D., Seng, J., "Storageless Value Prediction Using Prior Register Values," *Proc. ISCA-99*, Atlanta, May 1999, pp. 270–279.

[Tyson95]       Tyson, G., Farrens, M., Matthews, J., Pleszkun, A. R., "A Modified Approach to Data Cache Management," *Proc. 28th Annual Int. Symp. Microarchitecture*, Dec. 1995, pp. 93–103.

[Vajapeyam97]    Vajapeyam, S., Mitra, T., "Improving Superscalar Instruction Dispatch and Issue by Exploiting Dynamic Code Sequences," *Proc. ISCA-24*, Denver, June 1997, pp. 1–12.

[Vajapeyam99]    Vajapeyam, S., Joseph, P., Mitra, T., "Dynamic Vectorization: A Mechanism for Exploiting Far-Flung ILP in Ordinary Programs," *Proc. ISCA-99*, Atlanta, May 1999, pp. 16–27.

[Villasenor97]    Villasenor, J., Mangione-Smith, W. H., "Configurable Computing," *Sci. Am.* (May 1997).

[Vuletic97]    Vuletic, M., Ristic-Djurovic, J., Aleksic, M., Milutinović, V., Flynn, M., "Per Window Switching of Window Characteristics: Wave Pipelining vs. Classical Design," *IEEE TCCA Newsl.*, 1–6 (Sept. 1997).

[Wang97]    Wang, K., Franklin, M., "Highly Accurate Data Value Prediction Using Hybrid Predictors," *Proc. 30th Annual Symp. Microarchitecture—MICRO-30*, Research Triangle Park, NC, Dec. 1997, pp. 281–290.

[Wilson94]    Wilson, A., LaRowe, R., Teller, M., "Hardware Assist for Distributed Shared Memory," *Proc. 13th Int. Conf. Distributed Computing Systems*, May 1993, pp. 246–255.

[Wilson96]    Wilson, K. M., Olokotun, K., Rosenblum, M., "Increasing Cache Port Efficiency for Dynamic Superscalar Microprocessors," *Proc. ISCA-96*, Philadelphia, May 1996, pp. 147–157.

[Woo95]    Woo, S. C., Ohara, M., Torrie, E., Singh, J. P., Gupta, A., "The SPLASH-2 Programs: Characterization and Methodological Considerations," *Proc. ISCA-95*, Santa Margherita Ligure, Italy, June 1995, pp. 24–36.

[Yoaz99]    Yoaz, A., Erez, M., Ronen, R., Jourdan, S., "Speculation Techniques for Improving Load Related Instruction Scheduling," *Proc. ISCA-99*, Atlanta, May 1999, pp. 42–53.

[Zhang99]    Zhang, Z., Cintra, M., Torrellas, J., "Excel-NUMA: Toward Programmability, Simplicity, and High Performance," *IEEE Trans. Comput.* **48**(2), 256–264 (Feb. 1999).

[Zhou90]    Zhou, S., Stumm, M., McInerney, T., "Extending Distributed Shared Memory to Heterogeneous Environments," *Proc. 10th Int. Conf. Distributed Computing Systems*, May–June 1990, pp. 30–37; **30**(9), 46–49 (Sept. 1997).

# APPENDIXES

# A Reflective Memory System for PCs

This appendix* presents the experiences from a project run as a cooperative venture between a university research team from Europe and a leading industry from the United States, specializing in the area of reflective memory systems (RMSs). Reflective memory architecture represents a class of distributed shared memory (DSM) architectures, characterized by its nondemand update-based writethrough consistency mechanism applied to shared data replicas. The main goals of this project were to (1) design a cost-effective board that connects a personal computer as a node in an RMS system and (2) come up with some new ideas and their evaluation, in order to increase the efficiency of the basic RMS solution. Both goals were achieved through a synergistic interaction between the well-composed team from the university and the experienced researchers from industry. As a side effect, some new avenues for research in the more general area of DSM were open, which resulted in several new projects and published books. This text describes the methodology and development strategy utilized in the project, as well as particular challenges and generated results, putting the stress of presentation on lessons learned and their discussion.

## A.1. INTRODUCTION

Distributed shared memory (DSM) architecture represents a successful hybrid of shared memory multiprocessors (SMPs) and distributed systems (DSs), that has the advantages of both architectures [Protić et al. 1996a]. All processors in a DSM system access a shared logical address space, which simplifies the programming model and makes many applications portable from SMP systems. However, the underlying architecture is typically a DS, so the logically shared memory is physically distributed, which means that, with an appropriate interconnection network, the system has good potential for scalability.

---

*Prepared by Veljko Milutinović, Milo Tomašević, Jelica Protić, Savo Savic, Milan Jovanovic, and Aleksandra Grujic.

A DSM system is composed of nodes connected by some kind of interconnection network (ring, mesh, hypercube, LAN, bus hierarchy, etc.). Each node can be a single-processor system, or even an SMP (as in the DASH system [Lenoski et al. 1992]). In both cases, it contains a portion of local memory mapped to the unique shared address space of DSM. In general, shared data stored in local memories can migrate from one node to another, depending on the access pattern of an application. It can also have multiple copies replicated in local memories of several nodes, which requires that all copies have to be kept up to date according to some consistency rules defined by a memory consistency model (MCM). There are a variety of algorithms for consistency maintenance [Stumm and Zhou 1990], and the majority of them can be classified as multiple reader, single writer (MRSW) and multiple reader, multiple writer (MRMW). Consistency-related information is stored in system tables or a directory, which can be centralized or distributed, corresponding to the responsibility for management of the DSM system.

The meaning of the word "consistency" has been the most variable aspect of DSM systems from early efforts in this area, in the mid 1980 up to their current maturity. The memory consistency model (MCM) determines its meaning by defining the legal ordering of memory references issued by some processor, as observed by other processors in the system [Gharachorloo et al. 1990, Protić et al. 1996b]. While the strongest MCMs, *sequential* and *processor* consistency, do not distinguish between ordinary shared data accesses and synchronization accesses, more sophisticated models such as *weak*, *release*, *lazy release*, and *entry consistency* take advantage of the fact that shared data accesses are typically protected by some synchronization operations. Therefore, the points of enter/exit of critical sections and barriers are used as suitable moments when some consistency-related actions should take place. The global trend in most sophisticated MCMs [Bershad et al. 1993, Keleher et al. 1994] is to minimize the necessary communication between nodes, paid for by making the programming model more complex. This trend distances DSM systems from their original goal (simplification of the programming model) but greatly improves their performance by reducing and/or hiding the communication latency.

The mechanism that provides the illusion of shared memory in a physically distributed memory system can be implemented in software, in hardware, or in cooperation of hardware and software, which results in a hybrid implementation. The first DSM systems were implemented on the underlying network of workstations, typically connected by Ethernet [Fleisch et al. 1989]. They were implemented in software, as run-time systems and/or modifications of the original operating system. Some of the systems developed more recently also included (or were primary-based) compiler modifications [Bershad et al. 1993]. On the other hand, a class of DSM systems implements the DSM mechanism in hardware, which brings full transparency to all software layers, besides their performance advantages. The best potential for further improvements in the field of DSM is in combining a software approach, which can better explore the characteristics of the application known by the programmer, with the hardware approach, which helps in performing critical operations more efficiently. Therefore, the majority of the most recent DSM systems can be classified as hybrid implementations.

Reflective memory systems (RMSs) [Lucci et al. 1995] belong to the class of hardware-implemented DSM. Their main characteristic is the nondemand, writethrough update consistency mechanism. The majority of DSM systems generally keep track of valid copies of shared data present in local memory of the node and, if data are not present in the valid state, fetch it from another node(s) on demand. Unlike these systems, an RMS performs all necessary updates automatically, using broadcast over the RM bus as an appropriate mechanism. Each node maps some segments in its local reflective memory to the shared address space, treating these segments as open "windows." Each write operation to an open window is immediately propagated and received by all other nodes that also have opened windows containing that particular address. This method makes the write operation costly, but the read operation is not delayed, since the valid data are always present in the local reflective memory.

## A.2. WHY REFLECTIVE MEMORY SYSTEMS?

Although the first RM systems [Encore 1990] appeared in 1990, they suddenly gained emerging popularity in the late 1990s. In the meantime, many variants of demand-based DSM systems were built, proving the validity of the concept, but few of them became commercial products. The reason for that can be found in the high overhead of the operating system and the considerable time consumption of the layered protocol software. In addition, communication latency in such a system is hard to predict, which is not tolerable for real-time applications. All these problems are avoided in RM systems, with their simple and efficient hardware mechanisms for consistency maintenance. The communication in these systems is usually overlapped with computation, and the shared memory access time is predictable and relatively low. The simplicity of the architecture and the design that is typically based on the off-the-shelf components also helped initial RMS systems to stay on the market for almost a decade.

There are multiple types of systems based on nondemand writethrough update, some of them referred to as *mirror memory systems* [Maples and Witte 1990] or replicated memory systems [Oguchi et al. 1995]. However, the term *reflective memory* is most frequently used. This class of systems includes some commercial products—Modcomp [Furht 1997, Systran (1995), and VMIC (1995)]—targeting the real-time market, as well as ATC's ring-based approach [Ramanjuan et al. 1995] and DEC's memory channel [Gillett 1996]. The SHRIMP project of Princeton University [Blumrich et al. 1994] also belongs to this class. An almost exhaustive survey [Jovanovic and Milutinović

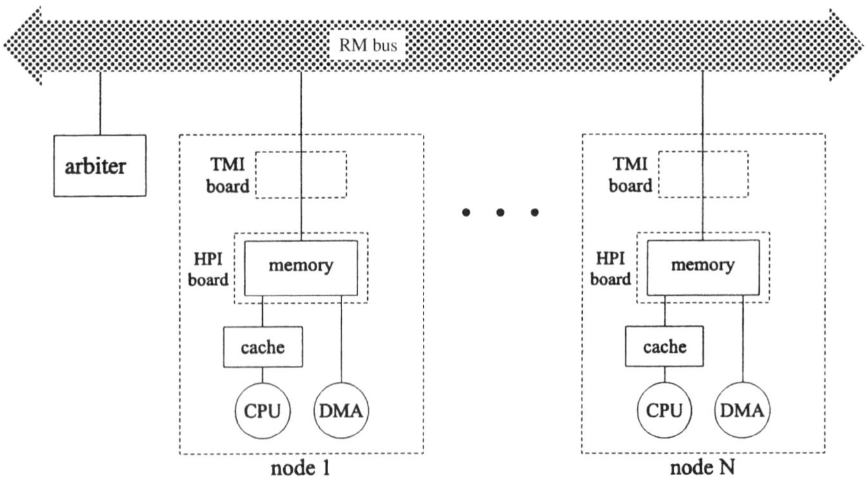

**Figure A.1.** Architecture of the reflective memory system (RMS) (CPU—central processor unit; DMA—direct memory access unit).

1998] covers the majority of RMS systems. The first RM system was made and patented by Gould Electronics in 1985. Encore acquired Gould in 1989 and updated the system, which later evolved into a number of RMS variants. The latest developments resulted in passing the technology from Encore to DEC Corporation, and its promotion of the DEC memory channel. This appendix focuses on Encore's development of RMS.

In the Encore RMS [Gertner 1993] of each node connected to the RM bus consists of a processor with cache, local memory, DMA device, and the RM board with memory and interface to the RM bus (see Fig. A.1). Bus arbitration is centralized, using a round-robin synchronous arbitration algorithm. The processor (responsible for word transfers) is connected to the RM board via the local processor/memory and/or the host system bus. The DMA unit (responsible for block transfers) is attached via the host system bus. Transmit/receive mapping tables are used to configure local memory pages as reflective (shared) or private (nonshared), as well as for address mapping, since copies of a shared page could be at different addresses on different nodes. On each write to the local RM memory, a check is made in the transmit mapping table if the address belongs to an open window, and, if so, the data word, together with the translated address in reflected shared space, is put into the transmit FIFO buffer. The node with a pending write requests the RM bus from the RM bus controller. When the node obtains the bus grant, the bus write operation (address + data) is initiated on the RM bus. All other nodes check if the address falls into a local open receive window. In that case, the data and translated receive address are written into the receive FIFO. Finally, data from the receive FIFO are forwarded to the local RM memory.

The state-of-the-art among RM systems is RM/MC (reflective memory/memory channel), which supports blocked updates. The RM/MC system is based on the multiplexed synchronous 64-bit wide bus, connecting up to eight processing nodes, with the 15,000-ft maximal distance (for a fiberoptic cable). RM/MC supports both single-word and block update transfers. Each node in the system is composed of two boards: transition module interface (TMI) for the RM/MC bus and host port interface (HPI) for the host system bus. There are implementations for the VME bus, the EISA bus, and the PCI bus.

Since RMS and similar systems are based on asynchronous, noninstantaneous communication, they typically cannot guarantee strict consistency. However, if all writes are seen by all processors in the system in the same order (which is true for centralized arbitration, and no bypassing in transmit/receive FIFO buffers), sequential consistency can be achieved. More relaxed consistency models can be implemented if we allow some bypassing and/or deleting in transmit/receive FIFO buffers. However, the latest developments in this area introduced some hybrid solutions [Denton and Johnson 1996]. Encore proposed LAM—a software-based consistency maintenance for RM/MC underlying hardware. The LAM uses library routines to manage allocation of the RM space and to synchronize the accesses to this space, providing the programmer with the means of entry consistency like MCM.

## A.3. REQUIREMENTS FOR EFFICIENT RMS IN PC ENVIRONMENT AND DEVELOPMENT STRATEGY

The earlier RMS systems were usually implemented in a minicomputer environment and used for a wide range of applications, from real-time to online transaction processing. However, in the early 1990s the expansion of the DSM concept to a network of workstations (NOW) and a nctwork of personal computers (NOPC) became one of the most important development trends in this field [Li 1995]. Encore was one of the first to understand this trend and concluded that using a standard and relatively powerful machine such as a PC, as a node in the RM system, could lead to a cost-effective system with high configuration flexibility. That was the motivation to start this project, with the prospective to build a system that costs less, connects more nodes, and supports more traffic.

The first technical goal of the project was the design of a fully operational RMS board, which plugs into a PC and turns it into a node within the bus-based DSM system, using the reflective memory approach. The board had to be compatible with the RM bus interconnect (Encore's proprietary bus) and also had to provide a larger amount of local memory. In addition to the requested performance characteristics superior to existing competitor solutions, the profit margin had to be built in the predefined value ($999).

Besides the strictly technical goal of designing the board, Encore was also interested in some theoretical research that would go beyond the currently operational systems and result in the development of the know-how, which could create new breakthroughs in the near future. The requests for this part of the project were not so strict; while the management staff of our research sponsor insisted on a solution for a 64- or 128-node RM system, the engineering staff was skeptical and suggested the generation and elaboration of several ideas that would probably improve the performance of the system and their detailed simulation analysis. The essence of this strategy is to achieve a relatively large global performance improvement through a series of small incremental improvements. This was probably the only way to go, since the RM concept existed for years, and a considerable advance could be obtained only through a superposition of relatively small advances in different directions.

In addition to the technical and theoretical goals, Encore also had in mind particular business goals when starting this project. Some time before the RM technology was passed to DEC, in order to serve as a basis for their later development of its memory channel, DEC invested 10% of its overall profit in the area of networking, to the development of Encore's RM. Technical solutions of RM probably influenced DEC's MC; however, it remains a secret for us which particular results of our research were built in DEC's final product.

In order to achieve the aforementioned goals, it was important to properly compose the research team. The perfect number was six:

1. A Ph.D. student to survey the field of DSM and come up with some fresh ideas that could open new horizons and bring breath and depth to the research

2. A senior, experienced Ph.D. researcher, to pass the knowledge in the area of multiprocessors to the younger members and help them solve all kinds of problems on the day-to-day basis

3. Two postgraduate students, to specialize in simulations and the logic design of the proposed improvements

4. The postgraduate student with the highest GPA, to specialize in the board design and testing, and cooperate with the engineers associated with the research sponsor in the final stages of the project—making sure that the board is fully operational

5. Last, but not least, the project leader, to guide the research strategically, encourage and motivate the team, and organize the logistics tasks and the communication with the sponsor based on weekly reports

## A.4. DESIGN TRADE-OFFS

The first goal of the project was to design a board that interfaces a personal computer to the RM bus. Since the interface between the local RM memory and the RM bus is similar for different computers, it is moved into a separate hardware module called transition module interface (TMI). The TMI is connected with the host RM board by a special cable with multiplexed address and data lines. The host board (Fig. A.2) contains the system bus interface, local RM memory, and memory arbiter and control logic. The memory on the host board and all data paths are 32 bits wide. There are no FIFO buffers between the memory and the system bus, since the onboard memory cycle is synchronized with the system bus cycle. The system bus burst cycles are also supported. The board is designed to have 16 or 64 MB of DRAM organized in four standard 72-pin 36-bit SIMMs. The system base address of the memory is programmable in 16/64-MB quantums. The memory arbiter gives the highest priority to the DRAM refresh logic, then to the system bus interface, and at last to the receive FIFO (data coming from the

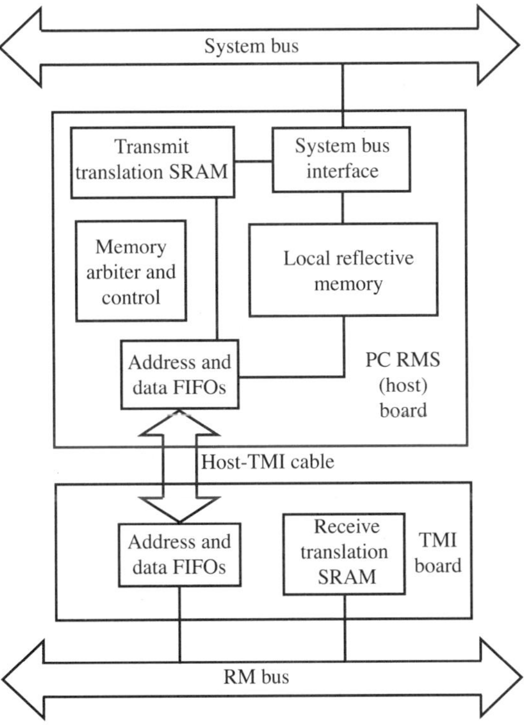


**Figure A.2.** Implementation scheme of the PC-RMS (TMI—transition module interface).

RM bus). The priority can be changed when the receive FIFO is almost full, or on the system bus timeout.

The host board also contains transmit translation SRAM. This mapping table keeps the shared/private indication bit for each 8-kB segment of the memory and a global RMS address where a block should be mapped if it is shared. From the memory side, address and data FIFOs are accessed by separate buses, while on the host TMI cable side address and data lines are multiplexed to reduce the bulk of cabling. The TMI demultiplexes the address and data lines, puts the address and data into separate FIFOs, and connects them to the RM bus. The receive translation SRAM is placed on the TMI to keep the information about mapping from the RMS address space to the local RM memory. If an address belongs to a block (that is reflected), data and the translated address are written into the receive FIFOs on the TMI board, then multiplexed, sent to the FIFOs on the host board, and finally written to the memory.

After the design of the PC-RMS board was finished, it was carefully analyzed, in order to note the drawbacks of the RM concept and its implementation and then, on the basis of its criticism several improvements were proposed [Savic et al. 1995]. Each proposed solution was separately simulated, in order to examine its potential effects and evaluate its impact in comparison with other proposals. After that, a specific decision-making methodology (modified for DSM) [e.g., Hoevel and Milutinović 1991] was applied, in order to choose a subset of proposals that would give the solution with a relatively large global performance improvement, achieved through a series of small incremental improvements. However, the final decision about the incorporation of proposed improvements into the final product was left to the research sponsor; our task was only to give recommendations, having in mind results of particular simulations. Therefore, simulation of the overall system, implementing all proposed improvements, was not done, because of time limitations and the fact that overall effects will be measurable on the implemented system. Here we present four suggested improvements that we found the most interesting:

*Improvement 1: Caching of Shared RM Regions.* In the existing RM systems, the shared reflected regions of the RM memory are not cached. The RM board appears as a memory slave on the host system bus and cannot initiate any action on this bus when a write from the RM bus occurs. Since the effects of caching to the overall system performance are well known [Tomašević and Milutinović 1993], we suggested some changes that make caching possible in RM systems. Each block in the RM memory should be associated with 1 bit, which denotes its caching status (C bit). When a block is cached, its C bit should be set. After that, when a write from the RM bus to a shared location arrives, the C bit will be checked, and if it is set, invalidation will be sent to that cache, and the C bit will be reset. Only one additional bit per block,

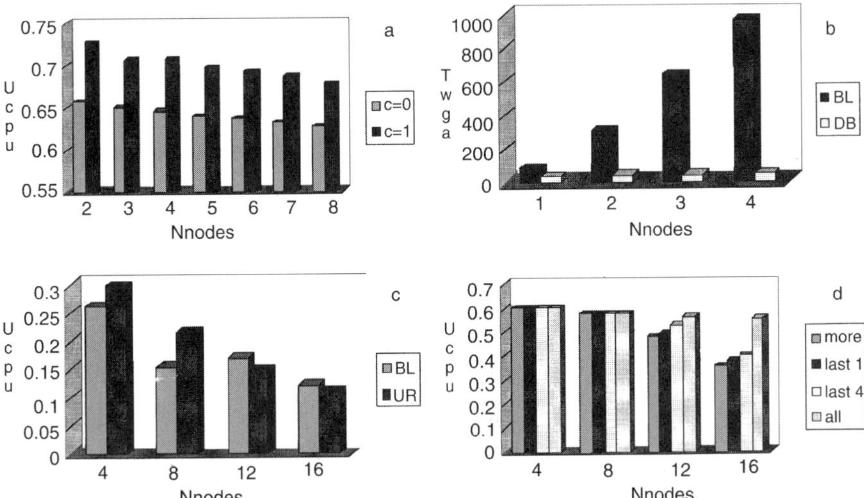

**Figure A.3.** Performance results for four proposed improvements (Ucpu—processor utilization; Nnodes—number of the RM nodes; Twga—average waiting time for the RM bus grant; C = 0—without caching; C = 1—with caching; BL—baseline system with single bus and round-robin policy; DB—double bus system; UR—urgent request arbitration). Panel *a* shows the comparison of processor utilization with and without caching. Panel *b* shows the comparison of average waiting time for the bus grant in a single (BL) and double (DB) bus RM systems. Panel *c* shows comparison of processor utilization for a system with the round-robin policy (BL) and the dynamic priority policy with urgent request arbitration (UR). Panel *d* shows the impact of four write-filtering variants on the processor utilization.

together with the very simple comparator logic, represents reasonable complexity paid for significant performance gain (Fig. A.3*a*). The simulation shows that caching of the shared memory segments results in an improvement of about 10% for the CPU utilization. The improvement is better for smaller node counts, since for larger node counts, the contention on the RM bus caused by multiple invalidations becomes significant. The benefits of caching should have been even more pronounced if the system follows writeback policy.

*Improvement 2: Double RM Bus.* Since the shared bus represents a notorious bottleneck in bus-based systems, the natural solution for the restricted scalability is to employ multiple buses. We have analyzed the solution with a double RM bus. When the need for accessing the RM bus arises, the bus request is issued on both buses at the same time, yet only one bus will be granted that transfer (whichever grant comes first). The proposed solution can be efficient in cases of both transient and permanent bus overloads, in terms of traffic of writes to the shared memory regions. The overall effect is that the average time to obtain the bus is decreased, which results in the higher system throughput. This also offers the potential for increased system reliability. However, the cost of this approach is significant because of the introduction of the second bus, and the hardware complexity of the bus interface logic is increased. This complexity increase can be justified only by a satisfactory performance improvement (Fig. A.3*b*). Under conditions of saturated bus traffic in the single-bus RM system, the average waiting time for the bus grant increases linearly, but in the double bus system this time is insignificant and practically insensitive to the system size.

*Improvement 3: Improved RM Bus Arbitration.* The RM bus controller assumes equal priority for all requests coming from each node. However, the overload of the nodes can be different, so some FIFOs can be full while FIFOs in other nodes are empty. The solution for this problem is to give a higher priority to the overloaded node, from the moment it reaches the "almost full" condition. When the number of items held in the transmit FIFO buffer reaches a certain threshold, the "urgent need" for the RM bus will be signaled. Then, the RM bus arbiter temporarily departs from the round-robin policy and gives the upper hand to that node. This priority will be assigned temporarily (for only one transfer on the RM bus). When no urgent-need signals are asserted, the arbiter is back to the round-robin arbitration. For the purpose of announcing the priority request, bus lines reserved for the node ID can be used, since they are free during the RM bus transfer. This approach introduces no additional bus lines but requires very careful timing. The simulation shows that under conditions of increased bus traffic, this method improves processor utilization, but only for a lower number of nodes, while in the hypothetical system with 12 or 16 nodes it has no advantages (see Fig. A.3*c*).

*Improvement 4: Write Filtering.* The essence of this improvement is the filtering of subsequent write requests issued to the same address by the same node, while these requests are sitting in the transmit FIFO of the writer. This method helps in reducing the traffic on the RM bus. According to the number of previous writes in the FIFO that should be compared to the issued write operation, we have proposed three solutions: (1) variant "all"—comparison with the entire contents of the transmit FIFO, (2) variant "last 4"—comparison with the last four writes, and (3) variant "last 1"—comparison with the previous write only. Performance improvements for all these variants are given in Figure A.3d. The best effects of this solution can be expected for applications with high processor locality of memory references (long write runs). When applying this improvement, one must be careful about the consistency model assumed by the programmer, since the loss of some write operations is allowed only for relaxed consistency models. Simulations were performed for no write filtering and for variants "last 1," "last 4," and "all." Processor utilization is almost the same for all four variants in the systems with less then eight nodes; however, in systems with larger node counts, the difference becomes significant, in favor of write filtering applied to the longer sequences in the transmit FIFO buffer.

## A.5. FINAL RESULTS AND LESSONS LEARNED

Designing a board to match the vague specification was not only a technical challenge but also an organizational one. Being the leading manufacturer of RM systems for a long time, Encore was eager to innovate their product by implementing research results generated in the academic environment. Entrusting the project to the university 7000 miles away was a bit risky, but due to the benefits of the information superhighway, it was necessary to travel from one continent to another only three times. In spite of the cultural mismatch and all the organizational differences, the development and basic testing were done at the university, and the detailed testing was finished at Encore, with the assistance of their engineers. The result was the fully operational board (see Fig. A.4). This technology transfer showed that globalization is a promising trend, and that the cooperation of industry and research institutions can be really fruitful.

However, the dark side of the industrial–academic cooperation was unveiled when the time came to publish the research results. Because of the proprietary limitations, Encore was not willing to provide confidential data generated by experiments and performance measurements on the designed board. Besides that, the traces used as input of our simulations were not generated by some standard benchmark applications, but they were produced by the team selected by the research sponsor. They provided us with files containing address traces that they believed to be representative of the end-user programs and applications relevant to Encore. Furthermore, the

**Figure A.4.** Photograph of the PC-RMS board.

sponsor was not cooperative in providing detailed characterization of these applications.

The lack of detailed characterization of traces used in simulations, as well as measurement results, made performance figures we obtained incomparable to the similar research efforts in the field. Although not representative of all realistic applications, standard benchmarks such as SPLASH-2 seem to be still inevitable in the research work that pretends to be published in major journals. The attitude of the reviewers toward our research papers was generally positive: they did not want to reject our manuscripts, but requested "minor" revisions, claiming that any hiding of information about conditions and assumptions of presented experiments is not tolerable by the research community. In this way, we were faced with a problem we could not solve, but at least we learned a lesson; when cooperating with sponsors from industry, one has to be precocious and make a subcontract concerning conditions for publishing of research results. Otherwise, the only solution is to publish articles in company journals, which certainly have reduced visibility, compared to IEEE or ACM publications.

Fortunately, hard work always leads to results. While exploring the possible enhancements of the RM concept, in order to achieve the theoretical goal of the project, after the incubation period has passed, we came up with some general ideas, applicable in the wider area of DSM systems, as well as some solutions in the field of cache consistency in multiprocessor systems. Even two years after the project was finalized, these ideas were applicable in other projects, and resulted in several thesis and research papers, summarized in Milutinović (1996). The most important new concept was the STS idea, which suggests the separation of data that expresses predominantly spatial locality from the data that expresses predominantly temporal locality, and different treatment of these two data spaces [Milutinović 1996]. The continued efforts inspired by the write filtering idea presented here resulted in even more sophisticated improvements of the RM system [Jovanovic et al. 1995], while the application of some concepts found in RM to the cached data in multiprocessor systems resulted in a new solution for lazy prefetching [Milenković and Milutinović 1998].

Finally, was it worth the effort? Instead of the research paper published in a respected journal (which is probably what every researcher dreams of when starting a new project), we ended up with a working board (to the best of our knowledge, the first implemented multiprocessor system with reflective memory DSM mechanism, using the PC-based computer nodes) and a lot of experience and accumulated knowledge, which resulted in a mature view of the field, presented in Milutinović (1999). The detailed survey of DSM concepts and systems that we did, inspired by this project, and the great collection of papers we read, did not finish in our archive; instead, we also published a tutorial book with the state-of-the-art works in the field of DSM [Protić et al. 1998].

## A.6. CONCLUSION

This project was a pioneer effort from two points of view—it was organized in cooperation between a university and an industrial company located on two distant continents before the concept of globalization and virtual project teams became commonplace, and it resulted in the first commercial DSM system based on a network of PCs. Projects like this bring great benefits to the industry, allowing efficient outsourcing of some research and development activities to the appropriate research centers. Thanks to the information superhighway, geographic limitations no longer apply when choosing the most suitable partners. On the other hand, university researchers get the opportunity to cope with real-world problems and to incorporate their knowledge into technology development. However, the obstacles to the process of globalization and sharing of results with the research community are industrial secrets and proprietary limitations, which slow down the free flow of information. The avenue we opened with this project has proved to be a prospective one; in the years after we completed our research, several other approaches in the field of DSM for PCs were initiated and proved to be successful.

### Acknowledgments

The authors are grateful to their colleagues from Encore (now Compaq): Z. Aral, I. Gertner, M. Natale, A. Gupta, N. Tran, and B. Grant for their contribution to the project presented in this appendix. A. Milenković and D. Raskovic from the University of Belgrade, who took part in the follow-up projects, also joined many of our discussions and helped us to understand better some important issues. The research work of Dr. V. Milutinović was partially supported by Purdue University, under grant 750-1285-0055. The entire research was inspired by Encore Computer Systems, after grant 200-3/4.

### REFERENCES

Bershad, B. N., Zekauskas, M. J., Sawdon, W. A., "The Midway Distributed Shared Memory System," *Proceedings of the COMPCON 93*, Feb. 1993, pp. 528–537.

Blumrich, M., et al., "Virtual Memory Mapped Network Interface for the SHRIMP Multicomputer," *Proc. 21st Annual Int. Symp. Computer Architecture*, IEEE Computer Society Press, Los Alamitos, CA, 1994, pp. 142–153.

Denton, R., Johnson, T., *Distributed Shared Memory Using Reflective Memory: The LAM System*, Univ. Florida Technical Report (TR96-021), Gainesville, FL, July 1996.

Encore Computer Corp., *RMS Functional Specification*, Encore Computer Corporation, Fort Lauderdale, FL, 1990.

Fleisch, B., Popek, G., "Mirage: A Coherent Distributed Shared Memory Design," *Proc. 14th ACM Symp. Operating System Principles*, ACM, New York, 1989, pp. 211–223.

Furht, B., "Architecture and Performance Evaluation of the MMM," *IEEE TC Comput. Architecture Newsl.*, 66–75, (Mar. 1997).

Gertner, I., *The Reflective Memory/Memory Channel System Overview*, Encore Computer Systems, Fort Lauderdale, FL, 1993.

Gharachorloo, K., et al., "Memory Consistency and Event-Ordering in Scalable Shared-Memory Multiprocessors," *Proceedings of the 17th International Symposium on Computer Architecture*, 1990, pp. 15–26.

Gillett, R., "Memory Channel Network for PCI," *IEEE Micro* **16**(1), 12–18 (Feb. 1996).

Hoevel, L., Milutinović, V., "Terminology Risks with the RISC Concept in the Risky RISC Arena," *IEEE Comput.*, 136 (Dec. 1991).

Jovanovic, M., Tomašević, M., Milutinović, V., "A Simulation-Based Comparison of Two Reflective Memory Approaches," *Proc. 28th Hawaii Int. Conf. System Sciences* (HICSS-95), IEEE Computer Society Press, Los Alamitos, CA, 1995, pp. 140–149.

Jovanovic, M., Milutinović, V., "An Overview of Reflective Memory Systems," *IEEE Concurrency* (in press).

Keleher, P., Cox, A. L., Dwarkadas, S., Zwaenepoel, W., "TreadMarks: Distributed Shared Memory on Standard Workstations and Operating Systems," *Proc. USENIX Winter 1994 Conf.*, Jan. 1994, pp. 115–132.

Lenoski, D., Laudon, J., et al., "The Stanford DASH Multiprocessor, *IEEE Comput.* **25**(3), 63–79 (Mar. 1992).

Li, K., "Development Trends in Distributed Shared Memory," *Panel of 22nd Int. Symp. Comput. Architecture*, Santa Margherita Ligure, Italy, June 1995.

Lucci, S., Gertner, I., Gupta, A., Hedge, U., "Reflective-Memory Multiprocessor," *Proc. 28th IEEE/ACM Hawaii Int. Conf. System Sciences*, Maui, Hawaii, Jan. 1995, pp. 85–94.

Maples, C., Wittie, L., "Merlin: A Superglue for Multicomputer Systems," *COMPCON 90*, Mar. 1990, pp.73–81.

Milenković, A., Milutinović, V., "Lazy Prefetching," *Proc. HICSS-98*, Mauna Lani, Hawaii, Jan. 1998.

Milutinović, V., "Some Solutions for Critical Problems in the Theory and Practice of DSM," *IEEE TC Comput. Newsl.*, 7–12 (Sept. 1996).

Milutinović, V., Tomašević, M., Markovic, B., Tremblay, M., "The Split Temporal/Spatial Cache: Initial Performance Analysis," *Proc. SCIzzL-5*, Santa Clara, CA, March 26, 1996.

Milutinović, V., *Surviving the Design of Microprocessors and Multiprocessor Systems: Lessons Learned*, IEEE Computer Society Press, Los Alamitos, CA, 1999.

Oguchi, M., Aida, H., Saito T., "A Proposal for a DSM Architecture Suitable for a Widely Distributed Environment and its Evaluation," *Proc. 4th IEEE Int. Symp. High Performance Distributed Computing* (HPDC-4), Washington, DC, Aug. 1995, pp. 32–39.

Protić, J., Tomašević, M., Milutinović, V., "Distributed Shared Memory: Concepts and Systems," *IEEE Parallel Distrib. Technol.*, 63–79 (1996a).

Protić, J., Tartalja, I., Tomašević, M., "Memory Consistency Models for Shared Memory Multiprocessors and DSM Systems," *Proc. 8th Mediterranean Electrotechnical Conf. Melecon 96*, Bari, Italy, May 1996, pp. 1112–1115.

Protić, J., Tomašević, M., Milutinović, V., eds., *Distributed Shared Memory: Concepts and Systems*, IEEE Computer Society Press, Los Alamitos, CA, 1998.

Ramanjuan, R., Bonney, J., Thurber K., "Network Shared Memory: A New Approach for Clustering Workstations for Parallel Processing," *Proc. 4th IEEE Int. Symp. High Performance Distributed Computing* (HPDC-4), Washington, DC, Aug. 1995, pp. 48–56.

Savic, S., Tomašević, M., Milutinović, V., "Improved RMS for the PC Environment," *Microproc. Syst.* **19**(10), 609–619 (Dec. 1995).

Stumm, M., Zhou, S., "Algorithms Implementing Distributed Shared Memory," *IEEE Comput.* **23**(5), 54–64 (May 1990).

Systran, *SCRAMNet + Protocol Description*, Systran Corporation, Dayton, OH, 1995.

Tomašević, M., Milutinović, V., eds., *Cache Coherence Problem in Shared Memory Multiprocessors: Hardware Solutions*, IEEE Computer Society Press, Los Alamitos, CA 1993.

VMIC, *VMIC's Reflective Memory Network*, VME Microsystems International Corporation, Huntsville, Alabama, 1995

# Limes: A Multiprocessor Simulation Environment for PC Platforms

This appendix* presents a multiprocessor simulation environment, developed to facilitate research on multiprocessor systems using widely available hardware platforms. It comprises simulation tools, including both an execution-driven and a trace-driven simulator, applicable for memory architecture studies of shared address space multiprocessors. It also includes a detailed model of a bus-based cache-coherent symmetric multiprocessor system. The execution-driven simulator has a scheduling algorithm specially optimized for speed. It can run parallel applications based on the ANL programming paradigm, such as those found in the SPLASH-2 suite. The package also includes an extension for writing object-oriented parallel applications in the Java-like manner. Trace-driven simulation is based on a new concept of ideal traces. The trace-driven simulator supports an original technique for abstraction of events that can introduce timing dependencies in a trace, which, in turn, enables accurate simulation. Both execution-driven and trace-driven simulators can work using the same memory architecture simulators. The environment provides a simple general interface that allows for the hardware lying underneath the simulated processors to be modeled using object-oriented programming. The package currently runs on PC platforms with Pentium or a newer processor under the Linux operating system.

## B.1. INTRODUCTION

Simulation plays a vital role in multiprocessor studies. In a variety of simulation techniques, ranging from analytical modeling—which is often inadequate to model complex multiprocessor interactions—to hardware prototyping, which is costly and inflexible, software simulation has become relatively popular. Software simulation has certain benefits that make it the dominant method for validating proposed architectures. Software simulators

---

*Prepared by Igor Ikodinovic, Zoran Dimitrijevic, Davor Magdic, Aleksandar Milenković, Jelica Protić, and Veljko Milutinović.

are easier to develop, are less expensive than their hardware counterparts, and are able to perform simulations with a high level of accuracy. They are also more flexible, allowing frequent changes of simulation parameters and easy changes in the simulated architecture; this is significant because details of the architecture under evaluation may frequently need readjusting, according to the simulation results. In addition, as many simulations as there are available, host machines can be run simultaneously, which is important considering the number of experiments that usually have to be carried out.

The world of software simulation comprehends several different simulation methods, with a number of tools that follow them [Uhlig and Mudge 1987]. Certain trade-offs between accuracy, speed, flexibility, expense, portability, and ease of use are present in every simulation method. These issues should carefully be considered when evaluating simulation techniques or comparing them with each other. A rapid development in the computer field can also change conditions, making one method the best at some point in time but leading toward the introduction of new methods or the reemergence of some old ideas, so occasional reevaluation of the simulation techniques is necessary.

Currently, one of the most popular methods for simulation of multiprocessors [Goldschmidt 1993] is an execution-driven simulation, due to its speed and accuracy. This method has been used in a number of popular simulation tools. However, this method does not enable OS references to be included in the simulation. Trace-driven simulation* is another method that was widely used in the past but was replaced by other techniques, mostly due to its need for large disk space, problems with low disk transfer rates, and inability to accurately simulate complex interprocess interactions. Yet, with the rapid development of technology, disks with capacities of 10 GB or more and with transfer rates over 10 MB/s have become widely accessible, eliminating some of the drawbacks of this method. Trace-driven simulation can be performed with traces that can contain memory references from any source, including those from OS. Limes benefits from using both of these two simulation methods.

Limes consists of two simulators (an execution-driven and a trace-driven simulator) and a modifiable software representation of a realistic multiprocessor system. It currently runs on PC platforms with Pentium or newer processors under a Linux operating system.

The following sections will describe Limes. In Section B.2 we explain the goals that authors sought to satisfy with Limes and other requirements set before it. In Section B.3 we discuss the existing simulation tools and why

---

*The term *trace-driven simulation* sometimes pertains to all types of software simulations, meaning that a stream of memory references constitutes a trace no matter where it comes from (from direct execution or from a file). Here, by *execution-driven simulation* we mean that memory references come from direct execution of the instrumented code, and by *trace-driven simulation* that memory references come from a file residing on a disk.

Limes is different from them. Section B.4 presents the Limes structure in detail, including the execution-driven and trace-driven simulators and their internal algorithms, and gives insight into one realistic memory simulator on an SMP system example. Section B.5 deals with the Limes complexity and performance. Section B.6 gives the brief installation guide for installing and using Limes. Section B.7 gives some examples of projects where Limes was used. We conclude with Section B.8.


## B.2. GOALS AND DEMANDS

The nature of research that had to be performed, mainly involving shared memory multiprocessor studies, imposed a specific set of demands for a simulation environment that would be considered appropriate.

1. It had to be executable on PC platforms, because of the estimated number of experiments, the availability of such platforms in the environment, and the general inaccessibility and lower number of high-end machines. Still, we wanted to maintain the possibility of porting the environment to other platforms as needed.

2. Simulations were meant to be driven by realistic workload—the kind of workload that would execute on the proposed hardware platform once it comes to life. The character of the workload greatly influences the performance indicators, and using inadequate workload can often be misleading. For the execution-driven simulator we had to choose an appropriate set of applications it can run, in order to give the simulation results the necessary validity. The SPLASH-2 application suite [Woo et al. 1995] was considered a preferred workload, since this set of benchmarks became a de facto standard among the researchers involved in multiprocessor studies. The trace-driven simulator was meant to be able to work with traces generated in our environment, or elsewhere, as needed. Having both types of simulators would enable simulations with a wide range of different workloads.

3. The simulators were to deliver a high level of accuracy, keeping in mind the character of the studies, like simulations of bus-based cache coherence protocols [Tomašević and Milutinović 1992]. Memory architectures that are studied often have certain similarities in their structure, and the subtle differences that exist in their internal organization dictate not only that the simulation be precise, in order to accurately measure the impact of these differences on performance, but also that the simulator can easily be changed and adapted, so that little effort is spent when changing details in their structure. These and other requirements suggested the use of object-oriented programming (OOP) techniques in the building of the memory simulators. Writing a simula-

tor using an OO language (such as C++ ) allows great freedom for the writer of the simulator. The OO approach is also good when considering the desired level of efficiency. Having in mind the number and volume of experiments that need to be performed and a need for frequent changes of simulation parameters and memory architecture details, the advantages of the OOP approach become fully apparent.

Certain other conditions had to be fulfilled, such as the need to develop memory simulators independently from the simulation kernel, which would give an opportunity for development and testing of systems using parallel traces that come from different sources (i.e., from direct execution or from a trace residing on a disk). A part of the simulator that handles the instrumentation of the application assembly code was also meant to be written to be independent, which would allow us to change it easily when switching to different platforms and thus requiring no other changes in the rest of the environment in that process.

## B.3. EXISTING SOLUTIONS

The vast majority of existing simulation tools is developed for high-end multiprocessor computers. If made for uniprocessor machines, it is almost exclusively for platforms with RISC processors (mainly MIPSes and SPARCs). Sophisticated tools such as SimOS [Rosenblum et al. 1995] (runs on a MIPS-based SGI multiprocessor) and SimICS [Magnusson et al. 1998] (runs on SPARC machines) can simulate target architectures with a high level of accuracy using instruction set emulation. They are able to simulate the execution of an entire realistic operating system on a target machine, including complete simulation of the I/O subsystem. Besides the operating system itself, all kinds of applications can be used as a realistic workload for the simulations as well. The only thing that does not allow these tools to be considered perfect, except that they work only on RISC platforms, is the speed of simulation, which is still slower than that of execution-driven simulation.

Tools such as Tango [Davis et al. 1990] and its successor TangoLite [Herrod 1993], or CacheMire [Brorrson et al. 1993], are widely used execution-driven simulators; but again, they work only on RISC platforms. They cannot ensure a satisfying level of accuracy if ported to a different hardware platform, such as a CISC uniprocessor. Augmint [Sharma et al. 1996] is the only such tool that does in fact run on a PC (with a minor drawback of omitting iAPx86 string instructions and standard library routines from instrumentation). Augmint has an advantage that it also runs under Windows NT, as well as under a Solaris operating system on SPARC machines (portability was not of primary importance for Limes at the time, but we do consider porting it to other platforms now). However, a trace-driven simulator is not included in the Augmint environment, making it virtually impossible for OS references to be used as a workload in the simulations. TangoLite and CacheMire also lack the possibility of trace-driven simulation. They can only produce traces.

Keeping in mind insufficiencies of the existing tools regarding trace-driven simulation, nonexistence of such tools for PC platforms and for CISCs in general (Augmint was also just being developed at the time), and an awareness that every tool almost invariably requires modifications in order to allow for particular effects to be simulated and measuring techniques to be added, we felt that it would be worth the effort to invest some time in developing a simulator that would be well suited for our research goals, rather than to modify the existing tools. However, our demands were so widely set that the environment we wanted to develop actually represented a general tool for simulations of all shared memory multiprocessor architectures.

## B.4. LIMES STRUCTURE

The Limes simulation environment comprises both an execution-driven simulator and a trace-driven simulator. They enable a multiprocessor system to be simulated on a uniprocessor host machine, in our case a PC based on Pentium or a newer processor. The simulator of the target system* is devised to be independent from the source of memory references. This enables both execution-driven and trace-driven simulators to use the same memory system simulators. Memory/synchronization references come either from the execution streams of the parallel application's threads or from a trace residing on a disk, depending on the type of simulation. This is shown in Figure B.1. In the current version of Limes (v2.0) only one thread can be assigned to one processor.

The environment also includes an extension of the ANL programming paradigm, so that users can develop, test, and use as workload object-oriented parallel applications written in C++, but in a manner very similar to that employed by the Java programming language. The whole system is rather small, built on top of Limes. It can be used on any real machine that supports operations defined in ANL macros. The system implements only two classes—threads and monitors—and programs written using them are more readable and understandable than those using raw LOCKs and UNLOCKs. This can be of further value for researchers who may not necessarily be concerned with architecture studies but are focused primarily on investigating parallel algorithms.

---

*The simulator of the target system actually simulates the behavior of the target memory system. The memory system, for example, can include caches, TLBs, interconnection networks, or global memory. Simulation of other parts of the system (e.g., I/O system) is not considered here. These are the reasons why this simulator is also called a *memory system simulator*.

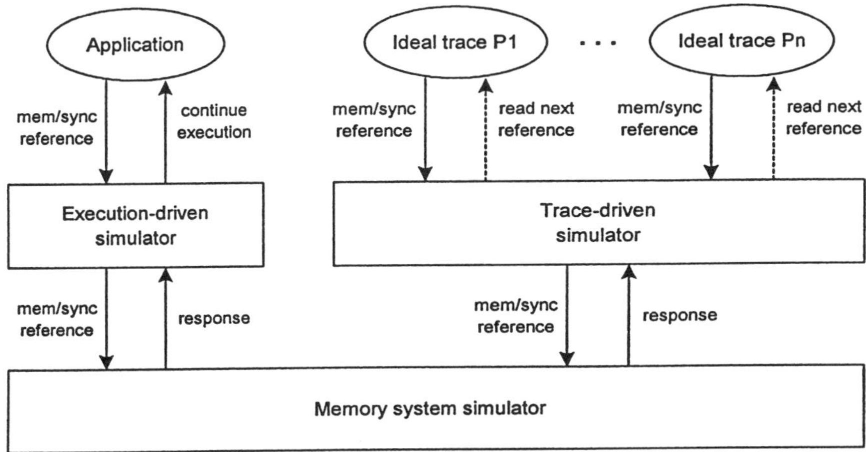


**Figure B.1.** Simulation in the Limes environment.

## B.4.1. Execution-Driven Simulator

Execution-driven simulation has two major parts. The first part is "static"—building an executable simulation. The second part is 'dynamic'—executing the simulation. Building the executable simulation will be discussed first. Details of the simulation will be described after that. At the end, it is explained how the optimized algorithm for thread scheduling works.

### Building an Executable

Applications that can be used to drive this simulator are like those from the SPLASH-2 suite, or more generally, those that use ANL (Argonne National Lab) macros for expressing parallelism. Application assembly code (created by the compiler) is instrumented with kernel callouts at compile time; the source code does not require any alterations for that. The simulator allows three levels of assembly code instrumentation, depending on what type of events should be instrumented. Level 0 instruments only synchronization primitives and user-defined machine instructions (which are also possible to define in Limes—e.g., prefetch, forward, etc.), level 1 instruments shared reads and writes as well, and level 2 instruments all other memory references, including local ones. The simulation speed is reciprocal to the instrumentation level. Level 0 may be appropriate for investigation of parallel algorithms; level 1, for shared memory architecture evaluations; and level 2, for simulations of caches.

Most C/C++ programs call standard library functions, and so do the parallel applications. Limes by default instruments all standard library functions whose argument can be a pointer, as they might read or write global memory without the control of the simulator. Other library functions are not instrumented (which has virtually no impact on the simulation accuracy), but can be, if needed; the instrumentation tool is open for additions. To avoid collision with library functions that are called by the simulator itself, instrumented functions are prefixed, and a set of macros is defined that enables redirection of calls to the instrumented functions instead of calling the originals. TangoLite, for example, can instrument only those portions of the run-time libraries that are not used by its own run-time system.

What makes the whole instrumentation process hard is that the CISC (unlike RISC)* instruction set is relatively complex, containing many instructions, which are frequently nonuniform, so that many addressing combinations are possible.

The instrumented application code, the simulator, and the memory system simulator are finally compiled and linked together into a single executable. The process of compilation and linking is controlled via a set of make files. If the application code is compiled with an appropriate option, the compiler produces some debugging information that the instrumentation tool can understand. If the application crashes, the simulator will use those pieces of information and print the source line that the offending thread was executing at that moment. It is worthy to know that the instrumentation does not prevent the application to be debugged with a standard debugger.

### *Simulation*

The whole simulation (application threads, simulator kernel, and memory simulator) executes in the context of a single UNIX process. Basically, during the execution the simulator scheduler acts like a layer between the parallel application and the simulated memory system. Parallel application executes its native machine code, one thread at a time, until it encounters an event of interest, such as a memory/synchronization reference (read/write, lock acquire/release, or some user-defined instruction). Then it gets rescheduled, and the actual operation is deferred until its timestamp reaches the global

---

*One important issue deserves attention here. Knowing that the simulator will execute applications that are compiled for a CISC processor, the question is whether the simulation results are bound to the behavior of such a processor. Can it be used for simulating some future RISC multiprocessor, too? It is hard to give an exact answer, but our results show that it is possible. One real SPLASH-2 application (FFT) was executed on a MIPS R2000 DEC station and one on a Pentium-based PC, with the same parameters (65536 complex doubles—a realistic problem size). The first simulation was performed using TangoLite, and the second one using Limes. MIPS generated 18.393 millions of shared reads and 12.908 millions of shared writes, while Pentium generated 18.552 millions of shared reads and 12.782 of shared writes during the execution of the application. The results indicate that for a real application, a RISC such as MIPS R2000 generates approximately the same number of shared reads and writes as Pentium does (in both cases the discrepancy is less than 1%). CISCs certainly generate more private references, but these references are generally of little importance to multiprocessor studies; shared references are what determine the behavior and performance of multiprocessor systems.

simulation time. The simulator is responsible for multiplexing threads and for scheduling of their memory requests at proper times, in correct order. Context switching that occurs immediately before an instruction of interest executes is realized through the replacement of such instructions with kernel callouts at compile time. The programming model supported is the *lightweight threads* model (all processes share the address space with the master process, except for the stack area, which is privately reserved for each process). Instruction execution times are calculated at the end of each basic block (a basic block ends with a branch, a label, or a memory reference). They are calculated by adding (a) the time needed for an instruction with a memory reference to complete and (b) the execution times of other instructions in the basic block. At compile time, each instruction is associated with a number that represents the number of cycles it takes to execute before the execution of the next instruction begins. For most instructions this number is one. By changing these values, it is possible to model a faster or a slower processor. For example, by decreasing their value, we can roughly simulate a super-pipelined processor. Simulation time is expressed in processor cycles of the *target* multiprocessor system.

If the simulation is run with an appropriate option, during the simulation process the simulator will produce a trace that contains references according to the level of instrumentation, along with additional information that can be used to support accurate trace-driven simulation; file format is open for the user to change it.

### *Optimized Scheduling Algorithm*

Scheduling is necessary during the execution of the simulation to maintain correct interleaving of memory activities of the application threads. The scheduling algorithm of the simulator can substantially influence the performance of the simulation, because scheduling activities are very frequent during the simulation. It is important to keep this overhead as low as possible. Other sources of overhead are related to the instrumentation process and to the memory system simulator.

The relative influence of the scheduling overhead on the simulation performance depends on the amount of overhead introduced by the other two factors. Having in mind the frequency of scheduling activities, it is certainly worthwhile to reduce it as much as possible.

Accurate simulation can be accomplished if the memory simulator was called after every cycle in the simulation, regardless of whether any requests exist in that cycle. However, it would slow down the simulation significantly. Another, equally correct, approach is chosen in which the scheduler calls the memory simulator in subsequent simulated cycles only if there is at least one new request or at least one stalled thread; idle cycles can be skipped. The other optimization regards the execution stream of the nonglobal instructions. The scheduling algorithm will force the continuous execution of nonglobal instructions as much as possible, thus saving the time needed for frequent invocations of the scheduler. The scheduler will call the memory simulator only if there are no threads that can continue execution, because they all wait for a memory operation to complete. An additional optimization regards shared writes. Threads are allowed to continue execution without first waiting for shared writes to be completely simulated, deferring that job as long as possible and thus saving the time for frequent calls to the memory simulator.

The scheduler works as follows. When a memory reference is met in a thread's execution stream, the scheduler is invoked. The scheduler then does not invoke the memory system simulator right away, but instead checks to see if there are any threads that have become free to continue execution, because their memory requests have been satisfied in the meanwhile. If there are such threads, then one of them with minimal time (there may be more than one) is scheduled to continue its execution, and the scheduler will wait for some other thread to call it again. If there are no threads that can continue execution, that is, if all of them are waiting for some memory request to complete, then the scheduler enters a loop while calling the memory simulator in every cycle, until some thread gets its request satisfied and enables it to continue execution. Such a scheduling algorithm imposes that the memory simulator must be called with two parameters: time_now and last_simulation_time. The memory simulator then performs simulation between these two moments. If time_now > last_simulation_time is called, this means that at last_simulation_time all threads' requests were satisfied. However, it does not mean that the memory simulator can simply skip this time. This is due to the optimization that is performed when simulating shared writes. When a shared write occurs, the memory simulator will return a *satisfy* signal right away, even if the actual write was not simulated. This way a thread can continue execution without delay. The actual shared write will be simulated later, on some other call to the simulator.

The memory simulator determines how many shared writes it can buffer before a coming write must wait for the first one in the buffer to be actually simulated. This is the reason why, on every call, the memory simulator first checks to see if there are any unfinished shared writes, and simulates them first if there are some. After the simulator establishes that there are no more unsimulated events, it can freely skip the rest of the time, all up to time_now.

## B.4.2. Trace-Driven Simulator

Making a trace of some program's execution ensures that a simulation can be split into two phases: local work and a memory system simulation. Local work is not of importance for multiprocessor studies. Trace-driven simulation performs only the memory simulation part, while local work is already built into a trace via timestamps. If multiple simulations are performed with the same trace, then trace-driven simulation can potentially bring speed-up over multiple execution-driven simulation runs. However, because of disk transfer rate limitations, this is not always possible. Multiple runs can yield different results when using execution-driven simulation if nondeterministic scheduling is performed. This is possible even with static scheduling of workload. This is not the case with trace-driven simulation. Trace-driven simulation also enables the use of traces that contain OS references, while execution-driven simulations cannot include OS references.

There is one more benefit when using traces as a workload. When comparison of two architectures is needed, and their simulators are not working on identical platforms, or even with the same simulators, trace-driven simulation enables completely accurate comparison (only an instruction interpretation method enables that, too). Using execution-driven simulation, for example, on one RISC and one CISC platform, would give completely different results for the same workload. The same traces, on the other hand, contain the same local work, so the comparison is reduced only to the memory systems that are both equally simulated.

The Limes' trace-driven simulator uses as input a trace generated by the execution-driven simulator from the Limes environment, or from some other source. Limes is very flexible concerning the trace file format. Its current version supports two formats: a text trace file format that can be viewed with a simple text editor, and a binary trace file format that is significantly shorter and used for storing large traces. Other formats can easily be implemented by altering the appropriate module of the trace-driven simulator (and the module for trace generation of the execution-driven simulator, if traces are produced with it). Every format, however, should be able to support *abstraction* by allowing certain additional information for each reference to be stored in a trace.

Limes enables accurate trace-driven simulation, where possible, by using the method of abstraction to eliminate timing dependencies. The technique used to support abstraction in Limes is original. The simulator currently supports the abstraction of shared reads/writes, locks/unlocks, user-defined instructions (prefetch, forward, etc.), barriers, and thread creation events. It is, of course, open for changes and ready to support abstraction of other timing-dependent operations, and of different implementations of currently supported primitives.

### Accuracy Issues

The execution path of a multiprocessor workload depends on the ordering of events in a system, which in turn depends on the machine memory system timings. When timing dependencies are present, a small change to the memory system architecture can induce numerous changes to the execution path of a program and cause inaccurate simulation. This happens because the trace itself seldom contains any information about the program's execution path.

This information is implicitly built into a trace via timestamps and is valid only while the memory system is not changed. If the memory system is changed, timings will change as well, and timestamps will no longer contain correct information about the program's execution path.

Holliday and Ellis (1990) find that traditional traces are not adequate for accurate trace-driven simulation, and they propose *intrinsic traces* as an alternative (an intrinsic trace consists of the control-flow graph of the workload plus timing and address data for each basic block). They argue that accurate trace-driven simulation without partial reexecution of the program is possible only for so-called graph-traceable programs (those are programs where all addresses can be determined during simulation, based only on the information gathered from the trace). In reality, the range of programs for which accurate trace-driven simulation can be obtained at a reasonable cost reduces to a class of programs whose threads have execution and data paths that cannot be influenced by other threads. Most applications comply with this condition.

As discussed in Goldschmidt and Hennessy (1992), accurate trace-driven simulation can be obtained (where possible) only by eliminating timing dependencies from the trace by abstraction of the operations that cause them. To eliminate timing dependencies by means of abstraction, we must first ensure that all necessary information is recorded in a trace, and then to support the abstraction on the side of the simulator (at the expense of additional simulator complexity).

Koldingr et al. (1991) discussed whether traces generated from multiple runs of the same program will yield the same results, and if tracing-induced dilation affects simulation accuracy. As for the first issue, the trace we once obtain can be used for simulation of different memory architectures, so we do not need multiple simulation runs to obtain different traces for different architectures. We can thus achieve accurate simulation, if the trace contains all the necessary information. As for the second issue, Limes produces traces without the time dilation effect, but we cannot guarantee that trace collection techniques used elsewhere [Stunkel et al. 1991] will not introduce that same effect. For example, like Limes, TangoLite, Augmint, and other execution-driven simulators that are used for trace collection do not introduce time dilation. Traces generated by SimOS or SimICS also do not suffer from the time dilation effect. This is because they all perform tracing on the *simulated* architectures. Microcode modification used in ATUM [Agarwal et al. 1986] and a technique of inline tracing used in MPtrace [Eggers et al. 1990] and TRAPEDS [Stunkel and Fuchs 1989] does introduce time dilation, as they are used to trace programs on a *host* machine. The dilation effect, however, is often negligible.

### *Abstraction*

Abstraction of operations that can influence a program's execution path means that some information about them is stored in a trace, so that they can be correctly redone during the simulation. That way the simulator will no longer be bound to the timestamps when the execution path is concerned, but will be able to maintain the correct ordering by redoing the critical timing-dependent operations.

Timing-dependent operations include:

- Shared memory references (reads and writes)
- Synchronization operations (such as locks and barriers)
- Operations for dynamic scheduling of workloads (e.g., allocation of tasks, loop iterations, or memory)
- Other timing-dependent operations such as thread creation

Timing-dependent operations other than those that can influence the program execution path (like those involving real-time clock)* can also be abstracted.

We divide timing-dependent operations into two groups: elementary and complex. They are abstracted in a different way. Elementary operations are those that are realized in the simulated hardware as primitives. Complex operations are realized through the use of elementary operations. In our case elementary operations are shared reads/writes, lock acquires/releases, and user-defined instructions, while barriers are realized as complex operations.

Timing dependencies can be eliminated from elementary timing-dependent operations by recording their completion time. In that way the simulator knows the amount of "pollution" introduced by the architecture on which the trace was generated and eliminates it.

Complex operations are much harder to abstract, depending on how "complex" they are. Barrier implementation via locks, used in Limes, is an example of a complex operation that needs significant support on the side of the trace-driven simulator. Apart from that, the trace itself must contain rather detailed information on each barrier that has to be provided by the trace generation environment.

In general, it is up to the trace-collecting tool to save all the necessary information in the trace. The abstraction technique we use is responsible for performing an accurate simulation based on this information.

### Concept of an Ideal Trace

The abstraction of elementary timing-dependent operations can be reduced to simply decreasing their timestamp values for the amount of their completion time. In other words, timestamps, which implicitly contain information about the execution path, are reduced to contain *only* that information and

---

*Operations involving a real-time clock are timing-dependent because the timings of other operations influence the time that the clock shows at a certain point.

*no* information on the memory system. The abstraction can be done this way by postprocessing the trace after it was generated, and before the trace is used for the simulations, which is much more efficient than if the whole procedure were done by the simulator at run time. It also means that no additional information on those operations is needed in a trace after postprocessing, which reduces its size. Complex timing-dependent operations, on the other hand, must be handled completely by the simulator at run time.

We based our simulator on *ideal traces*. An ideal trace is a trace where all elementary timing-dependent operations have already been freed of timing dependencies by adjusting their timestamp values (so that no additional data are needed by the simulator), and where complex timing-dependent operations have been properly abstracted.

The easiest way to produce it is to run a simulation using an ideal memory simulator, which makes it possible to complete every memory request in a single cycle. However, lock acquire operations will not be abstracted that way (for they must preserve the correct global ordering). We fight this inconvenience by forcing every attempt for gaining a lock to appear in a trace until it is finally gained. After the simulation is over, during the postprocessing phase, the superfluous lock appearances are eliminated, while the timestamps of all the references that follow are corrected for the amount of time the lock had to wait to gain it. This procedure assures a fast way to get an ideal trace, as an ideal memory simulator adds very little time to the simulation overhead.

### Simulation Using Tracer

Tracer is the Limes trace-driven simulation tool. By default, Tracer supposes that a trace it will use as an input is a single file that contains all processors' references. This format is chosen because it is more efficient regarding storage space than having a separate trace file for each processor. It allows one timestamp to be used for a group of operations that are done in the same cycle on different processors. It is more efficient to have the same number of 1-byte processor labels than 4-byte timestamp integers. Such a trace is also convenient for visual inspection if it is in textual format. However, although this format is good for keeping traces, it is not good when considering simulation efficiency. That is why Tracer first invokes a parser that makes ideal traces for each processor, which are then used as an input to the simulator (see Fig. B.1). At the same time the parser eliminates superfluous lock requests and corrects timestamps (as a part of the postprocessing phase of the lock abstraction process) and extracts in the special data structure information needed by the simulator to support abstraction of barriers. This process can be done only once, and the ideal traces obtained can then be used for multiple simulations.

The simulation starts by reading the first reference of each processor's thread and by scheduling the one with the smallest timestamp, and continues with scheduling requests until the last one is simulated. The process is quite straightforward, except for the barriers. In the current implementation each barrier in a trace is just a set of reads, writes, locks, and unlocks. There would be no way to tell if they functionally execute this synchronization primitive, if they weren't previously annotated during the trace generation phase. Each barrier is annotated with markers at certain points, which are recognized by the scheduler, and using that information it can perform the abstraction. The abstraction principle is quite general and can be used for handling different implementations of barriers, as well as other complex timing-dependent operations. Another event that must be recorded and abstracted is the creation of a child thread. It is necessary to abstract those events, to prevent scheduling of child threads before they were actually created by the parent.

This set of abstracted operations is sufficient to support all statically scheduled workloads. Abstraction of dynamically scheduled workloads can be supported by extending the abstraction paradigm used for statically scheduled programs.

### B.4.3. Memory System Simulator

The memory system simulator is completely independent from the simulation kernel and from the applications (traces) used to drive the simulations. As mentioned before, it can be used by both the execution-driven and the trace-driven simulator.

The simulator provides the memory system simulator with a stream of requests. They use a simple interface to communicate, and the same interface should be used when building up a simulator for any kind of shared address space multiprocessor system. How the memory simulator is realized is of no importance as long as it uses that interface.

### *SMP Memory Simulator*

A model of a bus-based cache coherent symmetrical multiprocessor system (SMP) comes with Limes. It is a system composed of $N$ identical processors, with an on-chip (L1) cache implementing one of the five supported coherency protocols. The processors are interconnected via bus, and all the communication through bus signals is adequately mimicked. The main memory module is also connected to the bus. This organization is represented in Figure B.2.

Each hardware unit is programmatically represented with a module (a C++ class), and each module is unaware of the others. That is, the modules do not communicate directly; rather, they are organized as isolated units, which communicate with the outer world through input/output *ports*: a module reads its input ports, performs the operation that depends on both the information on the input ports and the internal state of the module, and leaves the result on its output ports. These ports are shown in Figure B.3.

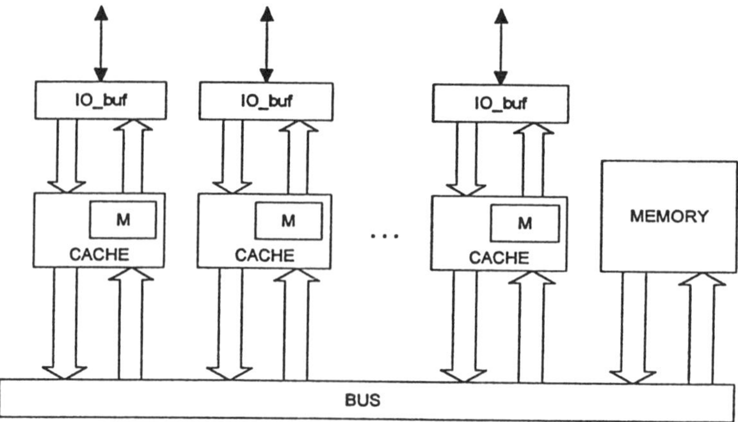

**Figure B.2.** Interconnection of modules in a bus-based cache coherent symmetrical multiprocessor system (CACHE—cache controller module; M—cache memory module; IO_buf—simple abstract modules that communicate with a simulation kernel; MEMORY—main memory module; BUS—the bus).

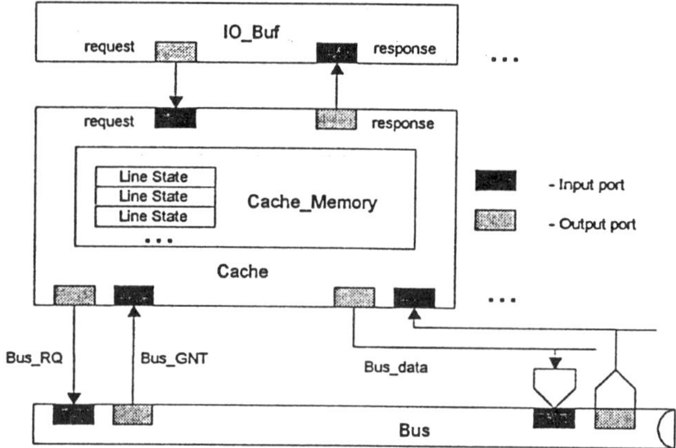

**Figure B.3.** Cache module and input/output ports.

Only one (abstract) module in the system is aware of the system topology, and it is not visible in Figure B.3. This module reads output ports from the modules and writes the information found there into the input ports of the modules to which they are connected, following a certain order and simulating signal flow; it also polls the modules for execution at proper times. This organization would allow one, for example, to easily add the TLB modules between the cache controllers and the processors without changing other modules (except the topology module, of course).

This design philosophy greatly resembles the one employed by VHDL. It allowed us to retain the desired level of accuracy, while preserving the simulation speed and the ability to change the model easily. This only shows that various approaches are possible using OOP (object-oriented programming); the freedom in writing a memory system simulator, however, should be complete. A guide to designing simulators is a part of the Limes documentation.

The system currently includes detailed examples of five snoopy cache coherence protocols: writethrough invalidate (WTI), word invalidate (WIN), Berkeley, Dragon, and MESI. The protocols differ in the operation of their cache controller modules, while the rest of the modules are functionally equal. The cache controller module is represented as a finite-state machine. Since all modules operate on a cycle-by-cycle basis, they change their internal state according to the current state and the information on their input ports. However, these states do not have to be understood in a strict sense as defined by the finite automata theory. The operation of the module is represented as a mixture of a flowchart and a state diagram. Modules are easily modifiable and extendible since they are written in C++, employing the OOP style.

### *DSM Memory Simulator*

A simple model of a distributed shared memory (DSM) multiprocessor system has been developed for Limes. The system comprised of $N$ identical processors. Each processor has local memory that is a part of a global distributed shared memory. Several memory consistency models are implemented to simulate access to the DSM. The system currently includes examples of processor, sequential, release, and lazy release consistency. The underlying network delays can be simulated in detail.

In order to simulate a DSM multiprocessor system, a specific memory simulator is developed. A different memory allocation strategy is needed for DSM systems, and some parts of the old Limes kernel had to be changed. The user can easily change the strategy for memory allocation (G_MALLOC macro). The memory simulator consists of the following parts: distributed shared memory itself, directory simulator, lock simulator, and network latency simulator. The size of the distributed shared memory is statically defined and distributed between the nodes. Current implementation of memory allocation enables a user to specify the home node for memory block to be allocated by the G_MALLOC macro. The global and static variables are by default allocated in the DSM space local to the first node. If the home node is not explicitly specified in the call to the G_MALLOC macro, then it is assumed that the home node for the new block is the node that called the macro.

The directory simulator implements the memory consistency model for distributed shared memory access. This simulator is called by the kernel every time the read or write instruction occurs. The simulator is responsible for all local page or block caching and acquiring of nonlocal DSM data. For the implementation of new consistency models this part needs to be changed.

In DSM systems lock management is quite different from the one in bus-based systems, because there is no bus to make the serialization. There are several different approaches to implement locks in DSM systems. However, all implementations make data access to the memory storing the locks sequentially (or at least processor) consistent. This simulator is called by the kernel every time the lock-related operations occur (acquire or release). The network latency simulator can be used to simulate different types of interconnection networks in detail. However, this part can be as simple as making the different constant delays for access to nonlocal parts of the distributed shared memory. If the aim of the research is to simulate the specific network configuration, this is the only part of Limes that the researcher has to change.

### Virtual Address Space Layout

When multiple threads of execution are created (for simulation purposes by Limes, in reality by any of the newer Linux releases), they are defined as regular processes that share address space with their parent. More formally, page table entries for mapping virtual addresses to physical addresses have the same values for all the threads. Figure B.4 depicts this situation for an example of four threads.

All the threads share the code segment (of course) and the data segment, and each one gets its reasonably wide share (say, 256 kB) of the upper area for the stack. The tops of their stacks are different. Stacks are private only

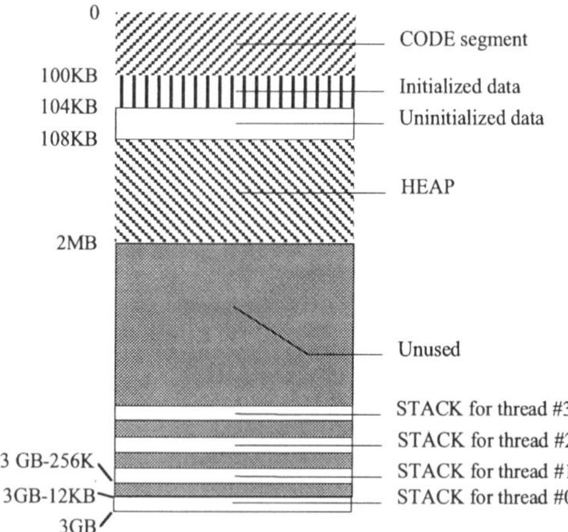


**Figure B.4.** Virtual address space layout for multiple threads (an example).

semantically: a thread *could* read and write another thread's stack, but never does so.

This brings us to the following point: Why do you not have to translate virtual addresses to physical addresses in simulation? The answer is that no two different virtual addresses point to the same physical address, and no two equal virtual addresses point to different physical addresses. So you can assume that these threads of the parallel application are the only set of processes that are executing on your simulated machine. And one can assume that the code segment was really loaded into memory at address zero, and so on.

There are certain leaks in this assumption. First, you do not have 3 GB of physical memory to assume that the stack is physically really placed at 3 GB minus something; and second, in reality, the lowest part of the physical memory would be assigned to system tables, and so on. But since all the allocations occur at page boundaries, it is only a matter of distribution of block tags in your associative memory in caches, which is unpredictable anyway, and has little impact on general behavior for two-way caches or caches with higher associativity, or larger caches. Still, it would be easy to create a translation function, should you need one.

Finally, a note on the ISSHARED(addr) macro: it returns true if (and only if) the address given belongs to the data segment. Formally, code segment is shared as well; but it can be read only and is therefore not important from the aspect of coherence maintenance. Note that reads from the code segment can occur, although Limes does not capture instruction fetches; this is because the application may read constants embedded in the code segment, mainly string constants, for various printf( )s that never occur in the parallel computation phase. If you really need to know whether an address belongs to the code segment, say, if (addr < DATA_SEGMENT_START).

### Virtual Address Space Layout in DSM Limes

The consistency models in DSM systems are more relaxed than in SMP systems, and the DSM memory is less consistent than the SMP memory. In order to accurately simulate the system, fetching of the nonlocal data is simulated in detail. A node in a DSM system holds a part of the distributed shared memory, and a part of the local, nonshared memory used for storing and caching of the nonlocal data and the directory data used to maintain the DSM consistency.

The virtual address space of the DSM simulator is shown in Figure B.5. The directory simulator dynamically allocates and frees the memory in the heap for the simulation of the MCM during the simulation. The main difference between the SMP and the DSM simulation virtual address space is that the space for the DSM is statically allocated at the beginning of the simulation and the G_MALLOC and G_FREE macros are reimplemented to correctly allocate and free the space in the DSM. This enables the user to implement new algorithms for memory allocation in DSM systems.

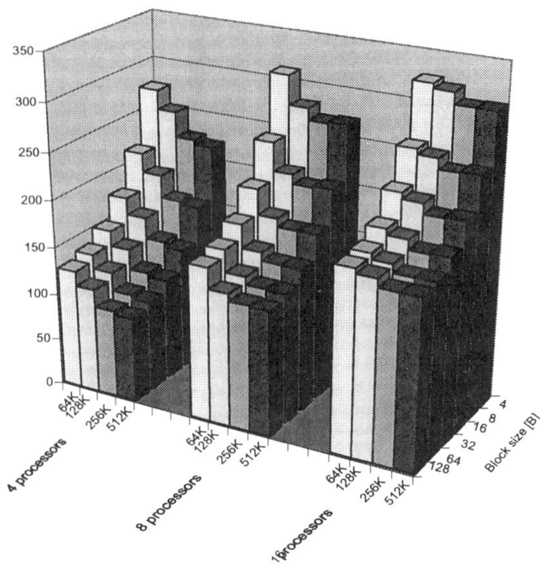

**Figure B.5.** Measuring bus traffic with Limes. This figure shows values (in MB/s) for FFT application.

## B.5. LIMES COMPLEXITY AND PERFORMANCE

The simulation kernel of the execution-driven simulator consists of nearly 1800 lines, and the tool for instrumentation of iAPx86 instructions contains an additional 1200 lines. The trace-driven simulator adds 1500 lines. The SMP model and five snoopy protocols comprise altogether some 4000 lines of code. The most complex classes—the detailed cache controllers—contain on average some 400 lines. The whole code is profusely commented. The complete Limes environment (including nine SPLASH-2 applications) takes about 820 kB (compressed).

The compilation process is rather quick, ranging from 1 s for simple memory models up to 15 s for the most complex ones, measured on a Pentium 133.

The Limes performance for execution-driven simulations is presented in Table B.1. The results show execution times and slowdowns for four SPLASH-2 programs. Simulations have been done for three different memory models and two instrumentation levels for each model. The abstract model does not invoke the scheduler or the memory simulator. It responds to the requests right away but still preserves global ordering. Ideal and MESI models both invoke the scheduler and the memory simulator. An ideal memory simulator returns a *satisfy* signal in a single cycle for every memory request (read/write), except for synchronization requests (lock/unlock).

**TABLE B.1. Execution-Driven Simulation Times and Slowdowns for Four SPLASH-2 Applications, for Various Memory Simulator Models**

| | OCEAN | | FFT | | LU | | RADIX | | Average |
| | | | | | | | | | |
| Simulator/Level | Time(s) | Slow-down | Time(s) | Slow-down | Time(s) | Slow-down | Time(s) | Slow-down | Slow-down |
| --- | --- | --- | --- | --- | --- | --- | --- | --- | --- |
| Uninstrumented | 8 | 1 | 2 | 1 | 16 | 1 | 7 | 1 | 1 |
| Abstract/level_1 | 312 | 39 | 65 | 32 | 863 | 54 | 77 | 11 | 34 |
| Abstract/level_2 | 684 | 85 | 94 | 47 | 1120 | 70 | 270 | 38 | 60 |
| Ideal/level_1 | 729 | 91 | 122 | 61 | 2438 | 152 | 132 | 19 | 81 |
| Ideal/level_2 | 1334 | 166 | 200 | 100 | 3045 | 190 | 698 | 100 | 139 |
| MESI/level_1 | 5273 | 660 | 850 | 425 | 10056 | 628 | 737 | 105 | 454 |
| MESI/level_2 | 5694 | 712 | 893 | 447 | 10297 | 644 | 1798 | 256 | 515 |

MESI is the most complex memory simulator in the current version of Limes, performing the simulation of a bus-based SMP with MESI cache coherence protocol. All simulations were performed for 16 processors, where each had a 64-kB large, two-way set-associative cache, with 16-byte cache lines.

Dimensions of the problems were appropriate: OCEAN worked on a $130 \times 130$ grid, FFT with 65536 complex doubles, LU with a $512 \times 512$ matrix, and RADIX with 262144 integers. Simulations were done on a 133-MHz Pentium PC.

Slowdowns for the abstract model indicate the instrumentation overhead introduced by the simulator, where correct global ordering is still retained. Results show that this is the biggest source of slowdown compared to other factors, introducing an average slowdown factor of 34 for level_1 and 60 for level_2 instrumentation. Slowdowns using an ideal memory simulator indicate the scheduling plus the memory simulator invocation overhead. It additionally slows down the simulation for a factor of 1.7–2.8 (2.4 in average) for level_1, and for a factor of 2–2.7 (2.3 in average) for level_2 instrumentation. Finally, simulation of a realistic bus-based SMP system with a MESI protocol indicates the memory simulator overhead. For level_1 simulations it brings a slowdown that goes from 9.5 to 16.9 (13.3 in average) and for level_2 simulations the slowdown ranges from 6.7 to 9.5 (8.6 in average) in addition to the instrumentation overhead. Pure memory simulator slowdowns, not counting the scheduling and the simulator invocation overhead, go from 4.1 to 7.9 (5.6 in average) for level_1, and from 2.6 to 4.5 (3.7 in average) for level_2 simulations.

These results are comparable to the TangoLite performance. TangoLite has an average slowdown of 45 for a simulation with no memory simulator, which compares to an average Limes slowdown of 34 in the abstract/level_1 case. An average memory simulator slowdown for Limes is 13.3 for MESI/level_1, and for a similar memory architecture TangoLite has a slowdown factor of 17. In total, Limes has a slowdown of 454 compared to the TangoLite average slowdown of 765.

**TABLE B.2. Sizes of Execution-Driven Simulation Executables and Memory Dilation Overheads for Four SPLASH-2 Applications, for Various Memory Simulator Models**

| Simulator/ Level | OCEAN Size (kB) | OCEAN Over-head (%) | FFT Size (kB) | FFT Over-head (%) | LU Size (kB) | LU Over-head (%) | RADIX Size (kB) | RADIX Over-head (%) | Average Over-head (%) |
|---|---|---|---|---|---|---|---|---|---|
| Uninstrumented | 167.8 | 0 | 121.5 | 0 | 125.1 | 0 | 120.7 | 0 | 0 |
| Abstract/level_1 | 237.9 | 41.7 | 138.3 | 11.4 | 137.8 | 10.1 | 129.3 | 7.1 | 17.6 |
| Abstract/level_2 | 323.9 | 93.0 | 154.7 | 27.3 | 158.3 | 26.5 | 141.6 | 17.3 | 41.0 |
| Ideal/level_1 | 237.9 | 41.7 | 138.3 | 11.4 | 137.8 | 10.1 | 133.4 | 10.5 | 18.4 |
| Ideal/level_2 | 323.9 | 93.0 | 154.7 | 27.3 | 158.3 | 26.5 | 145.7 | 20.7 | 41.9 |
| MESI/level_1 | 357.5 | 113.9 | 257.9 | 121.2 | 257.4 | 105.7 | 248.9 | 106.2 | 111.5 |
| MESI/level_2 | 443.5 | 164.0 | 274.3 | 125.7 | 277.4 | 121.7 | 261.2 | 116.4 | 132.0 |

Instrumentation inevitably increases the size of the application. Table B.2 shows some values of memory overhead that Limes introduces. These results can be compared to the TangoLite overhead, where instrumentation typically increases application static size by a factor of 4, against the factor of 2.1–2.3 for Limes.

Results for trace-driven simulation are presented in Table B.3. They show the trace-driven simulator performance against the execution-driven simulator. Simulation times are obtained using small, but sufficient examples. OCEAN, FFT, LU, and RADIX binary traces were 8.7, 9.0, 18.3, and 11.7 MB long, respectively. The disk transfer rate was about 1.0 MB/s. Simulations were performed for a 2-kB on-chip cache, for eight processors.

It is obvious that disk transfer introduces a constant overhead in the simulation. It can be reduced substantially by using a faster disk. The rest of the time is spent by the simulator. The version of the trace-driven simulator used to obtain these results is not optimized for fast disk access. Trace-driven simulation can potentially be twice as fast as execution-driven simulation with optimal disk access policy.

## B.6. INSTALLATION GUIDE

You do not need any privileges to install and run Limes, modulo the installation of version 2.6.3 of the GCC compiler, if it is not installed on the system already. The procedure is described below.

**TABLE B.3. Speed of Trace-Driven Simulation for Four SPLASH-2 Applications Compared to the Execution-Driven Simulation[a]**

| Memory Simulator/ Type of Simulation | OCEAN Time(s) | OCEAN Ratio | FFT Time(s) | FFT Ratio | LU Time(s) | LU Ratio | RADIX Time(s) | RADIX Ratio | Average Ratio |
|---|---|---|---|---|---|---|---|---|---|
| MESI/trace-driven | 145 | 2.38 | 57 | 2.37 | 36 | 1.80 | 66 | 1.94 | 2.12 |
| MESI/exec-driven | 61 | | 24 | | 20 | | 34 | | |

[a]The table shows simulation times for both simulators (running the same workloads) and their ratio.

For purposes of simplicity, it will be assumed that the username is joe, and that the home directory is /home/joe.

### B.6.1. Unpacking the Archive

Limes comes in a single archive, limes∗.tgz, where ∗ stands for the current version. While this procedure describes how to install Limes in a per-user manner, you can also choose to install it on the system to be globally accessible.

To unpack the archive, simply position in your home directory and do

```
tar xfz limes*.tgz
```

After this, a directory tree will be created.

### B.6.2. Setting Up the Environment

The Limes scripts and makefiles must know where the whole tree is. To communicate this information, set the environment variable LIMESDIR to point to the root of the tree. In the example, do

```
export LIMESDIR=/home/joe/limes  # if you use bash, or

setenv LIMESDIR /home/joe/limes  #if you use tcsh
```

(You can put them in your .profile or .login for convenience.)

### B.6.3. Installing the GNU C v2.6.3 Compiler

Check the version of your GCC compiler with "gcc -v." If it reports 2.6.3 or lower, the installation procedure is over. Otherwise, you will probably have to be root (or to ask one) to copy the development environment for the GCC 2.6.3 version on your file system. *Note:* Do not really *install* it, for it will overwrite your current GCC! Just read below how to *copy* it onto your file system, in a separate, harmless directory! Limes will know how to find and execute it.

As explained, Limes does various things with the application's assembly code and is therefore strongly dependent on the version of the compiler. But the 2.6.3 version does not have to be the default compiler on your system, of course. It will reside in a subdirectory and be called only by Limes. Here is how to do it:

1. First, get the Development disk series of the Linux distribution that contains GCC 2.6.3. An example is Slackware 2.2.0.1.
2. You will need the following archives:
   ```
   gcc263.tgz, include.tgz, libc.tgz, libgxx.tgz,
   lx128_2.tgz, binutils.tgz.
   ```

3. Create a directory, say, /oldgcc. Change the ownership of this directory to make it belong to a nonprivileged user. Now login as that user and cd to /oldgcc.

4. From now on, everything will be done from the directory /oldgcc as the current, and all the paths will be relative to that directory.

5. Unpack the archive
```
d1/gcc263.tgz(like tar xfz /cdrom/slackware/d1/
gcc263.tgz)
```

6. Unpack d6/lx128_2.tgz. Do
```
(cd usr/src/linux-1.2.8/include; ln -sf asm-i386
asm)(cd usr/src; ln -sf linux-1.2.8 linux)
```

7. Unpack d2/include.tgz. Do
```
(cd usr/include; ln sf /oldgcc/usr/src/linux/
include/linux linux)
(cd usr/include; ln -sf /oldgcc/usr/src/linux/
include/asm asm)
```

8. Unpack d5/libc.tgz. Do
```
sh install/doinst.sh
```

9. Unpack d4/libgxx.tgz. Do
```
sh install/doinst.sh
```

10. Unpack d8/binutils.tgz.

There is one more step: edit the "limes/globals.make" file in your Limes tree, and replace "ifeq (0,1)" with "ifeq (1,1)." If you have chosen another directory for GCC2.6.3 instead of "/oldgcc," change that in the file, too. The procedure is now over. It will not affect the rest of your system, except that you will now have 11 MB of disk space less.

If you do not have access to a Linux distribution that contains GCC v.2.6.3 files, you can download it from the following URL: `http://galeb.etf.bg.ac.yu/~dsm/limes/oldgcc.html` (or you will find a pointer to a site that contains it). It is, in fact, somewhat reduced and requires only 7 MB of space (the archive itself is 2.4 MB long).

## B.7. EXPERIENCE WITH LIMES

Limes has been used for a number of studies in the domain of shared memory multiprocessors, including SMP and DSM [Protić et al. 1996] systems.

In one of the studies regarding characterization of parallel applications for DSM systems [Marinov et al. 1998], Limes was used as a simulation environment, and one originally developed characterization tool was used to obtain parameters of interest. The data obtained by Limes were then used in two other studies—as an input to the analytical models of DSM systems, and to assist researchers in enhancing performance of DSM systems by improving memory consistency protocols.

In another study regarding SMP systems, Limes was used to evaluate the influence of different architectural parameters on the overall system performance. In particular, bus traffic and miss rate were measured on a system with a MESI cache coherence protocol while varying the number of processors, cache size, and cache block size, and using different workloads. The results were then used to infer some conclusions regarding the architectural details. As an example, values for bus traffic are shown in Figure B.5 [for FFT (fast Fourier transform) application from the SPLASH-2 suite, assuming 100-MHz processors]. Bus traffic decreases as cache size grows, because there are fewer cache misses that induce bus traffic. Bus traffic also decreases with the growth of cache block size, because the FFT exhibits spatial locality, so there is no negative influence of false sharing. When the number of processors grows, bus traffic increases, as there are more requests on the bus.

In a recent study concerning implementation of the lazy release consistency model in SMP systems, Limes was used successfully to simulate the environment that supports the proposed improvement in hardware, and to show some quantitative results.


## B.8. CONCLUSIONS

The intentions behind the development of this tool were to facilitate the multiprocessor studies at the University of Belgrade, and to provide researchers with an environment that can easily be adapted to fulfill their particular demands in order to make their study more effective. The tool can be of benefit to all those who need realistic simulations of shared address space multiprocessors (for architecture evaluation, in investigating real-time systems [Stankovic and Ramamritham 1993], etc.), and to the researchers in the field of parallel algorithms.

Limes comprises two usable simulators and a complete model of an SMP system, offering fast and accurate simulation on today's popular PC platforms. It employs some new abstraction techniques for accurate trace-driven simulation, with a concept that can be extended even to nondeterministic workloads, if properly supported by the trace generation tool. On the other side, the execution-driven simulator offers respectable speed using a fully optimized scheduling algorithm. For those who are more interested in investigating parallel algorithms, Limes offers a relatively fast type of simulation that still preserves correct global ordering; also, a new paradigm for easy and comprehensible parallel programming is available.

There is, however, enough room for some improvement and future work. A trace-driven simulator should optimize its access to the trace references and achieve higher simulation speed, and trace compaction techniques [Samples 1989] may be used. It can also be extended to support abstraction of some other types of timing-dependent operations, including dynamically scheduled workloads. Simulators could be improved to support multithreading or thread migration. Simulators of some other shared memory systems are currently being developed and will be included in the environment. In the end, Limes should soon be ported to work on other platforms and operating systems.

## Acknowledgments and Availability

The authors greatly appreciate the help from Professor Milo Tomašević, Professor Igor Tartalja, as well as from Darko Marinov of MIT. Of course, credit also goes to users of the package at universities around the world, whose feedback aided the authors in improving the tool.

The whole Limes package is in the public domain and can be found at the following Internet address: `http://galeb.etf.bg.ac.yu/~dsm/limes`.

## REFERENCES


Agarwal, A., Sites, R. L., Horowitz, M., "ATUM: A New Technique for Capturing Address Traces Using Microcode," *Proc. 13th Int. Symp. Computer Architecture*, June 1986, pp. 119–127.

Brorrson, M., Dahlgren, F., Nilsson, H., Stenstrom, P., "The CacheMire Test Bench —a Flexible and Effective Approach for Simulation of Multiprocessors," *Proc. 26th Annual Simulation Symp.*, Arlington, VA, March 1993, pp. 41–49.

Davis, H., Goldschmidt, S. R., Hennessy, J., *Tango: A Multiprocessor Simulation and Tracing System*, Technical Report CSL-TR-90-439, Stanford Univ. Computer Systems Laboratory, July 1990.

Eggers, S. J., Keppel, D. R., Koldinger, E. J., Levy, H. M., "Techniques for Efficient Inline Tracing on a Shared-Memory Multiprocessor," *Proc. ACM SIGMETRICS Conf. Measurement and Modeling of Computer Systems* **18**(1), 37–47 (May 1990).

Goldschmidt, S., *Simulation of Multiprocessors: Accuracy and Performance*, Ph.D. thesis, Stanford Univ., Palo Alto, CA, June 1993.

Goldschmidt, S. R., Hennessy, J. L., *The Accuracy of Trace-Driven Simulations of Multiprocessors*, Stanford Univ., Technical Report CSL-TR-92-546, Sept. 1992.

Herrod, S. A., *TangoLite: Introduction and User's Guide*, Technical Report, Stanford Univ., Palo Alto, CA, Nov. 1993.

Holliday, M. A., Ellis, C. S., *Accuracy of Memory Reference Traces of Parallel Computations in Trace-Driven Simulation*, Duke Univ. Technical Report (CS-1990-08), July 1990.

Koldingr, E. J., Eggers, S. J., Levy, H. M., "On the Validity of Trace-Driven Simulation for Multiprocessors," *Conf. Proc. 18th Annual Int. Symp. Computer Architecture* **19**(3), 244–253 (May 1991).

Magnusson, P. S., Dahlgren, F., Grahn, H., Karlsson, M., Larsson, F., Lundholm, F., Moestedt, A., Nilsson, J., Stenström, P., Werner, B., "SimICS/sun4m: A Virtual Workstation," paper presented at Usenix Annual Technical Conf., New Orleans, June 15–18, 1998.

Marinov, D., Magdic, D., Milenković, A., Protić, J., Tartalja, I., Milutinović, V., "An Analysis of SPLASH-2 from the DSM Point of View," *Proc. Hawaii Int. Conf. System Sciences*, Mauna Lani, HI, Jan. 1998.

Milutinović, V., "The Best Method for Presentation of Research Results," *IEEE TCCA Newsl.*, 1–6 (Sept. 1996).

Milutinović, V., *Surviving the Design of Microprocessor and Multimicroprocessor Systems: Lessons Learned*, IEEE Computer Society Press, Los Alamitos, CA, 1998, `http://galeb.etf.bg.ac.yu/ ~ vm/books/books.htm`.

Protić, J., Tomašević, M., Milutinović, V., "Distributed Shared Memory: Concepts and Systems," *IEEE Parallel Distrib. Technol.* **5**(1), (1996).

Rosenblum, M., Herrod, S. A., Witchel, E., Gupta, A., "Complete Computer System Simulation: The SimOS Approach," *IEEE Parallel Distrib. Technol.*, 34–43 (1995).

Samples, A.D., "Mache: No-Loss Trace Compaction," *Perform. Eval. Rev.* **17**(1), 89–97 (May 1989).

Sharma, A., Nguyen, A. T., Torellas, J., *Augmint: A Multiprocessor Simulation Environment for Intel x86 Architectures*, CSRD Technical Report 1463, Univ. Illinois at Urbana–Champaign, Urbana, Mar. 1996.

Stankovic, J. A., Ramamritham, K., "Advances in Real-Time Systems," *IEEE Comput. Soc.* (Sept. 1993).

Stunkel, C. B., Janssens, B., Fuchs, W. K., "Address Tracing for Parallel Machines," *IEEE Comput.* **24**(1), 31–38 (Jan. 1991).

Stunkel, C. B., Fuchs, W. K., "TRAPEDS: Producing Traces for Multicomputers via Execution Driven Simulation," *Prod. ACM SIGMETRICS Int. Conf. Measurement and Modeling of Computer Systems*, Berkeley, CA, May 1989, pp. 70–78.

Tomašević, M., Milutinović, V., "A Simulation Study of Snoopy Cache Coherence Protocols," *Proc. Hawaii Int. Conf. System Sciences*, Koloa, Hawaii, Jan. 1992, pp. 426–436.

Uhlig, R. A., Mudge, T. N., "Trace-Driven Memory Simulation: A Survey," *ACM Comput. Surveys* **29**(2) (June 1997).

Woo, S. C., Ohara, M., Torrie, E., Singh, J. P., Gupta, A., "The SPLASH-2 Programs: Characterization and Methodological Considerations," *Proc. 22nd Int. Symp. Computer Architecture*, June 1995, pp. 24–36.

# The Scowl Tool for PC-Based Characterization of Parallel Applications

This appendix* concentrates on the problem of defining and measuring parameters that characterize the behavior of parallel applications targeted to distributed shared memory (DSM) systems. These parameters can be used as input to various models for DSM systems performance evaluation. Also, typical patterns can be recognized, which helps generate new ideas for improvements of memory consistency protocols by adapting them to specific application characteristics. There are many studies partially concerning this issue, but we are not aware of any that provides a systematic and exhaustive set of parameters. Our study encompasses a variety of parameters, such as frequency of operation of various access types (private read/writes, shared read/writes, lock operations, barrier operations), average number of blocks accessed per interval, and so on. Results presented here are based on the SPLASH-2 application suite, which is chosen for its wide popularity and suitability for DSM systems. The developed instrumentation tool Scowl, along with the applied simulation environment Limes, is publicly available and applicable for performing measurements on other parallel applications, as well.

## C.1. INTRODUCTION

Systems with multiple processors that share physical or virtual common address space became very attractive because of their relatively simple programming paradigm [Milutinović 2000]. Dozens of symmetric memory multiprocessors (referred to as SMP) and distributed (or virtual) shared memory systems (referred to as DSM or VSM) [Protić et al. 1996a] were developed during the 1990s.

*Prepared by Darko Marinov, Davor Magdic, Aleksandar Milenković, Jelica Protić, Igor Tartalja, and Veljko Milutinović.

In order to evaluate new ideas related to SMP and DSM, one can use analytical, simulation, or implementation methods [Milutinović 1997]. The analytical method assumes the derivation of a mathematical model that describes the real system with many approximations. Although it is not possible to give general conclusions based exclusively on the analytical modeling, it should be the starting point for evaluating any new idea. It is relatively easy to implement, does not require expensive equipment, and is less time consuming than any other method. Few papers present an analytical approach to DSM evaluation [Kessler and Livny 1989, Stumm and Zhou 1990]. The second step in an analysis should be simulation. It allows modeling of details that are difficult or even impossible to include in the analytical model. Simulation can be done with more or less approximations, but the good point compared to the analytical model is the possibility of using both synthetic workload and realistic traces. The implementation method assumes the development of a real system, where all measurements can be performed. It is the most expensive and the most time-consuming method, but provides environment for exact measurement of the realized system.

The analytical method and simulation with probabilistically synthesized workload enable evaluation of sensitivity of the proposed mechanism on fine changes of a particular parameter of interest, and that is why they are worth doing. Development of such a model requires precise definition of parameters that characterize workload. Experiments are usually conducted by fixing most of the parameters to their typical values, while one or two parameters vary in the range between minimum and maximum realistic values. Ranges of parameters can be obtained using real SMP/DSM systems or dedicated execution-driven simulation tools, with appropriate measurement instrumentation. A set of parameters that should be used as input to a model depends on the complexity of the model itself and on the specific idea that one wants to evaluate.

The issue of special importance in the field of SMP/DSM systems is the memory consistency model [Protić et al. 1996b]. Sequential and processor consistency do not differentiate between ordinary and synchronization variables. Weak and release consistency make that difference but do not care about the identity of synchronization variables. Finally, lazy release and entry consistency care about the identity of synchronization variables and their relations to ordinary variables. The last two models are the most promising and the most interesting for evaluation, but we are not aware of any paper that provides values of parameters necessary for the evaluation.

There are several tools used to create an illusion of a multiprocessor on a single processor machine, such as TangoLite [Herrod 1993] developed at Stanford University. One of them is Limes [Magdic 1997], developed at the University of Belgrade. Limes runs on a PC with an i486 or higher, on Linux operating system, and executes parallel applications written in C/C++, using specific macros to express parallelism. Limes is based on the execution-driven approach and can be used for architecture evaluation studies as well as parallel algorithms evaluation. The idea presented in this appendix is to extend Limes with Scowl, a measuring instrumentation tool capable of providing a wide set of parameters relevant for evaluating SMP/DSM systems with relaxed memory consistency models.

Applications intended for simulation in the Limes environment are to be coded using ANL (Argonne National Lab) macros. Written in such a manner, applications are portable to real multiprocessor machines, as well as to other simulators, such as TangoLite. ANL macros can be divided into several groups: (1) macros for thread control, (2) macros for shared memory allocation, (3) synchronization macros, and (4) environment macros. In this appendix we pay special attention to the macro for shared memory allocation (G_MALLOC) and the synchronization macros for locks (LOCK and UNLOCK, namely, acquire and release operations, used for entering and exiting critical sections, respectively) and for barriers (BARRIER).

This appendix is organized into nine sections. After the introduction, Section C.2 concentrates on the problem statement, the difference between the selected approach and the existing body of research. The importance of the problem is also emphasized. Section C.3 summarizes some of the previous research efforts in the field. Section C.4 focuses on the selected approach, based on building dedicated simulation tools, in order to measure characterization parameters divided into specific semantic groups. Details about semantic groups are given in Section C.5. Section C.6 describes Limes and Scowl. Section C.7 gives results obtained by this analysis, organized in semantic groups. Conclusions are presented in Section C.8.

## C.2. PROBLEM STATEMENT

At the starting point of any analytical analysis or simulation analysis with synthetic benchmarks, one is faced with a difficult problem of determining realistic values of input parameters of the model. Some authors mostly use educated guesses to choose appropriate values [Kessler and Livny 1989]; others concentrate on published values obtained by characterization of existing benchmarks such as Woo et al. (1995), while some combine both approaches [Tartalja and Milutinović 1992]. However, it seems that specific characteristics of applications that lead to innovative algorithms or architectures are not noticeable in published characterizations, but have to be discovered by the researcher personally in order to propose a particular new solution.

The development of a software tool for characterization of parallel applications has advantages over the usage of existing address traces stored in files. The approach using such a tool is not restricted to a specific set of applications, since any application written according to specified rules can be characterized in this way. On the other hand, it helps the researcher to choose and measure parameters relevant for the particular evaluation. Existing address traces are not sufficiently customized for studies of sophisticated memory consistency models. Therefore, it is useful to provide the software tool that enables execution-driven collecting of relevant data and/or generation of customized, reduced traces suited to specific needs. This software should be modular in order to enable easy widening of the set of parameters to be measured.

In addition to the obvious importance of quantitative results one can obtain and directly use as inputs for some evaluating models, work on characterization can help in better understanding how the applications behave, having in mind that they are very large programs, complicated for code inspection. Some results can also help us in understanding how representative the applications are; if we conclude that their characteristics are similar, we can save time for running all of them, knowing that simulation experiments are usually time consuming. Finally, the results obtained by characterization can inspire us to extend our benchmark suite with some applications that differ from the set we use.

## C.3. PREVIOUS RESEARCH EFFORTS

Parallel program behavior, characterized by memory-reference patterns, has already been investigated in several works in the past. For example, Agarwal and Gupta have analyzed memory access patterns of several parallel applications (as are logic simulation or VLSI wire routing) under a MACH operating system [Agarwal and Gupta 1988]. They did the measurements using the ATUM address tracing scheme [Agarwal et al. 1986] implemented on the four-processor DEC machine VAX8350.

Eggers and Katz have characterized access patterns of write shared data in parallel applications (e.g., simulated annealing algorithm for cell placement, logic verification, circuit simulation) [Eggers and Katz 1988]. They used two multiprocessor platforms scaling from 5 to 12 processors, a Sequent machine under the UNIX operating system, and an ELXSI 6400 machine under the Embos operating system.

Gupta and Weber have measured and analyzed cache invalidation patterns and message traffic in shared memory multiprocessors [Gupta and Weber 1992]. They have used the Tango reference generator [Davis and Goldschmidt 1990] and architectural simulator based on partitioned shared memory among the processing nodes, infinite caches, and an invalidation, directory-based, cache coherence scheme similar to that used by the Stanford DASH multiprocessor [Lenoski et al. 1992]. Four out of five applications used by the authors are from the SPLASH parallel application suite [Singh et al. 1992].

An interesting contribution to the benchmark characterization has been done by Conte and Hwu (1990). They have defined a way of measuring benchmark characteristics separate from system architecture and performance. The measures of temporal and spatial locality of references are particularly emphasized in their work.

The problem of choosing a set of parallel applications for the quantitative evaluation of ideas in architectural studies is very difficult. The set should be representative of all parallel applications that might be run on a scalable DSM system. Meanwhile, there are very few real applications available, and the existing applications are not always the best parallel implementations in terms of algorithms or coding. In addition, since the execution of the applications is simulated, and the execution time of each simulation run has to be finite, the runs have to use much smaller data sets and total run lengths than might be expected from runs on a real machine.

Early research in shared-address-space multiprocessing was based on small workloads consisting of a few simple programs. Using different programs and different problem sizes made comparisons across studies difficult. In a way, the problem was surpassed with SPLASH (Stanford ParalleL Applications for SHared Memory) [Sing et al. 1992]. Although SPLASH has provided a degree of consistency and comparability across studies, it has many limitations. First, SPLASH contains only a small number of programs; moreover, it does not cover all aspects of scientific and engineering computing. Also, the implementation of SPLASH programs is not optimized for modern memory system characteristics and for machines that scale beyond a relatively small number of processors.

To overcome these limitations the SPLASH suite has been expanded and modified to include several new programs as well as improved versions of the original SPLASH programs. The resulting SPLASH-2 suite [Woo et al. 1995] contains programs that (1) represent a wider range of computations in scientific, engineering, and graphics domains; (2) use better algorithms and implementations; and (3) are more architecturally aware. Applications from SPLASH-2 suite are briefly described in Section C.6.

## C.4. APPROACH OF THIS RESEARCH

The essence of our approach to characterization of parallel applications aimed at DSM/SMP systems is to (1) define a list of parameters important for analytical or simulation study of DSM/SMP systems, according to our experiences in analyzing such systems [Tartalja and Milutinović 1992, Protić and Milutinović 1997, Milutinović and Milenković 1997]; (2) classify the parameters according to some common characteristics into semantic groups; (3) design Scowl—a tool that extends Limes, an execution-driven simulator of parallel applications, in order to enable measurements of relevant parameters; and (4) perform experiments that measure relevant parameters on a subset of SPLASH-2 applications, using Limes/Scowl running on a Linux P5-based platform. Our approach is presented in Figure C.1.

Measurement of relevant parameters can be done by using address traces stored in files (static approach) or by performing execution-driven simulation, specifically, collecting in run time the values relevant for the determination of parameters (dynamic approach). Some parameters can be determined by

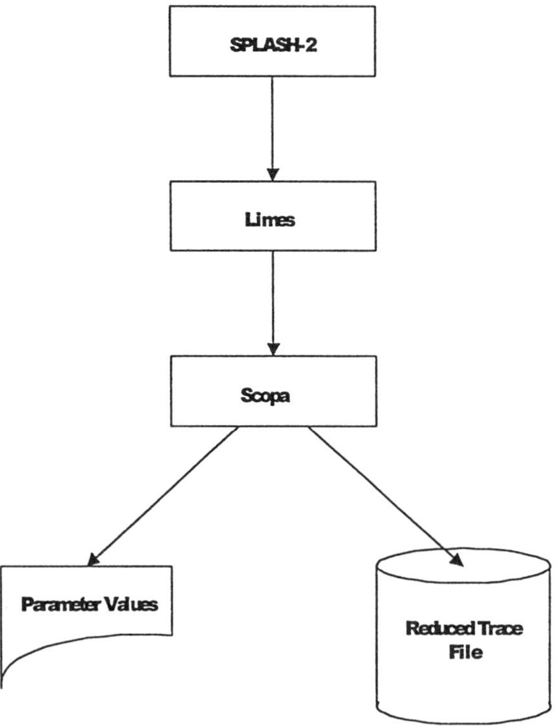



**Figure C.1.** The structure and dataflow of the designed measurement system. Simulation of a multiprocessor system is provided by Limes. An application from SPLASH-2 suite is linked together with Limes and Scowl to produce a single executable file. Limes supplies Scowl with shared memory operations (read, write, lock, unlock, barrier). Scowl represents the PRAM shared memory simulator capable of collecting statistics of these operations. The results are either stored in a reduced trace file or used for dynamic computation of parameters.

combining the two methods. In our approach, files with a kind of reduced address trace (each record represents aggregate data for multiple memory accesses) are generated, and some parameters are computed using data from those files; some other parameters are dynamically computed during the experiment. The organization of data in the files is convenient for obtaining a wide range of parameters (the format of a reduced trace file is presented in Section C.6). The use of these files is convenient because they can be analyzed using simple programs, without knowing the details of Limes and Scopa; this method is also less time consuming than dynamic simulations.

Parameters that we decided to measure were defined, having in mind relaxed memory consistency models, such as entry consistency and lazy release consistency. Special attention was paid to identifying particular application behavior inside and outside critical sections. Unlike other work that we are aware of, we also took care to identify particular locks protecting critical sections. We classified parameters that we measured according to their organization and semantics; parameters from the first group represent scalar values, the second group provides arrays of values indexed by the type of operation, and the last group represents arrays of values indexed by the size of the coherence unit. Consequently, the first group gives some general characteristics of applications, the second group concentrates on read/write private/shared ratios in the specified domain (in/out critical sections, between barriers etc.), and the third group actually examines the effects of coherence units granularity; these effects highly depend on locality behavior of applications. Details of these parameters and semantic groups are presented in Section C.5.

Our choice of applications to be characterized is based on the fact that most recent studies that compare DSM/SMP systems use a SPLASH-2 benchmark suite. However, it seems that this set of benchmarks is only partially representative for adequate simulation studies. Early research in the area of DSM was based on some well-known problems such as matrix multiplication, the traveling salesperson problem, and quicksort, where the researchers coded using an available programming language and environment. The simulation study of Midway, the first system implementing entry consistency, used such self-coded benchmarks instead of using SPLASH-2. Even some recent research is based on self-coded benchmarks according to their particular needs; in Ifto et al. (1996), the most interesting results are based on the authors' *N*-body kernel. Our approach was to perform measurements on applications from SPLASH-2, but making the measuring tool flexible enough and capable of characterization of other applications that may be needed.

In our work we use the same tools for characterization that provide parameters important for generation of probabilistic workload, and for actual dynamic simulations of architectures and protocols with realistic workload, so that results can be comparable. Our approach conducts all experiments (realistic and synthetic) in a single environment that is highly available—a low-cost PC running a Linux operating system.

## C.5. SPECIFICATION OF PARAMETERS AND MEASUREMENT CONDITIONS AND ASSUMPTIONS

The terms we will use in the rest of this appendix include the following:

*Epoch*—a sequence of instructions executed by a processor between two successive barrier operations.

*Interval*—a sequence of instructions executed by a processor between two synchronization points. There are two types of intervals: (1) *critical section interval* (CSI), a sequence of instructions executed by a processor between LOCK and UNLOCK primitives, and (2) *noncritical section interval* (NCSI), a sequence of instructions executed by a processor outside a critical section (between BARRIER and LOCK primitives, UNLOCK and LOCK primitives, and UNLOCK and BARRIER primitives). The assumption is that there are no explicitly nested critical sections, and it holds as a condition in SPLASH-2 programs.

*Segment*—a contiguous part of shared address space allocated statically by the compiler or dynamically where specified by the programmer using G_MALLOC (ANL macro for allocation of shared data).

*Subsegment*—a contiguous part of a segment accessed in an interval.

### C.5.1. Parameters

In this work we have measured a set of parameters that we divide into several semantic groups. The first group of parameters (Table C.1) contains some general behavioral scalar indicators such as the number of locks and barriers used on all processors or the frequency with which a given processor requires the same lock it has just released. The common characteristic of parameters in this group is that each parameter represents only one value per application. The second group of parameters (Table C.2) encompasses total and average numbers of instructions executed between particular synchronization points. The common characteristic of parameters in this group is that each parameter represents an array of separately measured values for each type of operation (no memory access, private/shared reads/writes of ordinary shared variables, and synchronization primitives). The third group of parameters (Table C.3) is devoted to parameters considering physical coherence units of address space referred to as blocks (VSM pages or SMP cache lines). These parameters are measured for a wide range of block sizes and are given separately for CSI and NCSI.

Except for parameters denoted by par1.3 and par1.5, which had to be measured from the "lock's point of view," all the others were measured from the "processor's point of view"; specifically the parameter was measured on each processor, and then averaged per processor. For instance, the value of parameter par1.9 (frequency of NCSIs with no shared data writes) is obtained by counting the total number of NCSIs on a processor and the number of NCSI with no shared data writes. This statistic is measured for each processor separately, and the ratio representing the frequency is determined by dividing the two numbers thereafter. Finally, parameter par1.9 is calculated as the average value of the ratios on all processors.

### C.5.2. Conditions and Assumptions

Values presented in the tables are based on the measurements performed under the following conditions:

- The PRAM architectural model is simulated, as it best characterizes the inherent behavior of the applications.

```
L(L1)  P0:  accesses(I1)
U(L1)  P0:  accesses(I2)
L(L2)  P3:  accesses(I10)
U(L2)  P3:  accesses(I11)
L(L1)  P1:  accesses(I6)
L(L2)  P0:  accesses(I3)
U(L1)  P0:  accesses(I7)
L(L1)  P3:  accesses(I12)
U(L2)  P0:  accesses(I4)
U(L1)  P3:  accesses(I13)
B(B1)  P0:  accesses(I5)
B(B1)  P1:  accesses(I8)
B(B1)  P2:  accesses(I9)
B(B1)  P3:  accesses(I14)
```

*accesses(In) ::=* W: *list of segments* R: *list of segments*

*list of segments ::= {segment}*

*segment ::=* Si: *list of subsegments*

*list of subsegments ::= subsegment {subsegment}*

*subsegment ::=* ⟨sa,nosab⟩

**Figure C.3**. Part of the reduced trace file. (In—interval identifier; W—write; R—read; Si—segment identifier; sa—starting address; nosab—number of successive accessed blocks; for the rest, see Fig. C.2). This is a part of the reduced trace file corresponding to the execution shown in Figure C.2. Global order of executed synchronization operations is always preserved in the file; that is, for any execution such as the one shown in Figure C.2, the file will have the same structure. R/W accesses performed during a particular interval are listed according to the structure of accesses (In), which is informally described above. It is also possible to generate lists not just for each interval but for all NCSIs in an epoch. Such a feature of the simulator is to be used for kinds of analyses that do not need partial data by intervals.

- Applications operate on problem sizes that were suggested as suitable for an upto-64 multiprocessor system. This is further elaborated in Section C.7.

- The gcci-486 (GNU C compiler for Intel architectures) is used for compiling the SPLASH-2 applications. All parallel applications were compiled with -O2 switch for maximum optimization level.

- The resulting machine code is a mixture of both integer and floating-point machine instructions.

- One machine word is 32 bits wide. Memory operands are either 1, 2, 4, 8, or 10 bytes wide, whereas the former three are counted as one-word operands, while 10-byte operands are treated as being three-words long.

**TABLE C.1.  Scalar Parameters Group**

| | |
|---|---|
| par1.1 | Total number of locks used on all processors |
| par1.2 | Total number of barriers used on all processors |
| par1.3 | Average number of processors that access the same lock |
| par1.4 | Frequency that given processor requires the same lock that it has just released |
| par1.5 | Frequency that given lock is acquired again by the same processor that has last released this lock |
| par1.6 | Frequency that given epoch has no critical sections |
| par1.7 | Average number of CSI in epochs with at least one CSI |
| par1.8 | Frequency of CSI with no shared data writes |
| par1.9 | Frequency of NCSI with no shared data writes |
| par1.10 | Average number of cycles between two consecutive acquire operations for the same lock |

**TABLE C.2.  Access-Type-Indexed Array Parameters Group**

| | |
|---|---|
| par2.1 | Total number of instructions executed by all processors |
| par2.2 | Total number of instructions executed in all CSI on all processors |
| par2.3 | Average number of instructions executed during one CSI |
| par2.4 | Total number of instructions executed during all NCSI on all processors |
| par2.5 | Average number of instructions executed during one NCSI |
| par2.6 | Average number of instructions between two successive acquire operations for the same lock |

- Any memory reference is counted as one, regardless of the operand size.
- An operation that both reads and writes memory (such as incrementing a variable) is treated as a read instruction followed by a write instruction.
- An access is considered to be shared if it accesses memory within a shared segment. Shared segments are (1) static data segment and (2) parts of heap allocated with G_MALLOC. These shared segments are not coherence units, but rather logical units allocated by the programmer. Since the code segment is not considered to be shared, occasional reading of constants from it is neglected.
- All the simulations are conducted for a 32-processor system.

In addition, the following assumptions are made:

- Every instruction is executed within a single cycle. That gives, on average, the execution rate of one instruction per cycle.

- There are as many processors in the simulated system as there are threads in a parallel application; or rather, there are as many available processors in the simulated system as there are threads in an application.

- Threads are pinned to their processors; that is, no migration is modeled.

- No process other than the parallel application exists in the simulated system, and the activities of the operating system (i.e., its kernel) during the execution are negligible.

## C.6. MEASURING TOOLS

In this section we present the details of the tools used in this characterization: Limes (LInux MEmory Simulator) and Scowl (System for Characterization of Parallel Applications). More information and the sources for the tools (which are in the public domain) can be found at the following URL: `http://galeb.etf.bg.ac.yu/~dsm/`.

### C.6.1. Limes

The simulation environment we used to conduct this research consisted of Limes: a software tool for simulation of multiprocessors [Magdic 1997] and a P5-based PC, running under a Linux operating system. The simulation tool performs execution-driven simulation; the concepts are briefly described here.

In execution-driven simulation of multiprocessors, the parallel application and the simulation kernel are linked into a single executable. The host CPU executes machine instructions of the application as long as they do not attempt to access memory. The moment it happens, the execution of the native application instructions is stopped, and the control is transferred to the simulation kernel. Therefore, the simulation tool has complete control over every memory reference the application generates. In fact, memory references occur quite often, but the fact that the application instructions are executed natively by the host CPU saves us from the need to interpret them or even to be aware of their semantics. In addition, as the instructions are executed, the target system time is measured, in terms of processor cycles elapsed during the execution of these instructions. So immediately before each memory reference, the simulation kernel is entered with all the relevant parameters: address, data size, type of operation, and the operation time-stamp.

Simulation kernel calls are inserted through a process called *code instrumentation*. The C or C++ code of a parallel application is translated to assembly language by the compiler. The resulting assembly code is then instrumented for intercepting memory references and counting the time. In the instrumentation process, a kernel callout is inserted before each instruction that accesses memory (some special treatment is required for instructions that both read and write memory, but this treatment is simple). This ensures that the operation is not actually committed until the proper time. Elapsed cycles are counted by accumulating estimated execution time of parts of the basic blocks through which the control flow passes in its trip between two visits to the kernel. The term *estimated* indicates that the processor's pipeline is not accurately modeled; instead, it is assumed that, on average, one CISC instruction is executed per one cycle.

Apart from taking control over memory accesses that the application makes, the simulation tool has one more important role; it creates threads of the parallel application and assigns each one a distinct simulated processor and maintains a private timer for it. Logically, each thread executes on its own virtual processor. In simulation, the threads are multiplexed so that a thread executes on the only available host processor at a time, being preempted every time it attempts to reference memory. Upon each preemption, the thread with the shortest execution time is scheduled for execution—its context is restored and it returns to user mode. When it returns to the kernel, its timer will be updated by the time spent in user mode, another thread with the smallest time will be chosen for execution, and so on. In this manner, a correct order of memory operations is preserved, by postponing the execution of an operation until the time associated with it becomes the smallest of all the times associated with other events (i.e., other memory operations or synchronization operations). The net result is that the content of the global memory is as it would be if the application were executed on a multiprocessor system with processors having one instruction per cycle execution rate and zero memory latency.

The simulator allows simulation of memory systems other than the PRAM system, described above. It can be instructed to simulate a realistic memory system consisting of TLBs, caches, the interconnection network, and similar, keeping account of the latencies for memory operations that take longer than one cycle into execution time per processor. However, since this study deals only with inherent characteristics of the applications and not with its application to any particular system, we considered that detailed hardware simulation would be unnecessary, revealing no more information than is currently the case. In fact, in many parts the characterization can be carried out without instrumentation of private references since they, by definition, do not have global consequences, and therefore issuing a private reference does not impose a need for rescheduling. This can speed simulations up significantly, since private references are very frequent on CISC processors. The interesting thing is that the number of *shared* references is approximately equal to reported numbers for identical simulations conducted on RISC machines.

### C.6.2. Scowl

Scowl is realized as an extension of Limes. Basically, Limes provides for hardware simulations, so in order to facilitate a mostly software-based study, it was necessary to change the simulation kernel in addition to writing a simulator. It was achieved by introducing some new simulation system calls, which allow efficient control of synchronization primitives.

Two different simulators were written in building the tool capable of measuring specified groups of parameters. One simulator is used for parameters of groups 1 and 2 that can be measured in execution-driven simulation. The other simulator serves as a generator of reduced trace files needed for figuring out parameters of group 3. Certain postsimulation analysis using a specially written program is needed to actually compute them.

The first simulator works with both private and shared references (the way they are considered in Limes). Nevertheless, it has its own structure describing which parts of memory are shared. Except for the structure for shared data, the simulator manages information about critical section passes on each processor and critical section passes of each lock, which are at the end used to determine the average values. This simulator can easily be modified to measure some other set of parameters, such as more lock identity oriented parameters.

The second simulator deals only with shared data. Since there are still too many memory references, it is not intended to calculate the end value of parameters, but instead it is used to produce the reduced trace files (a kind

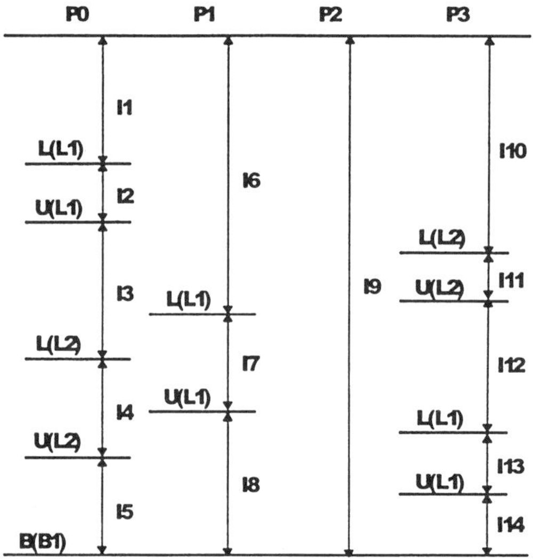



**Figure C.2.** An example of an epoch executed on FOUR processors. (P—processor; Li—lock identifier; Bi—barrier identifier; L( )—lock acquire; U( )—lock release; B( ) —barrier operation; In—interval identifier). This scenario shows a part of execution of an application on processors P0 to P3 between two barriers. The example shows the total of 14 intervals: 5 CSI (intervals I2, I4, I7, I11, and I13), and 9 NCSI (I1, I3, I5, I6, I8, I9, I10, I12, and I14). The reduced trace file is shown in Figure C.3.

**TABLE C.3. Block-Size-Indexed Array Parameters Group**

| | |
|---|---|
| par3.1 | Average number of blocks modified inside one CSI/NCSI with shared data writes |
| par3.2 | Maximum number of blocks modified inside one CSI/NCSI with shared data writes |
| par3.3 | Minimum number of blocks modified inside one CSI/NCSI with shared data writes |
| par3.4 | Average number of blocks only read inside one CSI/NCSI with at least one read-only block |
| par3.5 | Maximum number of blocks only read inside one CSI/NCSI with at least one read-only block |
| par3.6 | Minimum number of blocks only read inside one CSI/NCSI with at least one read-only block |
| par3.7 | Frequency that CSI/NCSI contains at least one read-only block |

of intermediate form). Output from this simulator is then used as input to the next pass, which calculates wanted parameters. For the time being, only one postsimulation analyzer exists for calculating parameters of group 3. However, the reduced traces can be used as input to other postsimulation analyzers that can generate some other set of parameters, for instance, parameters for epochs (periods between two barrier primitives) similar to the ones for critical section passes. The advantages of using reduced traces are that simulation becomes independent of the Limes simulation environment (even a Linux operating system, as a reduced trace is a plain ASCII file), and furthermore the process of obtaining parameters requires two orders of magnitude less time.

The format of reduced trace files is as follows. A reduced trace file consists of records (one per interval) showing addresses accessed during the interval. The interval is denoted by a synchronization operation that terminates the interval. Accessed addresses are given separately for both access types: writes and reads. For each of them, there is a list of segments that were accessed at least once during the interval. For each segment, a list of accessed subsegments is generated. For an example of execution of a program (shown in Figure C.2), the generated reduced trace file is presented in Figure C.3.

## C.7. RESULTS

This section presents the values obtained during the measurements. Applications used for measurements are listed in column "Code" of Table C.4. Nine applications from a SPLASH-2 benchmark suite were used. The whole suite consists of 12 applications, but three (Radiosity, Raytrace, and Volrend) were

**TABLE C.4.** **The First Group of Parameters for All Applications**

| 70Code | Problem size | par1.1 | par1.2 | par1.3 | par1.4 | par1.5 | par1.6 | par1.7 | par1.8 | par1.9 | par1.10 |
|---|---|---|---|---|---|---|---|---|---|---|---|
| Barnes | 4K particles | 87 | 4 | 16.05 | 65.62% | 47.91% | 47.06% | 62.21 | 3.47% | 8.33% | 172,400 |
| Cholesky | tk29.O | 67 | 1 | 14.76 | 26.09% | 68.95% | 50.00% | 1169.45 | 0.14% | 10.12% | 255,475 |
| FFT | 64K points | 1 | 1 | 32.00 | 0.00% | 0.00% | 85.71% | 1.00 | 0.00% | 65.63% | 3 |
| FMM | 4K particles | 347 | 1 | 4.58 | 38.62% | 56.51% | 38.24% | 21.36 | 0.41% | 35.52% | 2904,818 |
| LU | $512 \times 512$ matrix | 1 | 1 | 32.00 | 0.00% | 0.00% | 98.51% | 1.00 | 0.00% | 46.20% | 795,438 |
| OCEAN | $258 \times 258$ ocean | 4 | 14 | 32.00 | 92.75% | 0.00% | 76.89% | 1.00 | 90.05% | 47.64% | 16,051 |
| Radix | 256K integers | 33 | 2 | 32.00 | 0.00% | 0.00% | 62.50% | 21.67 | 0.00% | 90.41% | 9,139 |
| Waternsq | 512 molecules | 518 | 3 | 17.08 | 0.54% | 0.00% | 29.41% | 92.08 | 0.00% | 98.83% | 239,306 |
| Watersp | 512 molecules | 6 | 3 | 26.67 | 33.28% | 16.80% | 23.53% | 1.46 | 0.31% | 64.56% | 3,360,132 |

omitted since they are graphics oriented and are not ported to a Linux operating system. The applications used in characterization were taken "as are" (as can be found at the SPLASH-2 site: `http://www-flash.stanford.edu`), with some minor modifications of function declarations for a few of them, necessary to satisfy the GNU C compiler.

It should be noted that Water-Nsquared and Water-Spatial are treated as two different benchmarks although they solve the same problem. There are also two versions of programs for the LU and OCEAN problem included in the SPLASH-2 suite, but we used only the optimized ones. The measurements were done for the entire executions of the applications. As mentioned before, a 32-processor system employing the PRAM memory model is simulated.

Problem sizes used (given for each application in Table C.4) are "default problem sizes for up to 64 processor machines" as defined in README files of SPLASH-2 suite, except in two cases—Barnes and FM—where we had to decrease problem size from 16,000 particles to 4000 particles, due to limited resources of our simulations. The other parameters that can be changed for some applications were left at their default values, again bearing in mind that our intention was to characterize SPLASH-2 suite "as is." This implies executing FFT and LU (which are aware of cache) with cache of 64,000 lines of 16 bytes, and with cache of total size of 16 kB, respectively.

### C.7.1. Scalar Parameters Group

The first group of parameters is shown in Table C.4. It can be divided into several subgroups: par1.1 and par1.2 are basic indicators (number of synchronization variables); par1.3, par1.4, and par1.5 present the relations of processors and locks (critical sections); par1.6 and par1.7 present the relations of epochs (barriers) and critical sections (locks); par1.8 and par1.9 give information about shared data writes and intervals (of both type); and par1.10 is an advanced indicator of behavior of applications.

### C.7.2. Access-Type-Indexed Array Parameters Group

The second group of parameters for each application is shown in Tables C.5–C.13. Parameters in this group have one value for each type of operation: "na" (instructions that do not access memory), "rp" (reads of private data), "wp" (writes to private data), "rs" (reads of shared data), "ws" (writes to shared data), "la, lr" [lock acquire (LOCK) and lock release (UNLOCK)] operations, which execute the same number of times according to the way SPLASH-2 programs were written and "b" (barrier operations, BARRIER).

A memory reference is treated as private or shared according to the definition in Section C.5.2; in short, only semantically shared data is treated as shared—for example, it excludes the data allocated on heap with malloc and not G_MALLOC.

**TABLE C.5.  Parameters of Group 2 for Barnes**

|        | na | rp | wp | rs | ws | la, lr | b |
|--------|----|----|----|----|----|--------|---|
| par2.1 | 24,148,652 | 87,626,723 | 72,806,593 | 103,861,978 | 45,688,170 | 17,916 | 544 |
| par2.2 | 1,552,861 | 473,515 | 319,922 | 359,641 | 294,367 | 0 | 0 |
| par2.3 | 86 | 25 | 17 | 19 | 15 | 0 | 0 |
| par2.4 | 239,933,391 | 87,153,208 | 72,486,671 | 103,502,337 | 45,393,803 | 17,916 | 544 |
| par2.5 | 13,151 | 4780 | 3975 | 5669 | 2486 | 0 | 0 |
| par2.6 | 232,474 | 83,910 | 69,279 | 98,495 | 43,147 | 280 | |

**TABLE C.6.  Parameters of Group 2 for Cholesky**

|        | na | rp | wp | rs | ws | la, lr | b |
|--------|----|----|----|----|----|--------|---|
| par2.1 | 1,557,304,548 | 399,444,691 | 121,049,661 | 526,099,764 | 75,499,107 | 74,845 | 128 |
| par2.2 | 1,403,877 | 260,442 | 251,249 | 486,944 | 147,308 | 0 | 0 |
| par2.3 | 18 | 3 | 2 | 5 | 1 | 0 | 0 |
| par2.4 | 1,555,900,671 | 399,184,249 | 120,798,412 | 525,612,820 | 75,351,799 | 74,845 | 128 |
| par2.5 | 20,730 | 4789 | 1414 | 7012 | 908 | 0 | 0 |
| par2.6 | 1,022,310 | 337,356 | 97,565 | 315,989 | 66,707 | 62 | 0 |

**TABLE C.7.  Parameters of Group 2 for FFT**

|        | na | rp | wp | rs | ws | la, lr | b |
|--------|----|----|----|----|----|--------|---|
| par2.1 | 26,820,319 | 7,381,269 | 3,125,744 | 4,076,518 | 3,646,009 | 32 | 224 |
| par2.2 | 32 | 0 | 32 | 128 | 32 | 0 | 0 |
| par2.3 | 1 | 0 | 1 | 4 | 1 | 0 | 0 |
| par2.4 | 26,820,287 | 7,381,269 | 3,125,712 | 4,076,390 | 3,645,977 | 32 | 224 |
| par2.5 | 93,125 | 25,628 | 10,852 | 14,153 | 12,659 | 0 | 0 |
| par2.6 | 0 | 0 | 0 | 0 | 0 | 0 | 0 |

**TABLE C.8.  Parameters of Group 2 for FMM**

|        | na | rp | wp | rs | ws | la, lr | b |
|--------|----|----|----|----|----|--------|---|
| par2.1 | 324,987,222 | 349,606,033 | 249,675,692 | 101,314,505 | 4,067,746 | 14,354 | 1088 |
| par2.2 | 640,045 | 507,692 | 14,487 | 378,905 | 435,386 | 0 | 0 |
| par2.3 | 43 | 34 | 1 | 25 | 29 | 0 | 0 |
| par2.4 | 324,347,177 | 349,098,341 | 24,9661,205 | 100,935,600 | 3,632,360 | 14,354 | 1088 |
| par2.5 | 21,926 | 23,523 | 16,876 | 6900 | 238 | 0 | 0 |
| par2.6 | 608,550 | 657,156 | 469,903 | 189,624 | 5825 | 20 | 0 |

**TABLE C.9. Parameters of Group 2 for LU**

|        | na          | rp         | wp         | rs         | ws         | la, lr | b    |
|--------|-------------|------------|------------|------------|------------|--------|------|
| par2.1 | 282,285,138 | 51,034,105 | 22,925,180 | 99,261,147 | 45,267,071 | 32     | 2144 |
| par2.2 | 32          | 0          | 32         | 128        | 32         | 0      | 0    |
| par2.3 | 1           | 0          | 1          | 4          | 1          | 0      | 0    |
| par2.4 | 282,285,106 | 51,034,105 | 22,925,148 | 99,261,019 | 45,267,039 | 32     | 2144 |
| par2.5 | 127,846     | 23,112     | 10,382     | 44,954     | 20,500     | 0      | 0    |
| par2.6 | 0           | 0          | 0          | 0          | 0          | 0      | 0    |

**TABLE C.10. Parameters of Group 2 for Ocean**

|        | na          | rp          | wp         | rs          | ws         | la, lr | b      |
|--------|-------------|-------------|------------|-------------|------------|--------|--------|
| par2.1 | 223,098,081 | 144,857,134 | 35,963,631 | 156,781,581 | 34,775,267 | 6656   | 28,800 |
| par2.2 | 38,624      | 6624        | 6688       | 20,000      | 662        | 0      | 0      |
| par2.3 | 5           | 0           | 1          | 3           | 0          | 0      | 0      |
| par2.4 | 223,059,457 | 144,850,510 | 359,569,43 | 156,761,581 | 34,774,605 | 6656   | 28,800 |
| par2.5 | 6285        | 4081        | 1012       | 4416        | 979        | 0      | 0      |
| par2.6 | 432,117     | 287,897     | 64,554     | 341,317     | 74,836     | 9      | 42     |

**TABLE C.11. Parameters of Group 2 for Radix**

|        | na         | rp         | wp         | rs        | ws        | la, lr | b   |
|--------|------------|------------|------------|-----------|-----------|--------|-----|
| par2.1 | 59,783,120 | 36,261,715 | 24,052,299 | 6,936,718 | 3,115,127 | 2080   | 256 |
| par2.2 | 276,512    | 204,800    | 69,696     | 141,440   | 65,568    | 0      | 0   |
| par2.3 | 132        | 98         | 33         | 68        | 31        | 0      | 0   |
| par2.4 | 59,506,608 | 36,056,915 | 23,982,603 | 6,795,278 | 3,049,559 | 2080   | 256 |
| par2.5 | 25,129     | 15,226     | 10,127     | 2869      | 1287      | 0      | 0   |
| par2.6 | 138,827    | 107,530    | 27,411     | 104,271   | 44574     | 32     | 3   |

**TABLE C.12. Parameters of Group 2 for Waternsq**

|        | na          | rp         | wp         | rs         | ws        | la, lr | b   |
|--------|-------------|------------|------------|------------|-----------|--------|-----|
| par2.1 | 461,974,496 | 91,201,574 | 73,030,090 | 80,059,649 | 2,574,947 | 35,360 | 544 |
| par2.2 | 938,913     | 974,048    | 209,122    | 1,912,704  | 313,439   | 0      | 0   |
| par2.3 | 26          | 27         | 5          | 54         | 8         | 0      | 0   |
| par2.4 | 461,035,583 | 90,227,526 | 72,820,968 | 78,146,945 | 2,261,508 | 35,360 | 544 |
| par2.5 | 12,827      | 2510       | 2025       | 2174       | 62        | 0      | 0   |
| par2.6 | 2,881,907   | 597,291    | 465,363    | 511,381    | 19,178    | 274    | 2   |

**TABLE C.13. Parameters of Group 2 for Watersp**

|        | na          | rp         | wp         | rs         | ws        | la, lr | b   |
|--------|------------|-----------|-----------|-----------|----------|-------|-----|
| par2.1 | 367,533,725 | 89,037,782 | 70,592,527 | 62,415,862 | 1,016,365 | 609   | 544 |
| par2.2 | 745         | 1130       | 627        | 1348       | 671       | 0     | 0   |
| par2.3 | 1           | 1          | 1          | 2          | 1         | 0     | 0   |
| par2.4 | 367,532,980 | 89,036,652 | 70,591,900 | 62,414,514 | 1,015,694 | 609   | 544 |
| par2.5 | 310,120     | 75,128     | 59,565     | 52,670     | 855       | 0     | 0   |
| par2.6 | 1,826,463   | 462,500    | 357,616    | 311,293    | 5869      | 3     | 2   |

All accesses to memory are counted as 1, regardless of the actual size of the data being accessed: mostly 4 bytes (long words), relatively often 8 bytes (double long words) in computation, and rarely 1, 2, or 10 bytes. As stated in Section C.5.2, all instructions are assumed to execute within a single cycle.

### C.7.3. Block-Size-Indexed Array Parameters Group

The third group of parameters for each application is shown in Tables C.14–C.22. Column "f" (frequency) corresponds to par3.7 defined in Section 5.1, while the rest of the parameters (par3.1 to par3.6) are named in a more semantic way (average/maximum/minimum write/read-only).

This group of parameters is dependent on the layout of memory being allocated from the heap. The results presented here were obtained for the layout generated by calling the function malloc, provided by the compiler environment that was used (gcc 2.6.3.). While this simple translation to malloc is the way G_MALLOC is likely to be implemented in SMP systems, it could be made somewhat different in DSM systems. For example, we also measured the same parameters for G_MALLOC, which allocated each segment on the block boundary, but the results turned out to be similar; actually, the difference was less than 1%. It should be noted that the

**TABLE C.14. Parameters of Group 3 for Barnes**

| Block Size (bytes) | CSI | | | | | | | NCSI | | | | | | |
|---|---|---|---|---|---|---|---|---|---|---|---|---|---|---|
| | Write | | | Read-only | | | | Write | | | Read-only | | | |
| | avg | max | min | avg | max | min | $f(\%)$ | avg | max | min | avg | max | min | $f(\%)$ |
| 4    | 14.97 | 204.84 | 1.00 | 7.98 | 79.00 | 1.00 | 100.00 | 32.42 | 5049.16 | 2.06 | 107.63 | 9640.91 | 1.03 | 100.00 |
| 8    | 9.74  | 115.56 | 1.00 | 6.34 | 50.38 | 1.00 | 100.00 | 18.28 | 2789.91 | 1.53 | 66.48  | 5307.88 | 1.03 | 100.00 |
| 16   | 7.03  | 71.19  | 1.00 | 5.12 | 30.41 | 1.00 | 100.00 | 11.20 | 1658.81 | 1.25 | 45.61  | 3144.66 | 1.03 | 100.00 |
| 32   | 5.42  | 48.28  | 1.00 | 4.14 | 18.22 | 1.00 | 100.00 | 7.06  | 958.78  | 1.12 | 33.96  | 2066.47 | 1.03 | 100.00 |
| 128  | 3.66  | 24.94  | 1.00 | 2.80 | 8.88  | 1.00 | 100.00 | 3.32  | 339.53  | 1.00 | 18.52  | 915.41  | 1.00 | 100.00 |
| 512  | 3.25  | 19.66  | 1.00 | 2.21 | 5.31  | 1.00 | 100.00 | 2.20  | 168.88  | 1.00 | 12.05  | 405.16  | 1.00 | 100.00 |
| 1024 | 3.18  | 18.25  | 1.00 | 2.10 | 4.25  | 1.00 | 100.00 | 1.95  | 129.03  | 1.00 | 10.71  | 251.28  | 1.00 | 100.00 |
| 2048 | 3.13  | 17.91  | 1.00 | 2.08 | 4.00  | 1.00 | 100.00 | 1.73  | 94.91   | 1.00 | 9.82   | 148.97  | 1.00 | 100.00 |
| 4096 | 3.08  | 17.31  | 1.00 | 2.08 | 3.62  | 1.00 | 100.00 | 1.51  | 62.97   | 1.00 | 9.27   | 92.91   | 1.00 | 100.00 |

**TABLE C.15.  Parameters of Group 3 for Barnes**

| Block Size (bytes) | CSI | | | | | | | NCSI | | | | | | |
|---|---|---|---|---|---|---|---|---|---|---|---|---|---|---|
| | Write | | | Read-only | | | | Write | | | Read-only | | | |
| | avg | max | min | avg | max | min | $f(\%)$ | avg | max | min | avg | max | min | $f(\%)$ |
| 4 | 1.95 | 6.88 | 1.00 | 45.57 | 192690.88 | 1.00 | 100.00 | 491.08 | 178446.16 | 1.00 | 757.15 | 333146.00 | 1.00 | 99.96 |
| 8 | 1.68 | 5.91 | 1.00 | 23.82 | 96346.44 | 1.00 | 100.00 | 246.73 | 89236.94 | 1.00 | 386.97 | 166573.50 | 1.00 | 99.96 |
| 16 | 1.68 | 4.94 | 1.00 | 13.05 | 48174.22 | 1.00 | 100.00 | 125.89 | 44657.28 | 1.00 | 201.74 | 83287.25 | 1.00 | 99.96 |
| 32 | 1.68 | 4.94 | 1.00 | 7.67 | 24088.59 | 1.00 | 100.00 | 65.33 | 22360.72 | 1.00 | 107.88 | 41644.12 | 1.00 | 99.96 |
| 128 | 1.68 | 4.06 | 1.00 | 3.45 | 6024.38 | 1.00 | 100.00 | 19.44 | 5627.31 | 1.00 | 34.45 | 10412.53 | 1.00 | 99.96 |
| 512 | 1.68 | 4.06 | 1.00 | 2.39 | 1506.88 | 1.00 | 100.00 | 7.05 | 1440.31 | 1.00 | 13.67 | 2604.62 | 1.00 | 99.95 |
| 1024 | 1.68 | 4.06 | 1.00 | 2.22 | 754.44 | 1.00 | 100.00 | 4.90 | 739.56 | 1.00 | 10.02 | 1303.31 | 1.00 | 99.95 |
| 2048 | 1.68 | 4.06 | 1.00 | 2.12 | 378.22 | 1.00 | 100.00 | 3.80 | 383.44 | 1.00 | 8.03 | 652.66 | 1.00 | 99.95 |
| 4096 | 1.68 | 4.03 | 1.00 | 2.08 | 190.12 | 1.00 | 100.00 | 3.24 | 202.09 | 1.00 | 6.97 | 327.31 | 1.00 | 99.95 |

**TABLE C.16.  Parameters of Group 3 for Barnes**

| Block Size (bytes) | CSI | | | | | | | NCSI | | | | | | |
|---|---|---|---|---|---|---|---|---|---|---|---|---|---|---|
| | Write | | | Read-only | | | | Write | | | Read-only | | | |
| | avg | max | min | avg | max | min | $f(\%)$ | avg | max | min | avg | max | min | $f(\%)$ |
| 4 | 1.00 | 1.00 | 1.00 | 1.00 | 1.00 | 1.00 | 100.00 | 11308.94 | 24352.47 | 7680.19 | 7299.43 | 25605.00 | 1.03 | 88.89 |
| 8 | 1.00 | 1.00 | 1.00 | 1.00 | 1.00 | 1.00 | 100.00 | 5654.50 | 12176.31 | 3840.16 | 3651.03 | 12805.00 | 1.03 | 88.89 |
| 16 | 1.00 | 1.00 | 1.00 | 1.00 | 1.00 | 1.00 | 100.00 | 2827.29 | 6088.28 | 1920.16 | 1826.77 | 6405.00 | 1.03 | 88.89 |
| 32 | 1.00 | 1.00 | 1.00 | 1.00 | 1.00 | 1.00 | 100.00 | 1413.68 | 3044.22 | 960.16 | 914.52 | 3204.00 | 1.03 | 88.89 |
| 128 | 1.00 | 1.00 | 1.00 | 1.00 | 1.00 | 1.00 | 100.00 | 353.46 | 761.12 | 240.16 | 230.01 | 802.00 | 1.00 | 88.89 |
| 512 | 1.00 | 1.00 | 1.00 | 1.00 | 1.00 | 1.00 | 100.00 | 88.40 | 190.38 | 60.12 | 130.96 | 321.00 | 1.03 | 88.54 |
| 1024 | 1.00 | 1.00 | 1.00 | 1.00 | 1.00 | 1.00 | 100.00 | 49.75 | 107.12 | 33.88 | 116.40 | 293.00 | 1.03 | 88.54 |
| 2048 | 1.00 | 1.00 | 1.00 | 1.00 | 1.00 | 1.00 | 100.00 | 24.89 | 53.59 | 17.00 | 107.10 | 275.00 | 1.03 | 88.54 |
| 4096 | 1.00 | 1.00 | 1.00 | 1.00 | 1.00 | 1.00 | 100.00 | 12.47 | 26.84 | 8.56 | 102.46 | 266.00 | 1.03 | 88.54 |

**TABLE C.17.  Parameters of Group 3 for Barnes**

| Block Size (bytes) | CSI | | | | | | | NCSI | | | | | | |
|---|---|---|---|---|---|---|---|---|---|---|---|---|---|---|
| | Write | | | Read-only | | | | Write | | | Read-only | | | |
| | avg | max | min | avg | max | min | $f(\%)$ | avg | max | min | avg | max | min | $f(\%)$ |
| 4 | 51.18 | 92.00 | 1.00 | 2.86 | 3.75 | 1.00 | 100.00 | 223.31 | 12519.06 | 2.00 | 1298.21 | 387476.03 | 1.00 | 99.78 |
| 8 | 26.23 | 47.00 | 1.00 | 2.83 | 3.75 | 1.00 | 100.00 | 128.80 | 6696.69 | 1.03 | 694.44 | 193739.53 | 1.00 | 99.78 |
| 16 | 13.72 | 24.00 | 1.00 | 2.71 | 3.75 | 1.00 | 100.00 | 74.62 | 3602.50 | 1.03 | 366.25 | 96871.25 | 1.00 | 99.78 |
| 32 | 7.45 | 13.00 | 1.00 | 2.57 | 3.75 | 1.00 | 100.00 | 40.92 | 1966.19 | 1.00 | 205.71 | 48437.12 | 1.00 | 99.78 |
| 128 | 2.66 | 4.00 | 1.00 | 2.50 | 3.75 | 1.00 | 100.00 | 22.32 | 603.88 | 1.00 | 77.23 | 12110.56 | 1.00 | 99.78 |
| 512 | 1.49 | 2.84 | 1.00 | 2.33 | 3.75 | 1.00 | 100.00 | 16.60 | 260.56 | 1.00 | 37.83 | 3029.41 | 1.00 | 99.78 |
| 1024 | 1.27 | 2.00 | 1.00 | 2.25 | 3.75 | 1.00 | 100.00 | 14.96 | 201.91 | 1.00 | 26.28 | 1516.22 | 1.00 | 99.78 |
| 2048 | 1.18 | 2.00 | 1.00 | 2.11 | 3.75 | 1.00 | 100.00 | 13.37 | 173.09 | 1.00 | 19.04 | 759.12 | 1.00 | 99.78 |
| 4096 | 1.13 | 2.00 | 1.00 | 1.94 | 3.59 | 1.00 | 100.00 | 10.91 | 97.06 | 1.00 | 13.50 | 380.59 | 1.00 | 99.78 |

### TABLE C.18. Parameters of Group 3 for Barnes

| Block Size (bytes) | CSI | | | | | | | NCSI | | | | | | |
|---|---|---|---|---|---|---|---|---|---|---|---|---|---|---|
| | Write | | | Read-only | | | | Write | | | Read-only | | | |
| | avg | max | min | avg | max | min | $f(\%)$ | avg | max | min | avg | max | min | $f(\%)$ |
| 4 | 1.00 | 1.00 | 1.00 | 1.00 | 1.00 | 1.00 | 100.00 | 5308.22 | 31313.50 | 450.44 | 1685.18 | 16422.34 | 1.03 | 98.64 |
| 8 | 1.00 | 1.00 | 1.00 | 1.00 | 1.00 | 1.00 | 100.00 | 2654.11 | 15656.78 | 225.25 | 848.52 | 8229.00 | 1.03 | 98.64 |
| 16 | 1.00 | 1.00 | 1.00 | 1.00 | 1.00 | 1.00 | 100.00 | 1327.06 | 7828.47 | 112.75 | 430.02 | 4132.00 | 1.00 | 98.64 |
| 32 | 1.00 | 1.00 | 1.00 | 1.00 | 1.00 | 1.00 | 100.00 | 663.54 | 3914.28 | 56.50 | 219.91 | 2083.00 | 1.00 | 98.64 |
| 128 | 1.00 | 1.00 | 1.00 | 1.00 | 1.00 | 1.00 | 100.00 | 165.89 | 978.66 | 14.28 | 59.00 | 523.00 | 1.00 | 98.64 |
| 512 | 1.00 | 1.00 | 1.00 | 1.00 | 1.00 | 1.00 | 100.00 | 41.58 | 244.78 | 3.91 | 19.03 | 139.00 | 1.03 | 98.60 |
| 1024 | 1.00 | 1.00 | 1.00 | 1.00 | 1.00 | 1.00 | 100.00 | 20.79 | 122.44 | 1.97 | 10.38 | 70.00 | 1.03 | 98.60 |
| 2048 | 1.00 | 1.00 | 1.00 | 1.00 | 1.00 | 1.00 | 100.00 | 10.40 | 61.28 | 1.00 | 6.51 | 36.00 | 1.03 | 98.60 |
| 4096 | 1.00 | 1.00 | 1.00 | 1.00 | 1.00 | 1.00 | 100.00 | 10.00 | 45.22 | 1.00 | 6.10 | 35.00 | 1.03 | 98.60 |

### TABLE C.19. Parameters of Group 3 for Barnes

| Block Size (bytes) | CSI | | | | | | | NCSI | | | | | | |
|---|---|---|---|---|---|---|---|---|---|---|---|---|---|---|
| | Write | | | Read-only | | | | Write | | | Read-only | | | |
| | avg | max | min | avg | max | min | $f(\%)$ | avg | max | min | avg | max | min | $f(\%)$ |
| 4 | 1.89 | 2.00 | 1.00 | 3.80 | 4.00 | 2.00 | 100.00 | 2872.03 | 33953.06 | 161.19 | 1784.28 | 27706.00 | 2.00 | 100.00 |
| 8 | 1.76 | 2.00 | 1.00 | 2.90 | 3.00 | 2.00 | 100.00 | 1436.02 | 16982.72 | 80.59 | 904.17 | 13885.50 | 2.00 | 100.00 |
| 16 | 1.65 | 2.00 | 1.00 | 2.86 | 3.00 | 1.00 | 100.00 | 922.51 | 8590.44 | 68.84 | 490.14 | 7079.75 | 2.00 | 100.00 |
| 32 | 1.00 | 1.00 | 1.00 | 2.86 | 3.00 | 1.00 | 100.00 | 473.40 | 4411.19 | 40.16 | 259.73 | 3624.81 | 2.00 | 100.00 |
| 128 | 1.00 | 1.00 | 1.00 | 1.90 | 2.00 | 1.00 | 100.00 | 125.76 | 1185.06 | 11.38 | 81.88 | 1030.94 | 2.00 | 100.00 |
| 512 | 1.00 | 1.00 | 1.00 | 1.90 | 2.00 | 1.00 | 100.00 | 37.68 | 438.59 | 3.62 | 35.25 | 369.78 | 2.00 | 100.00 |
| 1024 | 1.00 | 1.00 | 1.00 | 1.90 | 2.00 | 1.00 | 100.00 | 20.02 | 243.59 | 2.28 | 24.86 | 208.00 | 2.00 | 100.00 |
| 2048 | 1.00 | 1.00 | 1.00 | 1.90 | 2.00 | 1.00 | 100.00 | 10.58 | 137.38 | 1.66 | 17.27 | 117.50 | 2.00 | 100.00 |
| 4096 | 1.00 | 1.00 | 1.00 | 1.90 | 2.00 | 1.00 | 100.00 | 6.13 | 89.50 | 1.31 | 12.58 | 81.50 | 2.00 | 100.00 |

### TABLE C.20. Parameters of Group 3 for Barnes

| Block Size (bytes) | CSI | | | | | | | NCSI | | | | | | |
|---|---|---|---|---|---|---|---|---|---|---|---|---|---|---|
| | Write | | | Read-only | | | | Write | | | Read-only | | | |
| | avg | max | min | avg | max | min | $f(\%)$ | avg | max | min | avg | max | min | $f(\%)$ |
| 4 | 31.52 | 32.00 | 1.00 | 4.94 | 5.00 | 1.00 | 100.00 | 4263.61 | 9247.53 | 960.31 | 913.87 | 17812.16 | 2.00 | 98.73 |
| 8 | 15.77 | 16.00 | 1.00 | 4.45 | 5.00 | 1.00 | 100.00 | 2234.55 | 5135.50 | 480.19 | 458.04 | 8908.28 | 1.97 | 98.73 |
| 16 | 7.89 | 8.00 | 1.00 | 3.22 | 4.00 | 1.00 | 100.00 | 1220.00 | 3085.06 | 240.16 | 230.04 | 4455.69 | 1.97 | 98.73 |
| 32 | 3.95 | 4.00 | 1.00 | 2.11 | 3.00 | 1.00 | 100.00 | 711.53 | 2054.88 | 120.16 | 116.04 | 2228.91 | 1.97 | 98.73 |
| 128 | 1.00 | 1.00 | 1.00 | 2.02 | 3.00 | 1.00 | 100.00 | 329.20 | 1281.84 | 30.16 | 30.59 | 559.50 | 1.94 | 98.7 |
| 512 | 1.00 | 1.00 | 1.00 | 1.98 | 2.00 | 1.00 | 100.00 | 229.59 | 1088.41 | 7.62 | 9.24 | 142.47 | 1.97 | 98.69 |
| 1024 | 1.00 | 1.00 | 1.00 | 1.98 | 2.00 | 1.00 | 100.00 | 202.78 | 1014.59 | 3.84 | 5.68 | 72.97 | 1.97 | 98.69 |
| 2048 | 1.00 | 1.00 | 1.00 | 1.98 | 2.00 | 1.00 | 100.00 | 120.04 | 515.00 | 1.97 | 3.90 | 38.22 | 1.97 | 98.69 |
| 4096 | 1.00 | 1.00 | 1.00 | 1.98 | 2.00 | 1.00 | 100.00 | 72.02 | 258.00 | 1.00 | 3.00 | 20.88 | 1.97 | 98.69 |

**TABLE C.21.  Parameters of Group 3 for Barnes**

| Block Size (bytes) | CSI | | | | | | | NCSI | | | | | | |
| | Write | | | Read-only | | | | Write | | | Read-only | | | |
| | avg | max | min | avg | max | min | $f(\%)$ | avg | max | min | avg | max | min | $f(\%)$ |
|---|---|---|---|---|---|---|---|---|---|---|---|---|---|---|
| 4 | 17.73 | 18.00 | 1.03 | 25.57 | 26.00 | 1.00 | 100.00 | 4124.78 | 13000.84 | 93.06 | 40.33 | 8610.03 | 1.00 | 99.91 |
| 8 | 8.88 | 9.00 | 1.03 | 15.74 | 16.00 | 1.00 | 100.00 | 2077.48 | 6756.72 | 46.53 | 21.87 | 4564.50 | 1.00 | 99.91 |
| 16 | 4.94 | 5.00 | 1.00 | 11.81 | 12.00 | 1.00 | 100.00 | 1283.59 | 4130.69 | 31.03 | 11.81 | 2290.25 | 1.00 | 99.91 |
| 32 | 2.97 | 3.00 | 1.00 | 8.86 | 9.00 | 1.00 | 100.00 | 649.47 | 2081.50 | 15.53 | 6.79 | 1152.62 | 1.00 | 99.91 |
| 128 | 1.49 | 2.00 | 1.00 | 7.40 | 9.00 | 1.00 | 100.00 | 171.04 | 532.47 | 15.53 | 3.83 | 495.97 | 1.00 | 99.91 |
| 512 | 1.12 | 2.00 | 1.00 | 6.05 | 8.00 | 1.00 | 100.00 | 45.59 | 134.22 | 15.53 | 2.83 | 329.50 | 1.00 | 99.91 |
| 1024 | 1.06 | 2.00 | 1.00 | 5.99 | 8.00 | 1.00 | 100.00 | 23.42 | 67.16 | 10.22 | 2.19 | 192.25 | 1.00 | 99.91 |
| 2048 | 1.03 | 2.00 | 1.00 | 4.98 | 7.00 | 1.00 | 100.00 | 12.07 | 33.62 | 5.62 | 1.49 | 97.62 | 1.00 | 99.91 |
| 4096 | 1.02 | 2.00 | 1.00 | 4.95 | 6.00 | 1.00 | 100.00 | 6.39 | 16.84 | 3.31 | 1.24 | 50.31 | 1.00 | 99.91 |

**TABLE C.22.  Parameters of Group 3 for Barnes**

| Block Size (bytes) | CSI | | | | | | | NCSI | | | | | | |
| | Write | | | Read-only | | | | Write | | | Read-only | | | |
| | avg | max | min | avg | max | min | $f(\%)$ | avg | max | min | avg | max | min | $f(\%)$ |
|---|---|---|---|---|---|---|---|---|---|---|---|---|---|---|
| 4 | 2.16 | 6.00 | 1.03 | 1.16 | 3.94 | 1.00 | 100.00 | 1279.57 | 4281.62 | 93.06 | 1099.04 | 7303.03 | 1.78 | 97.30 |
| 8 | 2.05 | 4.00 | 1.03 | 1.08 | 2.50 | 1.00 | 100.00 | 640.38 | 2150.06 | 46.53 | 578.25 | 3818.00 | 1.78 | 97.30 |
| 16 | 1.69 | 3.00 | 1.00 | 1.04 | 1.75 | 1.00 | 100.00 | 333.51 | 1108.44 | 31.03 | 321.25 | 2081.00 | 1.78 | 97.30 |
| 32 | 1.47 | 2.00 | 1.00 | 1.02 | 1.41 | 1.00 | 100.00 | 177.52 | 587.12 | 15.53 | 192.63 | 1211.00 | 1.75 | 97.30 |
| 128 | 1.42 | 2.00 | 1.00 | 1.01 | 1.16 | 1.00 | 100.00 | 52.50 | 176.41 | 15.53 | 104.43 | 625.88 | 1.75 | 97.30 |
| 512 | 1.00 | 1.00 | 1.00 | 1.01 | 1.09 | 1.00 | 100.00 | 26.66 | 67.69 | 15.53 | 95.58 | 619.81 | 1.75 | 97.30 |
| 1024 | 1.00 | 1.00 | 1.00 | 1.01 | 1.09 | 1.00 | 100.00 | 17.09 | 34.88 | 15.53 | 49.79 | 319.12 | 1.75 | 97.30 |
| 2048 | 1.00 | 1.00 | 1.00 | 1.01 | 1.09 | 1.00 | 100.00 | 17.09 | 34.88 | 15.53 | 49.79 | 319.12 | 1.75 | 97.30 |
| 4096 | 1.00 | 1.00 | 1.00 | 1.00 | 1.06 | 1.00 | 100.00 | 8.39 | 14.34 | 7.78 | 18.50 | 111.00 | 1.75 | 97.30 |

simulation kernel and the simulator itself allocate space at a separated part of heap, so there is no interference of data allocated by the application and the simulation environment.

## C.8.  CONCLUSIONS

In this text we have presented our approach to the characterization of parallel applications and have provided values of parameters from an original set describing parallel applications behavior. We have developed a measuring tool called "Scowl" that cooperates with Limes, a previously developed execution-driven simulator of multiprocessors. Parameters that we measured were classified into semantic groups. We also generated reduced traces that can facilitate the analysis of application behavior.

The parameters obtained using Scowl and Limes are designed to assist people who devise new methods aimed at the improvement of DSM systems and who use an analytical approach for quantitative evaluation and comparison of existing systems, especially in the area of memory consistency models.

Reduced traces can be used to obtain values for user-defined parameters with less programming efforts and execution time.

The major contribution of this appendix is that it has provided an exhaustive set of valuable characterization-related details that can be used by researchers in the field, with two major goals: (1) to assign more appropriate explanations to existing solutions, based on the rationales standing out from the presented data; and (2) to study these details and to try to come up with new solutions for existing problems. Our future plans include the development of models for evaluation of algorithms and architectures in the field of DSM/SMP systems based on the parameter values obtained. We also plan to reevaluate memory consistency models and their implementations according to the results obtained in this study.

## REFERENCES


[Agarwal86]   Agarwal, A., Sites, R. L., Horowitz, M., "ATUM: A New Technique for Capturing Address Traces Using Microcode," *Proc. 13th Annual Int. Symp. Computer Architecture*, Tokyo, June 1986, pp. 119–127.

[Agarwal88]   Agarwal, A., Gupta, A., "Memory-Reference Characteristics of Multiprocessor Applications under MACH, *Proc. ACM SIGMETRICS Conf.*, May 1988, pp. 215–225.

[Conte90]   Conte, T. M., Hwu, W. W, "Benchmark Characterization for Experimental System Evaluation," *Proc. 23rd Annual Hawaii Int. Conf. System Sciences*, Kailua-Kona, Hawaii, Jan. 1990, Vol. 1, pp. 6–18.

[Davis90]   Davis, H., Goldschmidt, S., *Tango: A Multiprocessor Simulation and Tracing System*, Technical Report CSL-TR-90-439, Stanford Univ., Palo Alto, CA, July 1990.

[Eggers88]   Eggers, S. J., Katz, R. H., "A Characterization of Sharing in Parallel Programs and Its Application to Coherence Protocol Evaluation," *Proc. 15th Annual Int. Symp. Computer Architecture*, Honolulu, May 1988, pp. 373–382.

[Gupta92]   Gupta, A., Weber, W. D., "Cache Invalidation Patterns in Shared-Memory Multiprocessors," *IEEE Trans. Comput.* **41**(7), 794–810, (July 1992); Correction, *ibid.*, **41**(12), 1631–1632.

[Herrod93]   Herrod, S. A., *TangoLite: Introduction and User's Guide*, Technical Report, Stanford Univ., Palo Alto, CA, Nov. 1993.

[Iftode96]   Iftode, L., Singh, J. P., Li, K., "Scope Consistency: A Bridge Between Release Consistency and Entry Consistency," *Proc. 8th Annual Symp. Parallel Algorithms and Architectures*, Padua, Italy, June 1996, pp. 277–287.

[Kesser89]   Kessler, R. E., Livny, M., "An Analysis of Distributed Shared Memory Algorithms," *Proc. 9th Int. Conf. Distributed Computing Systems*, June 1989, pp. 498–505.

[Lenoski92]    Lenoski, D., Laudon, J., Joe, T., Nakahira, D., Stevens, L., Gupta, A., Hennessy, J., "The DASH Prototype: Implementation and Performance," *Proc. 19th Annual Int. Symp. Computer Architecture*, Gold Coast, Australia, May 1992, pp. 92–103.

[Magdic97]    Magdic, D., "Limes: A Multiprocessor Simulation Environment," *TCCA Newsl.*, 68–71 (Mar. 1997).

[Milutinovic97]    Milutinović, V., Milenković, A., "Cache Injection Control Architecture," *Proc. 5th MASCOTS*, Haifa, Israel, Jan. 1997.

[Milutinovic00]    Milutinović, V., *Surviving the Design of Microprocessor and Multimicroprocessor Systems*, IEEE Computer Society Press, Los Alamitos, CA, in press.

[Protic96a]    Protić, J., Tomašević, M., Milutinović, V., "Distributed Shared Memory: Concepts and Systems," *IEEE Parallel Distrib. Technol.* **5**(1), (1996a).

[Protic96b]    Protić, J., Tartalja, I., Tomašević, M., "Memory Consistency Models for Shared Memory Multiprocessors and DSM Systems," *Proc. 8th Mediterranean Electrotechnical Conf. Melecon 96*, Bari, Italy, May 1996b.

[Protic97]    Protić, J., Milutinović, V., "Entry Consistency Versus Lazy Release Consistency in DSM Systems: Analytical Comparison and a New Hybrid Solution," *Proc. 5th IEEE Workshop on Future Trends of Distributed Computing Systems FTDCS'97*, Tunis, Tunisia, Oct. 1997.

[Singh92]    Singh, J. P., Weber, W. D., Gupta, A., "SPLASH: Stanford Parallel Applications for Shared-Memory," *Comput. Architecture News*, **20**(1), 5–44 (Mar. 1992).

[Stumm90]    Stumm, M., Zhou, S., "Algorithms Implementing Distributed Shared Memory," *IEEE Comput.* **23**(5), 54–64 (May 1990).

[Tartalja92]    Tartalja, I., Milutinović, V., "An Approach to Dynamic Software Cache Consistency Maintenance Based on Conditional Invalidation," *Proc. 25th Annual Hawaii Int. Conf. System Sciences*, Kauai, Hawaii, Jan. 1992, pp. 457–466.

[Woo95]    Woo, S. C., Ohara, M., Torrie, E., Singh, J. P., Gupta, A., "The SPLASH-2 Programs: Characterization and Methodological Considerations," *Proc. 22nd Annual Int. Symp. Computer Architecture*, Santa Margherita Ligure, Italy, June 1995, pp. 24–36.

# MISD, SIMD, and MIMD System / Accelerator Designs

The following appendixes appear on the World Wide Web
`http://galeb.etf.bg.ac.yu/~vm/`:

- Design of an MISD System/Accelerator
- Design of an SIMD System/Accelerator
- Design of an MIMD System/Accelerator